The Moon in the Nautilus Shell

The Moon in the Nautilus Shell

DISCORDANT HARMONIES RECONSIDERED

FROM CLIMATE CHANGE TO SPECIES EXTINCTION, HOW LIFE PERSISTS IN AN EVER-CHANGING WORLD

Daniel B. Botkin

UNIVERSITY PRESS

OXFORD
UNIVERSITY PRESS

Oxford University Press publishes works that further
Oxford University's objective of excellence
in research, scholarship, and education.

Oxford   New York
Auckland   Cape Town   Dar es Salaam   Hong Kong   Karachi
Kuala Lumpur   Madrid   Melbourne   Mexico City   Nairobi
New Delhi   Shanghai   Taipei   Toronto

With offices in
Argentina   Austria   Brazil   Chile   Czech Republic   France   Greece
Guatemala   Hungary   Italy   Japan   Poland   Portugal   Singapore
South Korea   Switzerland   Thailand   Turkey   Ukraine   Vietnam

Copyright © 2012 by Daniel B. Botkin

Published by Oxford University Press
198 Madison Avenue, New York, New York 10016

www.oup.com

A previous version of this book was published in 1990 as *Discordant Harmonies*.
Oxford is a registered trademark of Oxford University Press

Library of Congress Cataloging-in-Publication Data
Botkin, Daniel B.
The moon in the nautilus shell : discordant harmonies reconsidered / Daniel B. Botkin.
     p.   cm.
Originally published as: Discordant harmonies, 1990.
Includes bibliographical references and index.
ISBN 978-0-19-991391-6 (hardcover : alk. paper) 1. Nature conservation. 2. Environmental protection.
3. Environmental policy. I. Botkin, Daniel B. Discordant harmonies. II. Title. III. Title:
Discordant harmonies reconsidered.
QH75.B67   2012
333.72—dc23   2012005799

9 8 7 6 5 4 3 2 1

Printed in the United States of America
on acid-free paper

This book is dedicated to my mother, Gertrude Fritz Botkin, who put her trust in knowledge, creativity, humanity, and a goal of leaving the world a better place than she found it.

CONTENTS

ACKNOWLEDGMENTS

I wish to acknowledge the help and comments of many friends and colleagues over the years in ways that led to *The Moon in the Nautilus Shell*. In particular, I wish to thank John R. Bockstoce, a colleague and friend since the 1970s who was among those who persuaded me to write a follow-on to *Discordant Harmonies*. Our discussions about ecology, environment, anthropology, and most other things about civilization, as well as about sailing and flying, have been of great value to me. Matthew J. Sobel, applied mathematician, engaged in many conversations that sharpened the concepts of stability and did his best to ensure that my thinking was clear and logical and that I understood the bases of economics as these apply to ecology. The historian Charles Beveridge introduced me to the works of George Perkins Marsh, Clarence Glacken, and Marjorie Nicolson. Ecologist Lawrence B. Slobodkin understood the issues I was thinking about and encouraged my pursuit of the answers. Richard Pfilf, career U.S. Forester, introduced me to many foresters and did his best to make sure my feet were on the ground, well connected to the realities of forests and their management. Benjamin A. Stout, who during his lifetime was a professional forest statistician, helped also with clear thinking, analysis of complex ecological data, and continual encouragement. Charles A. Sansone, whose friendship began when he was my junior high school teacher and continues today, has always provided honest commentary and encouragement. Kenneth L. Purdy, a high-school classmate and friend, read drafts of this manuscript and did his best to make sure it was readable to nontechnicians.

I also wish to thank Jim Welter, a lifelong fisherman from Gold Beach, Oregon, whose insights about salmon and their environments gave me faith that complex ecological problems were soluble; Fritz Maytag, whose conversations on many subjects continually stimulate my thinking and analysis, and Donald A. DeAngelis, Professor, Department. of Biology, University of Miami, Coral Gables, Florida, who read and critiqued drafts of the manuscript.

My wife, Diana Perez Botkin, the best professional editor I have ever met, played a crucial role in reviewing this book, editing it carefully. My sister, Dorothy Rosenthal, who introduced me to the writings of Joseph Conrad, Antoine de Saint-Exupéry, and many, many of the books that set the path for my life, has been one of my best advisers, helpers, and critics of my professional work. She, along with my wife, helped with line-by-line editing of the revised manuscript. Dan Tarlock also played a major role in encouraging me to revisit Discordant Harmonies.

In many conversations, a number of other colleagues have helped me understand the issues I discuss. These include Thomas Dunne, geomorphologist, University of

California, Santa Barbara; Kenneth Cummins, freshwater ecologist, Humboldt State University, Humboldt, California; Lee M. Talbot, wildlife ecologist and conservationist, George Mason University, Fairfax, Virginia; Edward A. Keller, geologist, University of California, Santa Barbara; Willie Soon, astrophysicist, Harvard University, Cambridge, Massachusetts; and Pierre Glynne, National Research Program, U.S. Geological Survey.

In addition, I wish to acknowledge the many friends and colleagues whose help with *Discordant Harmonies* continued to play an important role in *The Moon in the Nautilus Shell.* They include Harold J. Morowitz and Lucile Morowitz, who read some of the earliest drafts of the first book and encouraged me to continue; Thomas E. Lovejoy, ecologist, who reviewed the Wilson Center work; Roderick F. Nash, historian, with whom I discussed the history of ideas about nature; J. Baird Callicott, philosopher, whose knowledge of environmental ethics and of Aldo Leopold's career was invaluable.

I would especially like to thank Jeremy Lewis, my editor at Oxford University Press, for his careful review of this manuscript and his encouragement; and Jason Allen Ashlock, my literary agent, who often had brilliant insights into what was important to write and do in this transitional time for book publishing. I am also indebted to Paul B. Sears, conservationist, who got me started in ecology; Murray F. Buell, ecologist, who gave me inspiration as a scientist and teacher; and Heman L. Chase, New Hampshire surveyor, for the many hours we spent surveying in the woods of New Hampshire and Vermont.

Work that laid the basis for a good part of the book was begun under a fellowship from the Woodrow Wilson International Center for Scholars, Washington, DC, and a fellowship from the Rockefeller Foundation Bellagio Study and Conference Center, Bellagio, Italy. Aspects of the material in this new book are a result of research supported by the World Wildlife Fund and by several federal agencies, including the National Science Foundation, the National Oceanographic and Atmospheric Administration, the Marine Mammal Commission, and the National Aeronautic and Space Administration's Office of Life Sciences. The writing of *The Moon in the Nautilus Shell,* however, has received no financial support from any entity or individual.

Finally, I would like to acknowledge other ecologists' contributions in the past 20 years to the problems associated with the idea of the balance of nature, including in particular Stuart Pimm's book *The Balance of Nature?* (Chicago, University of Chicago Press, 1991) and his subsequent writings.

INTRODUCTION

More than twenty years have passed since I wrote a book called *Discordant Harmonies*. Much has changed in those twenty years, and much hasn't. The thesis of that book was that we had approached environmental problems from the wrong set of assumptions, assumptions deeply rooted in our civilization and culture. These assumptions, considered at the time to be scientific, were in fact heavily based on ancient pre-scientific myths about nature. Although that book has been influential, much that I hoped it would achieve remains undone. On the one hand, it is time to celebrate the advances that have been made in solving some environmental problems. On the other hand, I find it necessary to explain where we have continued to follow the wrong paths—wrong in premises and wrong in effects.

Thus I have written *The Moon in the Nautilus Shell*, which begins with the major points that I made in *Discordant Harmonies*, extends them in ways that I see now are necessary, and explains where the conventional ideas about nature—especially about people and nature—have continued down the wrong path. In this introduction I first explain what I intended in *Discordant Harmonies*, augmenting what I wrote in the hope that it will make those goals clearer than they were before.

We interact with nature in two ways: rationally and through an inner, personal, nonintellectual response. The inner, personal, nonintellectual includes, in the largest sense, all that is outside of rationality—our folkways, our myths, our spiritual feelings that arise from deep within us, our religious sensitivities. Both ways of interacting with nature are important. However, we get ourselves into trouble when we confuse the two, letting the inner personal determine what we tell ourselves are rational decisions and actions, and believing that rationality can replace the nonrational—that we moderns are so immersed in the rationality of science and its offspring, modern technology, that we don't need, don't even have, a nonrational side of our existence when it comes to nature or our connection with nature. Isn't the issue of global warming simply one of science and therefore of rationality? It seems to surprise us moderns when we discover there are debates about climate change that are charged with emotion, opinion, political and ideological biases. In the rise of environmentalism, the science of ecology, and other environmental sciences, we have often gotten the two aspects of our interaction with nature reversed. When we do so, it is to our and our environment's detriment. This new book helps to clarify the dilemma and suggests a path toward its resolution.

In *Discordant Harmonies*, I tried to create a book that expressed its message in story, imagery, metaphor, and analogy, as well as straightforward narrative. The message seemed complex and outside of our ordinary way of thinking, and I hoped these added features would increase the reader's understanding and response to that message. They have done so to some extent and for some readers; but not for all, nor as completely as I now want. Hence, this sequel, *The Moon in the Nautilus Shell*.

As I believed in the 1980s when I worked on *Discordant Harmonies* and still believe, there are ideas here that can be helpful to us in many ways. I hope you find them so.

Stability, Constancy, and Common Knowledge

One person who read *Discordant Harmonies* wrote that I exaggerated greatly when I said that ecologists once believed that nature undisturbed by people would never change. Others have made the same criticism. This is a misreading of my book. My point has been that models, theory, management policies, recommendations, etc., in ecology assume that ecosystems and species are in a steady state and will never change, period. Of course ecologists and other environmental scientists are aware of geological time and evolutionary processes; it's just that these get lost when scientists set down their ideas in ecological statements, theories, models, and assumptions. The way I put this to try to keep it simple is: If you ask ecologists whether nature is always constant, they will always say "No, of course not." But if you ask them to write down a policy for biological conservation or any kind of environmental management, they will almost always write down a steady-state solution.

Here is an example. In 1979, noted ecologist George M. Woodwell, geologist Gordon J. MacDonald, famous atmospheric scientist Roger Revelle, and atmospheric chemist David Keeling sent a report to the Federal Council on Environmental Quality. Entitled *The Carbon Dioxide Problem: Implications for Policy in the Management of Energy and Other Resources*, this report was considered of such historic importance that James Gustave Speth, at the time Dean of the Yale School of Forestry and Environmental Studies, had it reprinted in 2008 with a foreword that he wrote for an 80th-birthday symposium held at Yale University for Dr. Woodwell. In their report, Woodwell et al. had written that "the CO_2 problem is one of the most important contemporary environmental problems . . . [that] threatens the stability of climates worldwide." Clearly these authors had to know that the climate had changed in the past, and that it had always been changing, but they wrote to the executive branch of the U.S. government that the climate is "stable" and human actions threaten that stability.[1]

All the major computer models of climate, called general circulation models (often referred to by those who work with them as "GCMs"), are steady-state models. A user puts in a set of environmental conditions and the computer program runs until it achieves a steady state. As long as the input is the same, the end result will always be exactly the same. And when the model of climate reaches that steady state, it stays there. For those of you unfamiliar with how these models are used, climate forecasts are done by comparing two different runs of a model, one with what is defined as the "standard" climate—usually the climate for 1960 to 1980 or 1960 to 1990—and a "transitional" climate. Both simulations are run until they reach an equilibrium and then are compared.

Similarly, most of the computer models used to forecast the possible effects of global warming on living things, from specific species to biomes (kinds of ecosystems), are also steady-state. In fact, most are steady-state in several ways. The present geographic distribution of a species (or biome) is overlaid on a map of the present average temperature and precipitation. Then the output from a GCM creates a new map of the forecasted average temperature and precipitation, and the range of the species is mapped onto that new climate. Think of the distribution of the species (or biome) as a rectangle on a flat map of Earth. If the temperature is forecast to move north, then the rectangle moves exactly that much north.

This method assumes a steady state in several ways. The present distribution of a species is assumed to be in steady state with the present climate. The "standard" present climate is assumed to be steady-state except for human-induced changes in greenhouse gases. The future distribution of the species (or of a biome) is assumed to adjust instantaneously to the future climate and be in steady state immediately with it. The species in a biome are assumed to have no flexibility, no way that any adjustments can be made in their relationships with temperature and precipitation.

None of these steady-state assumptions are true. Species do not come into instant equilibrium with a new climate; they are always in the process of responding to previous environmental change. There are other factors that limit distribution that are changing over time. Individuals, populations, species have the capability to adjust to a changing environment; otherwise they all would have gone extinct in the past.

I agree with critics who say that the scientists who create these models must know that the climate has always been changing—they keep showing us graphs that demonstrate that. And the biologists who create the models of the response of biota to climate change must also know that species are not always (if ever) in steady state with their present climate. But their forecasting methods contradict what they must know.

This has been the general approach in ecology and related environmental sciences: Whatever the scientist's knowledge of the dynamic, changing properties of nature, the formal representations of these remove such considerations in most cases. In *The Moon in the Nautilus Shell*, I provide other examples of the same phenomenon—whether or not environmental scientists know about geological time and evolutionary biology, their policies ignore them.

It is strange, ironic, and contradictory. That was one of the reasons I wrote *Discordant Harmonies*, because I found that ecologists were giving advice that was contrary to data and knowledge that confronted them, and I wondered why. The explanation was in that book and is reiterated and extended in this new book.

For me, it all began with the passing of the U.S. Federal Marine Mammal Protection Act in 1973. This set up a Marine Mammal Commission, with commissioners, a staff, and a scientific advisory board. The Act stated that its primary goal was to protect the "health and stability of marine ecosystems," three terms that at the time had no specific scientific meaning. Second to that goal, the Act stated, was to achieve an "optimum sustainable population of marine mammals." In 1975, the commissioners got in touch with me and said they did not understand, either legally or ecologically, what was meant by "optimum sustainable population." They invited me to write a report that would explain it, in terms of the wording within the Act and my knowledge of ecology.[2]

So I read the Act. It was written with the assumption—stated explicitly in the Act—that marine mammal populations existed in a constant abundance except when harvested by people. This was then the standard mathematical theory about how populations grew in nature, from a small number to a constant "carrying capacity."

The request to write the report brought me into contact with a group of marine mammal scientists and fisheries biologists. It was clear that these scientists, those who had written the Act and those who were on the Commission's science advisory board, knew that these populations had varied in the past and were likely to vary in the future, with or without people's intervention. Some of the scientists had years of experience in the field observing wildlife populations. They were sincere and good scientists, doing their best, not intentionally telling the Commission or other scientists anything that they believed was untrue. Yet they were giving advice that contradicted facts that were available to them.

Trying to understand why they clung to this view set me on a quest to study how the idea of nature had been viewed throughout Western civilization's history. I spent a year at the Woodrow Wilson International Center for Scholars, reading everything I could about nature beliefs from the ancient Greeks to modern times, and this was the beginning of *Discordant Harmonies*.

Pandora's Box and the Acceptability of Change

As I will discuss later, a question that is asked almost every time I give a talk is "If you say one kind of change is okay, doesn't that mean every kind of change is okay, like killing off some endangered species [or whatever interests the questioner]?"

I reply first that the opposite is true: If we are convinced that nature never changes and should never change, then we have no way to compare one kind of change with another; all are "bad." Once we recognize the naturalness of many kinds of changes, then nature provides us with a metric to decide among them.

I give a mechanical analogy. "When you are driving your car, you certainly want to be able to change the direction the car is going; otherwise you would crash into something," I say, "but that doesn't mean you want to drive the car off a cliff. Similarly, there are kinds of changes that are natural in that they have been part of the environment for a long enough time for species to adapt to them, and many require these changes. If we take actions that lead to these kinds of changes and at rates and quantities that are natural in the sense I have just described, then these are likely to be benign. If we invent some novel change that species have not had a chance to evolve and adapt to, then those are more likely to lead to undesirable results, and we should be very cautious in using them."

Mythology, Science, and Religion

Another topic that I did not discuss directly in *Discordant Harmonies*, hoping instead that the content would communicate it to the reader, is the intersection between mythology, science, and religion. It seems clear to me now that I should explain these connections and why they are important to the fundamental message I was trying to convey, to the way we see ourselves within nature, how we perceive nature by itself, and how we act toward and within it. These in turn affect many aspects of our lives both rationally and nonrationally, those we are readily aware of and those we are not—the I and thou of people-and-nature touches every aspect of our lives.

Anthropologist Joseph Campbell wrote that every human society needs a mythology in the sense of an explanation of how the world came about and how it works. He wrote:

> No human society has yet been found in which . . . mythological motifs have not been rehearsed in liturgies; interpreted by seers, poets, theologians, or philosophers; presented in art; magnified in song. . . . Man, apparently, cannot maintain himself in the universe without belief in some arrangement of the general inheritance of myth. In fact, the fullness of his life would even seem to stand in a direct ratio to the depth and range not of his rational thought but of his local mythology.[3]

Campbell then asked:

> Why should it be that whenever men have looked for something solid on which to found their lives, they have chosen not the facts in which the world abounds, but the myths of an immemorial imagination?[4]

Perhaps the answer to Campbell's question lies in a brief statement by Charles MacKay in his remarkable 1841 book *Extraordinary Popular Delusions and the Madness of Crowds*, where he writes:

> Three causes especially have excited the discontent of mankind; and, by impelling us to seek for remedies for the irremediable, have bewildered us in a maze of madness and error. These are death, toil, and ignorance of the future—the doom of man upon this sphere, and for which he shews his antipathy by his love of life, his longing for abundance, and his craving curiosity to pierce the secrets of the days to come.[5]

While in our modern times one might think that these fears would lead only to rational action—research and technological development for better health care, better agriculture, invention of machines to toil for us, and such modern forecasting methods as computer programming—it may well be to the contrary: that these drive us, more deeply, to seek what feels to us (rather than simply appears to us) as firmer ground, some kind of universal explanation of how the world came about and how it must work.

Be that as it may, Campbell's point of view seems quite at variance with popular conceptions of myths, mythology, science, and religion. With the great success of science and its applications, it is easy to believe that modern people can exist without any mythology; science—that is, rational thought based on careful empirical observations—can tell us what the world is really like, how it really works, and how it came about. Campbell would say, certainly that could be true, but true or not, it could be our modern mythology.

Yes, in our scientific age, we look upon mythology as something of the past, beneath us and unnecessary, false stories that have become children's fairy tales and should be no more. Our scientific/technological age has given us the hubris to believe that we have risen above something that seems primitive to us, part of the lives of prehistoric human beings and of those ancient civilizations that came before modern science. We look down on what they believed to be the fundamentals of existence. We seem to think we don't have a mythology and don't need one as we text-message while we drive, play video games, fly high above Earth, ignoring the scene below as we watch a movie, build cities that seem to isolate us permanently above nature except when we want a little exercise and recreation. This technological life blinds us to our dependence on a fundamental worldview, to the realization that the opposite holds.

We are embedded deeply within our myths and our folkways. Those of us in Western civilization have two. One, the grand idea of a balance of nature, has its origins in the ancient world, having roots somewhere within the ancient Greek and Judaic societies, with influences beyond these. The other is modern science itself, when it serves as revealed truth in which we need only believe without question.

We need to see mythology—in the sense of a story about how the world came about and how it works—as still a necessary part of human existence. It is deep within us, like it or not; it is not a bad thing, it is just what we are. We need to confront the conflict between our old and new myths and work our way to understanding the role, utility, and value of our rationality within all of the human experience. If we do not, then we are bound to fall victim, as we have in the past and seem to be doing in the present, to irrationalities that do not serve us well. To put this most simply: Ironically, the more our science and technology seem to separate us from mythology, the more unknowingly dependent on that mythology we become.

This may seem preposterous to those brought up to believe that science is, without any doubt, our modern source of truth, and that scientists could not fall victim to what I have just described. Indeed, I became involved with this problem because I was confronted with it and couldn't quite believe it myself.

However rational we may believe we moderns are, no one can doubt that many of our decisions and actions are irrational. One of the causes of the 2008 economic meltdown, for example, was unwarranted faith in certain computer-based forecasting methods, used irrationally. The same problem plagues ecology and environmental sciences in general.

To sum up the underlying and fundamental message of my 1990 book *Discordant Harmonies*: Overwhelmingly we still believe in the nature as described by the ancient Greeks, which has come down to us through Judaism and Christianity. Rational science simultaneously stands in opposition to this first tradition and is also greatly altered by it. Our sciences of environment looked back to the ancient Greek myths, which they often simply repeated, in conflict with the facts that science's very rationality contradicted. With our modern hubris, we fail to recognize that we, too, in Joseph Campbell's words, "have chosen not the facts in which the world abounds, but the myths of an immemorial imagination" as the basis for our environmental policies. But the message in *Discordant Harmonies* was perhaps stated too subtly and I was too gentle in presenting it, so here it is, bold and direct. Keep this idea in mind as you read the rest of this book.

Science, Religion, and Folkways

From an anthropological point of view, mythologies—in the positive sense that I have been using the word—are intimately connected to religion. This has led to the now-familiar conflict between those of us who believe in biological evolution, as demonstrated by modern science, and those who prefer to take the Bible literally and believe in creationism. What has not come to the fore is the importance of the Judeo-Christian traditional beliefs in influencing how we, as ordinary citizens, think and feel about nature and the environment, and how environmental science has developed. This was a third underlying theme of *Discordant Harmonies:* that even today, in this age when we seem to have persuaded ourselves that we have risen above mythology, most environmental policies, laws, and ideologies are consistent with (to say the least) and arguably a restatement of the beliefs about nature in that Judeo-Christian tradition. This consistency can be either intended by a writer/scientist or inferred through his or her (inadvertent) use of imagery, metaphor, and analogy.

For example, when James Lovelock titled his 2006 book *The Revenge of Gaia*,[6] intentionally or unintentionally he was casting up from the sea of Judeo-Christian tradition the belief that man is a sinner, bound to sin; that most recently we have sinned against nature, and we are being punished for it (as we should be) by Mother Nature (you can substitute your personal preference of God or gods). Therefore we must do penance, suffer for our sins, which in this case means living minimally, using only enough energy to provide the bare necessities of life and disallowing us enough energy to be creative, to develop more science and technology and further exploit and damage Earth; nor should we, by implication (perhaps unintentionally) have enough energy to be otherwise creative, even enough energy for Lovelock to write his book.

If Joseph Campbell is correct that "whenever men have looked for something solid on which to found their lives, they have chosen not the facts in which the world abounds, but the myths of an immemorial imagination," then one could argue that a fundamental challenge before us is to find a way to integrate our two primary mythologies, to bring mainstream science and Judeo-Christian traditions together in our minds and hearts in a way that provides us with a consistent modern mythology, meant in the most positive sense, as Campbell uses it. The need and hope are that we find a path to bring us where these two ways of conceiving of nature become compatible. Philosophers of science tell us they should not, at least by necessity, directly conflict, because science deals with "how" and "when" and does not explain ultimate "whys." The conflict arises for us when we believe that religion can tell us the "how" and "when" and science can be a source of "why." These may seem obvious mistakes, easily avoided, but they are not; we seem to be continual victims of this confusion.

My recent experiences suggest some hope. In 2009 I was invited to speak at the annual conference of the Concordia Seminary in St. Louis, which the director told me was the more conservative of the midwestern Lutherans. I was concerned that I would find myself immersed in an anti-science debate, but instead the ministers and theology students asked questions that indicated they were looking for legitimate scientific answers.

I have to be careful here in phrasing this, because it is such a controversial and sensitive topic. At the extremes, fundamental interpretations of the Bible as literal truth about the "how" and "when" of nature conflict with modern scientific observations.

I focused on Western religious traditions in *Discordant Harmonies* in part because these have been the greatest direct influence on science since the beginning of the scientific-industrial age, and in part because it is the tradition I am familiar with. I didn't believe that I had the knowledge or background to reach beyond this, but I hope others will. All religions that I know of have deeply embedded beliefs about the character of nature and the relationships between people and nature. Japanese Shintoism is famous for its focus on human–nature contact, including the enhancement of natural beauty by people. Some, like Hinduism, do not have the same focus on the constancy and immutability of nature, and perhaps some non-Western religions can help us through this next phase in the development of civilization, all the more likely as civilization becomes more global. It is also my impression that since I wrote *Discordant Harmonies*, increasing numbers of people within Western societies have been exploring non-Western religions, sometimes combining elements of these with Western ones. These too might help in this next step in civilization.

Thinking over how we are persuaded even now that certain things about nature and environment must be true despite or without scientific evidence, I have come to realize that these older ways of thinking and of seeking understanding are still with us. As David Fischer explains in his landmark book *Albion's Seed: Four British Folkways in America*,[7] societies develop from their folkways, which he defines as "the normative structure of values, customs, and meanings that exist in any culture." Included in his list of folkways are speech ways, building ways, family ways, learning ways (how children are taught), religious ways, and magic ways (which include normative beliefs and practices concerning the supernatural).[8] To this I add knowledge ways—how a society perceives what knowledge is, how it is acquired, and how it is used. These folkways are persistent and dominate thinking and action today, just as they have over many generations. I will discuss this again in the first chapter.

The sensitivity of our modern society to the comparisons of science, religion, and folkways makes this a difficult subject to approach, let alone discuss. For that reason, among others, I avoided its direct discussion in my earlier book. I discuss it in this book because I realize it's absolutely necessary.

No one, to my knowledge, has yet tried to compose into a single picture the integration of the major, great myths of nature and the way that nature appears to us through modern science—especially, but not limited to, the science of ecology. Ecology is, after all, the scientific study of the relationship between living things and their environment. This is what nature—that which surrounds us, which molds and affects us, and which we in turn affect—has meant throughout human history. And this single picture was and is the goal of both *Discordant Harmonies* and *The Moon in the Nautilus Shell*.

What I Tried to Do in Writing this Book: Style and Substance

I have always had a love of literature and writing, and grew up in a house with my father's library of 12,500 books. He was a well-known folklorist and author, an English professor for many years, and wrote and published poetry as well as nonfiction. By the time I was fourteen, I had decided on a life goal: to become a scientist and then write about it, and in addition travel and explore nature throughout the world as much as time and funds would allow. I majored in physics as an undergraduate, but decided that I wanted to do science outside. Ironically, my science professors told me that there was no more science to do outside.

In college I wrote for the campus newspaper, as both the theater critic and science writer, trying out how well I could communicate complex scientific research to my fellow students. Since there wasn't going to be any science outside, I decided to make a career out of my second great interest and become a professional science writer. I was fortunate to obtain a fellowship in science writing from the University of Wisconsin, Madison, which made me into a reporter of scientific research for the university's news bureau. I took advantage of that situation to obtain a master's degree in English literature, pursuing creative writing, the history and development of the novel (I hoped to write one someday), and getting the opportunity to expand and deepen my knowledge of the greatest British literature. I read extensively about the art of writing, including the use of analogy, metaphor, imagery, symbolism, rhythm, and rhyme.

I wrote *Discordant Harmonies* with the impressions the written word and the beauty of language had left on me. I felt that the book's message would be better communicated if it was expressed not only as a scientific statement but also emotionally, so as to speak, through the use of metaphor, imagery, and symbols. I approached *The Moon in the Nautilus Shell* the same way.

For example: In *Discordant Harmonies* I made use of the classic contrast between light, representing knowledge, understanding, truth (in a classical sense), and darkness, representing ignorance, confusion, falsehoods. Where I help the reader discover the reality of nature, as I understood it, the stories take place in bright sunlight. Things that are unknown, poorly understood, or widely believed but I know are wrong, I tell about with stories that take place at twilight or at night. My search for wolves in a world that had randomness we were only beginning to understand, a world that therefore was poorly known, takes place in the wilderness forests at night at Isle Royale. It actually did, but the example was all the more meaningful for me stylistically because it happened at night. In discussing the realization that life has affected Earth's environment globally, I quote Isabella Bird's description of the incredible brightness and color of the surroundings as she climbed Mauna Loa, which she wrote about in her nineteenth-century classic, *Six Months in the Sandwich Islands*.[9] This introduced the reader to my trip up that mountain to the meteorological observatory at 11,500 feet, where Keeling made his excellent measurements of the atmosphere's chemistry. Knowledge and light went together.

Here's another example. One of my father's favorite poems was John Masefield's "Sea Fever," the first stanza of which is:

I must go down to the seas again, to the lonely sea and the sky,
And all I ask is a tall ship and a star to steer her by,
And the wheel's kick and the wind's song and the white sail's shaking,
And a grey mist on the sea's face, and a grey dawn breaking.[10]

He liked it especially because the rhythm of the words matched the rhythm of a boat moving with wind, tide, and currents at its dock. I liked this idea too, and when in the last chapter of *Discordant Harmonies* I described leaving Venice in a motorboat taxi, I made the rhythm of the words fit what I remembered to be the way those boats moved in the canals as their wakes and those of other boats pushed them turbulently along, with waves resounding back to them from the walls of buildings along the canal. When the boat reached the open water of the Venetian lagoon, the rhythm of the words became smooth, steady, and quick, like the movement of the water taxi.

Why Be a Scientist?

I believe there are three primary reasons that people choose to go into a scientific, and especially an academically scientific, career in one of the environmental sciences.

The first reason is that some of us are just naturally curious and have to know how everything around us works. It isn't a question of whether to go into one of these fields; it is inevitable, an intrinsic drive.

The second reason is that being a scientist is a well-respected profession that in academic life results in a safe and comfortable standard of living and a variety of career perks, including awards of various kinds.

The third reason is that some people have strong political, ethical, moral, or ideological convictions that they want others—often society at large—to accept. They want to change the world in specific ways that meet their beliefs. Science for them is not a goal but a method to achieve, even force, these beliefs and their consequent actions onto society.

Of course one could say that there may be a little bit of each within any environmental scientist. But I am mostly the first kind. I have always had to know how everything I saw around me worked. When I was a kid I took apart everything that I could get my hands on. I used to go around my neighborhood and knock on neighbors' doors and ask if they had any broken appliances or radios or other electronic devices that they didn't want and could give me so I could take them apart.

I approached ecology in much the same way. I loved nature and was surrounded by it and had to understand how it worked. There was no other choice for me. But I had to understand how it really worked, not impose on it some set of beliefs I already had.

Those who see science as a means to a political, ethical, moral, ideological end, a means to promote a preconceived set of beliefs, have not liked that my attempt to understand how nature works upsets their preconceived views of how it should work and how people should act in relation to it. When I wrote *Discordant*

Harmonies I was careful to avoid saying this directly and bluntly. I tried to draw the reader in with stories, metaphors, imagery, and writing style, so that the reader would be led inevitably to agree with me. At that time I was much more naive about the strength of political and ideological motivations that led some into environmental science. I thought everybody who became a scientist was driven by curiosity, satisfied by understanding.

Over the years, I have learned that this is not always true. What we have seen since the 1970s, and are seeing more so today, is the manipulation of science to push political and ideological ideas. This motivation has often existed among those fascinated by nature—you can see it in the debate between Gifford Pinchot and John Muir, between Thoreau and Agassiz, and among ancient Greek philosophers. But it has become a major political force in our societies, has led to misdirection of policies and actions, and has debased science. So I have seen no choice today but to point out the message of both *Discordant Harmonies* and *The Moon in the Nautilus Shell* as directly as possible, even at the risk in this introduction of deviating from the style and approach that I sought in my original writing. I hope you will forgive me for this stylistic aside.

May 2012

The Moon in the Nautilus Shell

PART ONE

The Current Dilemma

FIGURE 1.1 The church Santa Maria della Salute, in Venice, Italy, is said to be built on a foundation of more than 1 million tree trunks. (Photograph by the author.)

1

A View from a Marsh: Myths and Facts About Nature

In the universe the battle of conflicting elements springs from a single rational principle, so that it would be better for one to compare it to the melody which results from conflicting sounds.

—PLOTINUS, THE ENNEADS *(THIRD CENTURY CE)*

Our Current Dilemma

One of the famous and often-photographed sights in Venice is the baroque church of Santa Maria della Salute, known as La Salute, which decorates the outer Grand Canal, its graceful dome presenting an image of great solidity, of heavy but graceful architecture, set against the constant motion of the coastal waters. It is said that the building of that single church began with the driving of 1,106,657 trunks of alder, oak, and larch, trees once common in the region, into the muds of the lagoon.[1] Completely submerged, no longer exposed to the air, the wood was protected from decay and thus remains as the foundation for the church. So it is with the rest of the city. That most famous example of human artifact, the architecture of Venice, survives in a changing lagoon because of a foundation built of wood, a biological support structure, surprising in our modern age of steel and concrete. This image of architectural beauty within nature—beginning in the fifth century CE and continuing today—can be viewed as a metaphor for the overall message of this book, and this first chapter serves as an overview of the primary dilemmas we face and that I discuss throughout.

Venice functions as a city partly because of the natural ecosystems of the lagoon, which, among other things, decomposes the sewage and other organic wastes from the city. Venice also functions because of its connection to the tides and currents of the Adriatic Sea, which continuously remove sewage from the lagoon, transporting it to the open ocean. The human artifice of the city persists because of biological matter and ecological processes, in combination with the external environment beyond the direct reach of the Venetians. And these ecological processes occur over many different time periods, from the twice-daily tides to responses of the Earth's surface since the end of the last ice age 10,000 years ago.

Venice was founded in the fifth and sixth centuries CE as a refuge for people fleeing Germanic tribes, including the Lombards, who were destroying the Roman Empire.[2] Inhabitants of towns around the northern Adriatic fled to the marshes, from which they could more easily defend themselves. At first they returned to their home cities after a raid, but eventually they began to settle in the lagoon. The mud in the marshes was unstable and shifted continually; to create a city, it was necessary to stabilize the ground. The first Venetians did this by driving millions of saplings into the mud.

Venice's environment is not simply an out-there, and not simply something that Venetians have damaged only to their own disadvantage; the intimate involvement of people and environment made it possible to create this city and for it to persist for fifteen centuries. This tight integration of people and nature is the perception we need, but have not yet made our own. One of the purposes of this book is to help us make this transition—from seeing ourselves as outside of nature and therefore only harming it, to seeing ourselves as within and part of nature, and capable of developing a natural nurturing. As today's Venice illustrates, this isn't an easy transition, either as an idea or in deciding what actions to take.

In the early 1980s, I was in Venice to attend an international conference, "Man's Role in Changing the Global Environment."[3] That city was an intriguing place to discuss environment and people, because its history provides an image in miniature of our current situation. Standing across the Grand Canal from Santa Maria della Salute, I could not help but compare my view at the end of the twentieth century with the view that must have confronted the first settlers more than 1,000 years ago, a view of flat marshland that stretched dishearteningly as far as the eye could see. I imagined what it might be like to begin to build the foundation for a great city by driving saplings into the salt marsh to hold the mud in place. What ideas did that take? What views of nature, the "environment," and the relationship between people and nature?

Although impelled by necessity, the first Venetians did not go to those marshes so long ago empty-handed, without the benefit of some knowledge of technology or of natural history, without the benefit of civilization. They brought with them three things: ideas, technologies, and a perspective of the world—how nature works, how people might change nature, and how the world in the future might be different from the world they had known in the past. Today, our position in relation to the environment of our entire planet is similar to that of the ancient settlers of Venice in relation to the marshes of the Adriatic. We see problems shifting before us whose solutions are unclear. Some of us, fearing disasters, want to simply flee to the marshes, run away from the human enemy, ourselves, and do no more than stick our heads in the muds of forgetfulness. Some of us want only to wail and complain about that enemy. Others want to forge a new foundation for our lives, the equivalent of sinking millions of trees into the mud, to use our imagination, our inventiveness, and our ability to observe our surroundings carefully to improve both our lives and the condition of nature around us.

Modern Venice epitomizes many of the dilemmas we continue to face in the twenty-first century—a variety of environmental problems in which the role of people is sometimes obvious and sometimes undetermined, and even when the problems and their sources are recognized, solutions that would work and would be politically and societally acceptable elude us. The city is slowly sinking, so it suffers more and more frequently from fall to spring from high waters, the famous *aqua alta*, that flood the grand floors of many classic buildings and otherwise beautiful open spaces, like the famous Piazza San Marco. The lagoon's waters are polluted by sewage, because since its founding Venetians have simply dumped household wastes into the canals, and the city still lacks a centralized sewage treatment center. Thus Venice past and present illustrates our confusion about our role in nature.

Beginning in the nineteenth century with the writings of John Muir, Henry David Thoreau, and George Perkins Marsh, reemphasized in the first half of the twentieth century by Paul Sears in *Deserts on the March*, Aldo Leopold in *A Sand County Almanac and Sketches Here and There*, and Fairfield Osborn in *Our Plundered Planet*, and continuing through the 1980s, the nations of Western civilization, and especially Great Britain, the United States, and Canada, expressed increasing concern about the adverse effects of our technologies on local and regional environments, at the scale of a city like Venice and its environs, which include the lagoon and the farmlands and industries nearby. By the 1960s and 1970s, this had grown into a major political and ideological movement, today's environmentalism. New sciences developed, and earlier ones, such as ecology, blossomed in ways that its pioneers could never have imagined.

With the beginning of the space age and many new techniques for scientific research, the perspective broadened again, to a view of life as a planetary phenomenon and of people as capable of affecting the environment globally. By the 1980s we became aware that we were witnessing technological effects that might radically change the relationship between people and nature on a global scale and with greater potential power than ever before. But at the same time, some believed more hopefully that we could envisage not only the destructive effects that have received so much widespread publicity but also the possibility of constructive management that could achieve long-term uses of natural resources and enhance the environment in ways both pleasing to us and necessary for the survival of life on Earth.

For most of us, some environmental issues have become frightening and frustrating. This concern has reached a level unimagined by the early-twentieth-century writers about nature, going beyond ordinary interest and capturing the world's imagination. We are told repeatedly that we are witnessing technological effects that are radically changing nature, and the relationship between people and nature, on a global scale. Some environmental problems seem to be discussed in the media again and again, but given the level of disagreement and rancor among experts, we often don't know what to believe. Something said to be true in one decade is rejected in the next. Other ideas seem to recur every decade, at first eliciting loud calls of alarm and warnings of disaster from experts, but on further scientific inquiry appearing less dire and then temporarily fading away, only to return in a new form even more alarming than the last. Some examples of environmental problems that were thought not that long ago might lead to widespread disasters, but that may be unfamiliar to today's reader, include nuclear winter and acid rain damage to forests from eastern North America to Germany.

As the scale of these issues continues to increase, from the regionalized "silent spring" of Rachel Carson more than 50 years ago to concerns about global climate warming today, something is lacking in the way we deal with them. I wrote in 1989 that solving our environmental problems requires a new perspective that goes beyond science and has to do with the way everyone perceives the world. *That is still lacking—that new perspective has not taken hold.*

But we are not empty-handed; on the contrary, we are the beneficiaries of a rich history in recent centuries, in the development of science and ideas about conservation. We walk in the footprints of pioneers in biology geology, geography, and other Earth sciences, including Charles Darwin and Alexander von Humboldt, and the conservationists I previously mentioned. I wrote in 1989 that just as the first Venetians had a history of technology, knowledge of how to cut and use wood and stone, so we have a rich history in techniques, the great and powerful technologies of computers, of satellite remote sensing, and of modern chemicals, including DNA analysis. Just as the first Venetians were not the first to walk on the Adriatic shore, so we are not the first to understand that the environment is of importance.

It is more than 40 years since the phrase "spaceship Earth" was made popular[4] and more than 40 years since the Apollo astronauts took their famous photographs of Earth from space—a blue globe enveloped by swirling white clouds against a black background, creating an image of a small island of life in an ocean of empty space. With this view of our planet, we were, in the 1970s, like the first settlers in Venice standing on the Adriatic coast and seeing only flat marshland, unsure perhaps whether that vista was friendly or hostile. They began to use their knowledge and their technology to build a great city. So can we take the opportunity to build a new approach to our environment, but we have been repeatedly thwarted in this attempt.

Computer and satellite technologies have greatly changed many of the material aspects of our lives, including how much we walk, exercise, play video games, navigate; the reliability of automobiles; many kinds of medical diagnostic techniques—in fact, most material aspects of our lives. And since the 1980s, satellite observations have become a major way that we learn about the environment. The idea that life is connected globally and that people might affect the environment at a global level has become popular, fashionable. But ironically, to our detriment and that of the environment and our fellow creatures, the perception of our place in nature and what that nature is truly like has not changed. As a result, instead of viewing Earth as a life-supporting and life-containing planet, it is portrayed to us mostly as hostile and dangerous, a vast marshland that we can only view from afar and in fear, the topic of disaster movies like *The Day After Tomorrow* and scary television programs, deepening our gloom, further graying our skies. Nonetheless, it remains my hope that humanity, with its modern technologies, will come to see Earth as that beautiful blue sphere floating in space but ever changing, brilliant in the reflected light of its sun and filled with life, including and benefiting from the presence of our species, and persisting for several billions of years because it is never in steady state.

We tend to think that our actions are limited simply by tools and information. But it is not for the lack of a measuring tape or an account sheet of nature that we are unable to deal with the environment. The potential for us to make progress with environmental issues is limited by our basic assumptions about nature, the unspoken, often unrecognized perspective from which we view our environment. Our perspective, ironically in this scientific age, depends on ancient myths and deeply buried beliefs. To gain a new view, one necessary to deal with global environmental problems, we must break free of old assumptions and myths about nature and ourselves while building on the scientific and technical advances of the past.

This is still the heart of our dilemma, as true today as it was in 1960 and in 1990. The environmental movement of the past half-century is only the most recent emergence of older concerns about the relationship between us and our surroundings. Unfortunately, although we have made great progress in what we know about life on Earth, and in our technologies to monitor and change life's environment, we are still trapped by ancient ideas that prevent us from finding cures for our environmental deficiencies. Our rich experience in conservation, science, and technology presents opportunities to take positive approaches to environmental issues. Environmentalism of the 1960s and 1970s was essentially a disapproving and in this sense negative movement, focusing on aspects of our civilization that are bad for our environment. It played an important role by awakening people's consciousness, but it didn't provide many solutions to our environmental problems, or even viable approaches to solutions. That environmentalism was based on ideas of the industrial age—the machine age—ideas that developed in the eighteenth century and expanded in the nineteenth, ideas that I will argue in the rest of this book are outmoded.

That environmentalism has been perceived as opposing technological progress, but both those arguing for progress and those arguing for protection of the environment have shared a worldview, hidden assumptions, and myths about human beings and nature that dominated the industrial era. In the large, neither science nor environmentalism has gotten to the roots of the issues, which lie deep in our ideas and assumptions about science and technology, and go even deeper in myths and ancient worldviews. Only by exposing the roots will we be able to achieve a constructive approach to our environmental problems.

How Our Perception of Nature Must Change

The changes that must take place in our perspective are twofold: We must recognize the dynamic rather than static properties of Earth and its life-support system, and we must accept a global view of life on Earth. We have tended to view nature as a digital camera's still life, much like a tourist-guide illustration of La Salute; but nature with and without people is and always has been a moving-picture show, much like the continually changing and complex patterns of the waters in the Venetian lagoon.

These are changes that still need to happen. Since the publication of *Discordant Harmonies*, a small group of ecological scientists—in particular Stuart Pimm, Klaus Rohde, Stephen Hubbell, and Donald Strong—have added their voices to the need to accept the perception of nature as always changing.[5] But in the large, this idea gets lip service at best. For example, our laws, policies, beliefs, and actions continue to be primarily based on nature as a still life. This is all the more ironic in a society immersed in movies, television, and computer games that are dynamic, and cell phones that can take moving pictures. It is as if we watch but do not see; hear but do not listen; see the dance but do not feel the rhythm. Perhaps we need to go back and play once again Simon and Garfunkel's "The Sound of Silence" and this time listen.

In the past, *nature* has meant what was in our background. Today, Earth is our village, and in the past 40 years it has become popular, even conventional wisdom, to view it as a special place, a planet revealed by modern science to be strangely suited to support life. The ability of Earth's environment to support life was written about beautifully and precisely almost a century ago by Lawrence Henderson in his book *The Fitness of the Environment*.[6] Henderson was struck by the characteristics of Earth as a planet that together make life possible. For example, Earth is large enough to hold an atmosphere, but not so large that its atmosphere is like that of Jupiter, Neptune, and Uranus, toxic to life as we know it. It is close enough to the sun to be warm enough for living things, but not so close as to boil water everywhere, as on Mercury and Venus. And it rotates fast enough to make many rotations as it revolves around the sun; therefore, all of Earth's surface gets sunlight some of the time, unlike Mercury, where one side is continually heated and the other continually cooled.

Henderson was also struck by the chemical and physical characteristics of the simple compounds so important to and within living things and so common on Earth. He wrote at length about the importance of the high latent heat of water— water's capacity to store a lot of heat energy, a capacity greater than that of any other known small compound in the universe. This modulated Earth's temperature and that of individual organisms. Unlike most materials, water expands when it freezes, so solid water floats on liquid water. This, too, has important implications for life. Henderson recounted an old experiment that showed that if a vessel with ice at the bottom was filled with water, the water above the ice could be heated and boiled without melting the ice. If this were the case in nature, ice would gradually build up on the bottom of lakes and oceans until "eventually all or much of the body of water . . . would be turned to ice."

Since liquid water is necessary to sustain living cells, support the "meteorological" cycle, and stabilize the temperature of Earth, which in turn is important to life, then "no other known substance could be substituted for water as the material out of which oceans, lakes, and rivers are formed" or "as the substance which passes through the meteorological cycle, without radical sacrifice of some of the most vital features of existing conditions."[7]

Water is the "universal solvent," able to maintain many chemicals in solution so that the complex chemical processes of life can take place. Simply analyze human urine, Henderson suggested, within which there may be 80 dissolved compounds, to see that water is a wonderfully perfect liquid to support life.

All in all, water seemed a mysteriously perfect chemical for life and fortuitously abundant on Earth. Its odd combination of qualities, each unusual among similar small compounds and each making water peculiarly suitable to support life, perplexed Henderson. His insights have become common knowledge, and since the beginning of the space age—the 1960s—something new and important has been added to them: a growing understanding of the extent to which life has influenced the environment at the planetary scale over Earth's history, and a growing recognition that our planetary life-support and life-containing system, now called the *biosphere*, is deeply complex. The biosphere is unlike the mechanical devices of our own construction, and its analysis requires the development of new scientific approaches.

Mechanical devices are still our key metaphors, but more than figuratively. They are within the assumptions of most of the computer forecasting tools that dominate our planning for nature and ourselves and are at the heart of the actions we have taken in the past 50 years. Perhaps computers, so compact and still, with few moving parts, merely reinforce the image of nature as a still life, as fixed as the computer box on your table, no matter what flashes by us on the screen. Although much of the power and abilities of modern society arises from those still little boxes, there is nothing about them as moving, as powerful, as the sight and sounds of a steam train, a jet airplane, a Saturn V rocket leaving the Kennedy Space Center launch pad, or a group of race cars at a Formula One track of curves and straightaways. Nor, for that matter, is anything about them as moving as the sight and sounds and sweat of a racehorse passing within a few feet of us rounding the last pole, or of a leaping lion, a marching herd of elephants, a diving pelican. Perhaps this is why the intrinsic dynamism that computers can deal with is not coming through to us as new metaphors. One could hope that by the second decade of the twenty-first century the elaborate and very popular Xbox, Wii, and other computer-based and computer-made-possible games, which are highly dynamic and involve chance and risk, would be changing the way the younger generation views every-thing, including the environment. But if this is happening, it hasn't reached the level of policies, laws, or major ecological analyses.

Life Is a Planetary Phenomenon

We are accustomed to thinking of life as a characteristic of individual organisms. In-dividuals are alive, but an individual cannot sustain life. Life is sustained only by a group of organisms of many species—not simply a horde or mob, but a certain kind of system composed of many individuals of different species—and their environment. Together they form a network of living and nonliving parts that can maintain the flow of energy and the cycling of chemical elements that, in turn, support life.[8] A system that can do this is not only rare but also peculiar compared to mechanical systems, peculiar from the perspective that we have become accustomed to in our methods of analyzing and constructing the physical trappings of our modern civilization—automobiles, motorboats, radios, the devices of the industrial age.

What exactly is required to sustain life? We can imagine, and a few scientists have tried to make, very simple, closed systems that sustain life: an aster in a small glass vial with water, air, a little soil harboring a few species of bacteria or fungi. Such simple systems have been made, but have generally persisted for only a short time. In the 1970s, Clair Folsome of the University of Hawaii created the then longest-enduring of such systems from the muds and waters of Honolulu Bay; life in his sealed flasks survived for more than 20 years, undergoing occasional green booms and busts while resting in the quiet of a shelf in the north-facing window of a Honolulu labo-ratory. Even more impressive is the closed ecosystem made by Professor Bassett Maguire of the University of Texas, Austin. He took samples of water and its life from a small pond on a large bedrock structure called Enchanted Rock in central Texas and sealed these in a glass flask. Within the flask were small crustaceans called ostra-cods (of the genus *Cypris*). These are short-lived but have continued to reproduce, and their descendants continue to crawl within the flask, feeding on blue-green photosynthetic bacteria, since 1982. Some of these flasks still contained a living system.[9]

In striking contrast, life and its planetary support system, the biosphere, have persisted for more than 3 billion years, in spite of, or perhaps because of, the bio-sphere's immensely greater complexity. In the biosphere, some one and a half million species have been named, and my colleagues estimate that much greater numbers are as yet unnamed—somewhere between 5 million and 30 million, depending on who makes the estimate, dispersed in tens of thousands of local systems that we call ecological communities and ecosystems.[10]

These ecosystems are of perhaps 30 major kinds, which ecologists call *biomes*—such as the tropical rain forest, the coral reef, grasslands. We find that our planetary life-support system is complex at every scale and at every time. In the aggregate, the crowds of species make up a complex patchwork of subsystems at many different stages and states of development spread across Earth's surface. How this dense complexity persists and has persisted for so long is an intriguing and unsolved question. The answer holds a key to the survival of our own species, and is also an answer to the ancient question about the very character of nature. Our planetary life-support and life-containing system, the biosphere, is, after all, "nature" in its largest sense.

Current knowledge about the biosphere is ahead of—therefore out of step with—current beliefs, forecasting methods, and policies about nature. This is one of the main impediments to progress on environmental issues; it tends to blind us to the possibilities for constructive action. Our technology places before us a new vista, but our beliefs are forcing us to look backward. This confusion leaves us mired in a barren conceptual mud and tends to lead those concerned with conservation of the environment to emphasize the benefits of doing nothing and assuming that nature will know best.

Ancient Themes of Nature

Until the Industrial Revolution, there were two major beliefs about the character of nature: nature as organic and nature as divinely created. Divinely created nature was perceived as perfectly ordered and perfectly stable; it achieved constancy, and when disturbed, it returned to that constant condition, which was desirable and good.[11]

We find the idea of the perfect order of nature in the writings of the classical Greeks and Romans, and it was probably in the minds of people at the time of the founding of Venice. This apparent orderliness of nature was well expressed in the nineteenth century by George Perkins Marsh (1801–1882), the intellectual father of American conservation, in his 1864 book *Man and Nature*:

> In countries untrodden by man, the proportions and relative positions of land and water, the atmospheric precipitation and evaporation, the thermometric mean, and the distribution of vegetable and animal life, are subject to change only from geological influences so slow in their operation that the geographical conditions may be regarded as constant and immutable.[12]

In his own life, Marsh epitomized much of the Western intellectual history of ideas about nature. His was an amazing career.[13] Born in Vermont, he became the American ambassador to Italy and to Egypt. During the time he spent in those countries, he was struck by the way thousands of years of human settlement had affected the landscape in comparison with the relatively untouched forests of his native Vermont. Later, as the fish commissioner in Vermont, he gave remarkably insightful explanations for the disappearance of fish from Vermont streams, perceiving far beyond his time the complex interactions among groups of living things. He was no fool about nature; quite the contrary—he was one of environmentalism's first geniuses. That such a careful and thorough observer and thinker about nature's character could mistake nature's fundamental quality as constancy is something we must be alert to. It is not simply stupidity that leads us to think of nature in an imaginary and false way.

Marsh's assertion about the constancy and stability of nature is a theme that runs throughout Western history and remains the dominant perspective in our time. Why? For one thing, it is an easy and comfortable viewpoint to take. Imagine the first Venetians looking out from the marshes to the Adriatic coast, which stretched as far as they could see. Although the muds shifted within the marshes, the marshes and the coast themselves must have seemed as permanent as any rock or mountain. With our modern geological knowledge, we know now that this coastline has changed over time, that its present location is the product of the history of the great Pleistocene ice ages, and that the seas along the coast of Italy are the consequence of one of those slow but definite changes that are characteristic of nature.

The classical philosopher Lucretius saw nature very differently, not as constant but as organic, always changing. This idea, too, has a long history and has its adherents, although they have been in the minority. Marsh's idea of nature undisturbed by human influence continues to be the dominant point of view in ecology textbooks and the popular environmental literature. Perhaps even more significant, this idea of nature forms the foundation of twentieth-century scientific theory about populations, ecosystems, and large-scale environmental processes like the world's climate. It is the basis of most national laws and international agreements that control the use of wild lands and wild creatures, just as it was an essential part of the 1960s and 1970s mythology about conservation, environment, and nature. Although by 2012 one sees some movement away from this view, it is still true that the predominant theories in ecology either presume or have as a necessary corollary a very strict concept of a highly structured, ordered, and regulated, steady-state ecological system.[14]

Scientific evidence has shown for at least 40 years that this view is wrong at local and regional levels, whether for the condor and the whooping crane or for the farm and the forest woodlot—that is, at the levels of populations and ecosystems. Change now appears to be intrinsic and natural at many scales of time and space in the biosphere. Nature changes over essentially all time scales, and in at least some cases these changes are necessary for the persistence of life, because life is adapted to them and depends on them.

The quality of change is illustrated by the history of climate, which can be reconstructed with considerable accuracy. With all the interest and scientific focus on climate change in the past 20 years, there have been many reconstructions of past temperatures of Earth's surface, and the methods for doing these have improved. Arguably the most accurate, or at least one of the most accurate and carefully tested, has been made from ice cores taken out of Antarctica's

Dome C glacier, as part of the "European Project" (Figs. 1.2 and 1.3). The resulting patterns of temperature change during the past 800,000 years encompass a good part of the time that our own species, *Homo sapiens,* has been on Earth and therefore the period that should concern us most in our attempts to interpret the character of nature. These data continue to show, as reconstructions in the 1980s also showed, that variation is the rule, that the temperature is always changing, never fixed, never constant.

Figures 1.2 and 1.3 show changes in Earth's average surface temperature for four time scales. The temperature patterns from the beginning of the sixth century—about the time of the founding of Venice—to the present are shown in Figure 1.2. Since that time, including the ninth century, when the Franks stormed inland Italian cities and forced the inhabitants to seek permanent settlement in the lagoons of Venice, the temperature has varied without any obvious pattern, except that most of the time average temperatures have been colder than those of the twentieth century. (Note that this Antarctic Dome C ice core gives as its most recent temperature the year 1912, not later in the twentieth or into the twenty-first century. That is because it takes a while for the ice to form from snow and to become stable enough to provide a measureable date.) There are periods of small variation and periods of large variation, but there is no constancy or any simple pattern or simple regular cycle. Although we can calculate an average temperature during the past thousand years, as one can calculate an average for any set of numbers, there has not been an "average" temperature for Earth during the past millennium in the sense of a fixed average about which the temperature varied in a regular way.

Moving back in time and taking a longer view, we see in Figure 1.3(C) the change in average temperature for the past 18,000 years, a period reaching back before the origin of civilizations, to the time of the mammoth. We can discern a trend, a gradual warming, but not constancy and not even a regular cycle of temperatures.

Figure 1.3(B) shows the temperature patterns for the past 160,000 years. From this longer perspective, the warming trend of the past 30,000 years is merely one of many fluctuations.

Finally, looking back 800,000 years in Figure 1.3(A), we see for this largest scale of time a wandering of Earth's surface temperature—up and down, mostly colder than the present climate, but with periods of little variation and periods of great variation, times with apparent cycles and times without cycles. We see a particularly warm period, warmer than that of the most recent in the ice core, that occurred approximately between 150,000 and 125,000 years ago. These graphs illustrate that change appears intrinsic and natural at many scales of time and space in the biosphere, and that nature changes during all time scales.

The idea that change is natural and the failure to accept it have created problems in natural-resource management and have led to destructive, undesirable results. But how do you manage something that is always changing? There are several answers. The simplest and perhaps most helpful is to begin with an understanding

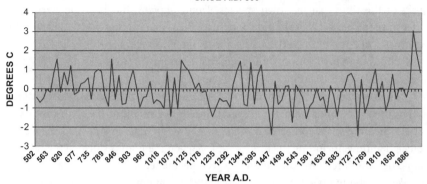

TEMPERATURE DIFFERENCE FROM 20TH CENTURY STANDARD SINCE A.D. 500

YEAR A.D.

FIGURE 1.2 Change in temperature since the sixth century, the time of the founding of Venice, Italy, as determined from ice cores taken from the Antarctic glacier known as Dome C, as part of the "European Project" published in 2007. This may or may not be representative of the entire Earth surface temperature, but it is as close as we can come at this time. (J. Jouzel et al. [2007]. Orbital and Millennial Antarctic Climate Variability over the Past 800,000 Years. *Science, 317,* 793–796. The abstract to this paper states that "high-resolution deuterium profile is now available along the entire European Project for Ice Coring in Antarctica Dome C ice core, extending this climate record back to marine isotope stage 20.2, ~800,000 years ago. Experiments performed with an atmospheric general circulation model including water isotopes support its temperature interpretation.")

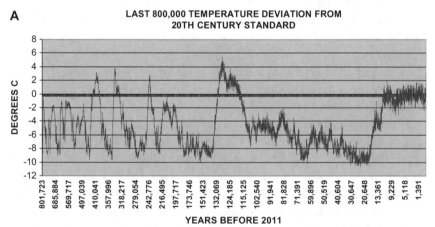

A

LAST 800,000 TEMPERATURE DEVIATION FROM 20TH CENTURY STANDARD

YEARS BEFORE 2011

FIGURE 1.3 Change in temperature over different periods of time during the past 800,000 years, as determined from ice cores taken from Antarctic glacier. (Same source as Figure 1.2.) Climatologists studying climate change prefer in general to look at the difference between temperatures at one time compared to another, rather than the actual temperature, for a variety of technical reasons

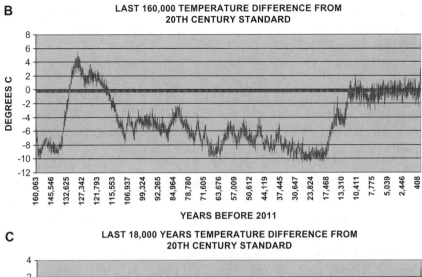

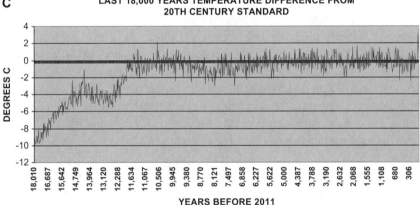

FIGURE 1.3 *(continued)*

of how nature really works and of our proper role in nature: people living within nature, neither poisoning it nor destroying its reproductive capabilities.

Of even more concern to many people is the possibility that by admitting to some kinds of change, we may have opened a Pandora's box of problems for environmentalists. Basically, once we acknowledge that some kinds of change are good, how can we argue against any alteration of the environment? The answer is that to accept certain kinds of change is not to accept all kinds of change. Moreover, we must focus on the rates at which changes occur, understanding that certain rates of change are natural, desirable, and acceptable, while others are not. As long as we refuse to admit that any change is natural, we cannot make this distinction and deal with its implications.

Ideas of Nature in Transition

Before the seventeenth century, an image of nature as divinely ordered seems to have dominated explanations about how nature worked, but the only metaphors that people had to describe Earth were organic, derived from animals and plants, from the phenomena of life with which people were familiar, including human life. Earth was described as a kind of fellow creature, and these metaphors were apparently taken quite literally.

Beginning in the seventeenth century, with the rise of Newtonian mechanics and the work of Galileo, Johannes Kepler, and other scientists, powerful new ideas developed. Along with the invention of machines such as the steam engine, they led to new metaphors, to the idea that Earth and the solar system operate like clockwork, like a machine.[15] Hence, from the human perspective, in the nineteenth century there was what Marjorie Hope Nicolson (1894–1981), an expert on literature and nature, has called the "death of the Earth"—the death of the idea of an organic Earth.[16] Although little discussed today, this change had a profound impact at the time. (Nicolson wrote some important and influential books about literature and nature, including my favorite, *Mountain Gloom and Mountain Glory: The Development of the Aesthetics of the Infinite*.)[17]

Today we analyze ecological systems looking backward, as though they were nineteenth-century machines, full of gears and wheels, for which our managerial goal, like that of any traditional engineer, is steady-state operation. To us, the mechanical view of Earth, nature as a machine, seems an old and permanent one. But it is not. The mechanical Earth is a seventeenth-century idea that developed in the eighteenth, blossomed in the nineteenth, and carried into the twentieth century. This concept of nature and of Earth would not have been familiar to the first Venetians. They would have known about and been comfortable with other, much more ancient ideas about nature, the idea either of an organic Earth or of nature as divinely created order.

These three images of nature—the machine, the creature, and the divine— dominate our thoughts about the environment, although we are not usually aware of them. It is important for us to examine these three images both as explanations and as metaphor. As explanations of how nature works, the divinely ordered image and the mechanical image have much in common. Both lead to the idea of nature as constant unless unwisely disturbed, and as stable, capable of returning to its constant state if disturbed. Both lead to the conviction that undisturbed nature, or perhaps nature with human beings playing their "natural" but fixed roles, is good, while a changing nature is bad. In contrast, the organic image as an explanation of how nature works focuses on change and on processes, with change seen as inevitable, to which, like it or not, human beings must yield.

To summarize since the middle of the twentieth century, we have been living through a time of change: a transition in our civilization from the mechanical age, that of gears and wheels, to a new era that appears to us as the space and computer age. We engineer genetics—even inventing digital machines to analyze human DNA quickly. We make robots that perform more and more tasks. We develop artificial limbs for people who have lost one of their own. We talk about the use of computers for "artificial intelligence." We are moving away from the mechanical view to a different set of perceptions in which the distinction between organic and inorganic is no longer very clear.[18] While this transition is taking place in our recreation rooms, in medical operating rooms, and in track and field games where people who have lost their legs are running on artificial ones, we have not settled on the right metaphors, images, and symbols that fit this transition in our perception of nature.

We interact with nature in two ways, rationally and non-intellectually—an inner, personal response. Both are important. We get ourselves into trouble when we confuse the two, letting inner personal determine what we tell ourselves are rational decisions and actions, and believing that rationality can substitute for the emotional. I mean "non-intellectual" and "inner personal" in the largest sense of all that is outside of rationality—our folkways, our myths, our spiritual feelings that arise from deep within us, our religious sensitivities. In the rise of environmentalism, the science of ecology and other environmental sciences, we have often gotten the two aspects of our interaction with nature reversed. When we do so, it is to our and our environment's detriment. The purpose of this book is to help clarify the dilemma and suggest a path toward its resolution.

A TRIAL OF INSECTS AND A WAY OF THINKING

When I first began work in ecology in the 1960s, the dominant belief among ecologists, foresters, and fisheries scientists was that nature, left alone, would achieve a constancy in form, structure, and dynamics, and that only human action forced nature away from this balance. But much has changed since then, so we need to look at nature from a new perspective.

To understand where we have come in our beliefs and thinking about nature, let's begin with what French philosopher Luc Ferry wrote in his amazing book *The New Ecological Order*. He tells stories about villages in medieval France that suffered from a plague of insects eating their crops.[19] In 1745 the farm fields around Saint-Julien were so infested that the farmers went to the town judge and asked that the insects be brought to trial for their deeds. The judge agreed on condition that the insects be represented by a lawyer. In the end, the insects won, on the argument that they were only doing God's work and were sent there to punish the villagers for their sins

From the thirteenth through the eighteenth century, before modern science and rationality dominated our way of thinking, such trials were common in Europe and included trials against rats, mice, leeches, and, in Marseilles, against dolphins. Sometimes the animals would win on the grounds that they were simply punishing people for their sins. The common pattern was that the judge would appoint a lawyer to represent and defend the accused creatures, and "a sergeant or court clerk" was "charged with loudly and clearly reading them [the accused creatures] the summons to appear, in person, at an appointed day and hour, before the court."[20] Sometimes it was decided that certain fields and forests would be set aside for the creatures to feed, and they would be legally excluded from others. They were told and expected to obey or pay the consequences.

These trials seem crazy to us because, Luc Ferry argues, we are products of the age of reason, of rationality, a belief system that seems entirely normal to us. But rationality is actually unusual, perhaps unique, in the history of civilizations. Throughout human history, more people have thought about nature and the environment the way the medieval villagers did than the way a nineteenth- or twentieth-century scientist did.

Ferry argues further that the age of reason separated people from nature, so that it seemed impossible that nonverbal, nonrational creatures could participate in civilized human activities, or that we could participate in theirs. Rationality has certainly helped in the advancement of science, but the separation of people from nature has had unfortunate consequences for our understanding of the environment.

THE DEATH OF RATIONALITY

Intense debates in the past 50 years about how to relate to nature suggest that the age of reason may be coming to an end. Underlying current debates about environmental issues are basic questions such as: How real is the concept of a balance of nature? What is the connection between people and nature? What are our roles in and obligations to nature—and, for those who see nature and its nonhuman creatures as able to make decisions, what are its to us?

The end of rationalism seems to offer two possible futures: a resurrection of the anti-rationality illustrated by the medieval trials of insects, or an integration of a kind of nature–humanity understanding within and as part of an extended rationality. One would hope for the latter. We do try to apply science to environment, but recent environmental debates seem to move us backward, to a reliance on nonrational, ideological beliefs instead of rationally derived facts in harmony with modern understanding of the environment.

In trying to apply rational analysis in the development of the modern science of ecology for almost half a century, I have repeatedly found such efforts overwhelmed by political, ideological, and wishful thinking that derives from and justifies itself on a thread of beliefs that traces back to the ancient Greeks and Romans.

The medieval insect trials stemmed from a way of thinking about environment that seems alien and impossible to us, but in fact similar kinds of prerationalist ideas still exist in our society, with belief coming first and dominating scientific findings. One of the most striking examples of this is the story of the crop circles in England. For thirteen years, circular patterns appeared mysteriously in grain fields in southern England. Proposed explanations included aliens, electromagnetic forces, whirlwinds, and pranksters. The mystery generated a journal and a research organization headed by a scientist, as well as a number of books, magazines, and clubs devoted solely to crop circles. Scientists from Great Britain and Japan brought in scientific equipment to study the strange patterns. Then, in September 1991, two men confessed to having created the circles by entering the fields along paths made by tractors (to disguise their footprints) and dragging planks through the fields. When they made their confession, they demonstrated their technique to reporters and some crop-circle experts.

In spite of their confession, some people continue to believe that the crop circles have some alternative causes. Crop-circle organizations still exist and now have Web sites. For example, a report published on the World Wide Web in 2003 stated that "strange orange lightning" was seen one evening and that crop circles appeared the next day.

How is it that so many people, including some scientists, still take those crop circles seen in England seriously? This is just one example among many demonstrating that some people want to believe in mysterious causes, perhaps finding them more exciting than scientific explanations. When this kind of thinking is limited to a fringe element or focused on a relatively unimportant issue, it is one thing. But when, as I have found, some degree of irrational thinking is more widespread, even among scientists, in connection with vital issues, such as climate change, it becomes an impediment to constructive action.

Thinking over how we are persuaded even now that certain things about nature and the environment must be true despite or without scientific evidence, I have come to realize that these older ways of thinking and of seeking understanding are still with us. As David Fischer explains in his landmark book *Albion's Seed: Four British Folkways in America*, societies develop from their folkways, which he defines as "the normative structure of values, customs, and meanings that exist in any culture." Included in his list of folkways are speech ways, building ways, family ways, learning ways (how children are taught), religious ways, and magic ways (which include normative beliefs and practices concerning the supernatural). To this I add knowledge ways—how a society perceives what knowledge is, how it is learned, and how it is used. These folkways are persistent and dominate thinking and action today, just as they have over many generations.

This, then, is the heart of the matter that confronts us. Like it or not, computers, along with modern science, have changed our view of life, and observations from space have changed our perceptions of our planet. We can no longer rely on nineteenth-century models of analysis for twenty-first-century problems. More than any other factor, confronting and recognizing these changes in our deep-seated assumptions are the major challenge that faces us in interpreting nature and in dealing with environmental issues. This was true in the last decades of the twentieth century and continues to be true today; to my surprise, much less has changed in twenty years than I expected.

As I suggested earlier, Venice of today epitomizes our modern dilemma. It is that combination of nature and human influences that I have been talking about— a city within what many see today as a beautiful lagoon with wildlife and vegetation lovely to travel through, as I have done. The city has been sinking—a foot in the past century. A study done by scientists at the IBM Thomas J. Watson Research Center in the 1970s showed that the primary cause was removal of groundwater under the city. That water filled in spaces between the sandy and muddy particles of the lagoon's basin, and as it was removed, the particles squeezed closer together because of the weight above. But the removal of groundwater was not due primarily to freshwater use by the Venetians; it was the use of water in agriculture and cities on the mainland that was sinking the city. In addition, the surface of our planet is still recovering from the last ice age; the sea level has been rising as glaciers melted since the peak of the ice extent about 10,000 years ago, and water expands as it warms.

The result is the famous *aqua alta*, flooding that occurs from the fall through the spring in Venice. The worst and most infamous happened in 1966, when winter high winds and storms at sea pushed Adriatic water into the lagoon. Waters in the city rose almost six feet (195 cm), flooding the ground floors of 16,000 houses and doing billions of dollars' worth of damage. Since then, *aqua alta* has become well known to tourists. *Aqua alta* flooding that has risen more than four feet in the city occurred in December 2008 and 2009. Arriving in the city during one of these events, visitors walk on temporary raised wooden platforms so they can visit St. Mark's Square and see La Salute and the city's many other elegant, elaborate, and highly decorated buildings. These floodwaters are not only high but also polluted. Unknown to most of the 50,000 tourists on a summer day, most of the sewage from the elegant Venetian buildings simply flows into the lagoon, as it has since the city was founded. The tides that move waters in and out of the lagoon twice a day take away much of the sewage into the Adriatic, and it is the tidal action that has kept Venice livable throughout the centuries.

Venice has therefore depended on environmental change on a twice-daily basis, and since tides are a global phenomenon, the city is connected to and dependent on dynamics of the Earth–moon system. The city's sinking results from events at a slower pace. Venetians have to find ways to survive environmental changes at a number of different levels of time and space, but try as the Venetians have been doing, doing has not always been done. Back in the 1980s when I was writing *Discordant Harmonies*, there had already been decades of discussions, plans, and government funds spent on designing some kind of gates that could be raised when high winds and storms were pushing Adriatic waters into the lagoon and causing *aqua alta*. Despite plan after plan, such a barrier has not yet been built. The current plan is called MOSES, in part to bring up the Biblical story of the parting of the Red Sea and the hope for a Venetian equivalent, and also because it is an abbreviation for *Modulo Sperimentale Elettromeccanico*. The MOSES project consists of 78 giant steel gates across three inlets that connect Venice to the waters of the Adriatic. When high waters begin to flow into the city, the gates are raised (air is pumped into them so that the upper part flows up) and in theory these will stop the *aqua alta*.

Permanent gates preventing an exchange of waters between the lagoon and the Adriatic are not possible, because of the sewage dumped directly into the lagoon. The entire lagoon would become a sticky and disease-promoting mess without that exchange. (The lagoon isn't a place you would want to swim. During the filming of the movie *Summertime* in the 1950s, actress Katherine Hepburn had to fall into the lagoon, and did so, contracting an eye infection that is said to have become a long-term problem for her.)

As the British newspaper *The Guardian* noted in 2006, the population of the city dropped from 121,000 in 1966 to 62,000, and if these trends continue, Venice might become well Italy's Disneyland.[21] That possibility was obvious to those of us working in Venice in the early 1980s, when the population had dropped well below 100,000. Many of the grand buildings that line Venice's canals are empty. Many of the people who serve tourists—restaurant waiters, hotel personnel— have their homes somewhere on the Italian mainland and either commute daily or share crowded rental apartments with others involved in the same work. They can't afford to live in Venice, and environmental change, both human-induced and beyond human influence, makes the city less and less desirable as a place for working people to live. This grand, elaborate, highly decorated, and famous city, this great tourist attraction, may have few residents in the future. And in a democratic society with Italy's political volatility, that very small number of citizens may not carry enough political weight to ensure the completion of movable gates to protect the city from *aqua alta* or the construction of a system to transport sewage to central treatment facilities.

With the rapid turnovers in the Italian government, it is never quite clear whether the next administration will continue to fund the MOSES project. Right now, work continues and it is said that the gates will be working by 2014. Whether politics and economics allow this and whether the gates work as they are supposed to, both in stopping *aqua alta* and allowing flushing of the lagoon's polluted waters, remain an unknown. The once great city-state, a center of commerce that impressed visitors with gold-plated buildings and elaborate ceremonies to compensate for its lack of land area, that paid its own way, today cannot afford to pay for the MOSES project and has to depend on the Italian government and the wishes of all the people of Italy. The interplay between Venetians and their lagoon-nature is made more complicated by the interaction of cultural history, tourism, and national economies, the last in turn affected by worldwide economic concerns. Perhaps only tourism will provide the driving force to sustain the city, thus truly turning it into an empty playland façade, no longer a vibrant international center for trade, commerce, and the arts.

We can look to ancient and modern Venice as models in miniature of what we need to face up to if we are going to find ways to solve environmental problems elsewhere. It therefore epitomizes our current dilemma: on the one hand the interaction between people and the rest of nature, which includes the interaction of changes at different rates and at different spatial scales, all happening at once and having to be dealt with together—those discordant harmonies of nature; on the other hand, the difficulty of human processes for solving Venice's problems given insufficient scientific and engineering knowledge, the variabilities of political will and just plain politics, and of economic conditions. Even if we recognize and accept that change is natural, Venice warns us of the difficulty of dealing with it if we want to sustain in one place a physical structure built by people and of continuing value to people, exactly where they have been and where they are now.

We may lack the political and economic will to solve our environmental problems. It will require that we first recognize, confront, and change our fundamental assumptions about the character of nature and the relationships between ourselves and nature. A more comfortable relationship with nature will enable us to proceed much more rapidly to develop a constructive approach to solving environmental problems. In this way, we can move on from the view of George Perkins Marsh, well aware of his warnings about the negative effects that our civilization can have on our surroundings, but also ready to look ahead. We can build a great civilization in which our role in the environment is a positive one, managing sustainable natural resources and enhancing the quality of our environment.

As I wrote in the Introduction, to do this, we have to embark on a journey through the ways that we have perceived and still perceive nature, then consider what paths are open to us to achieve not only a new perception of nature but a profoundly meaningful relationship between ourselves and nature, one that integrates our modern scientific knowledge and way of thinking with our humanity. The chapters that follow expand on the ideas introduced here. Let us hope that although the future of our global environment may sometimes seem as gloomy as an unending sea marsh on a gray day, we are really viewing the foundation of a new development, an advance in understanding our surroundings that leads to an advance in our civilization.

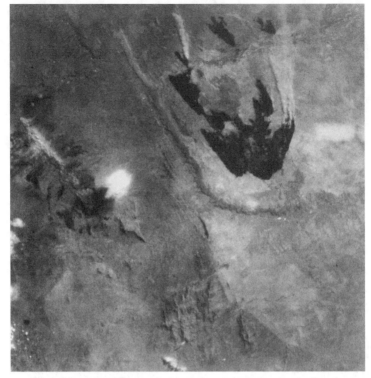

FIGURE 2.1 Tsavo (East) seen from space in false infrared, the darker the shading, the greater the amount of vegetation. The obtuse angled line separating the darker gray to the left and lighter gray to the right is the border of the park, the result of drought and the actions of people, and elephants. (NASA Landsat Photograph.)

FIGURE 2.2 The border of Tsavo National Park (East), photographed in 1977, shows the effect of drought and elephant die-off of the late 1960s. (Photograph by the author.)

2

Why the Elephants Died

BREAKDOWN IN THE MANAGEMENT OF LIVING RESOURCES

Animal populations must exist in a state of balance for they are otherwise inexplicable.

—ECOLOGIST A. J. NICOLSON (1933)[1]

The balance of nature does not exist, and perhaps has never existed. The numbers of wild animals are constantly varying to a greater or lesser extent, and the variations are usually irregular in period and always irregular in amplitude. Each variation in the numbers of one species causes direct and indirect repercussions on the numbers of the others, and since many of the latter are themselves independently varying in numbers, the resultant confusion is remarkable.

—ECOLOGIST CHARLES ELTON (1930)[2]

Views of Tsavo East

The Landsat satellite image (Figure 2.1), taken in the late 1970s from 140 miles above the Earth, over East Africa, shows a curious geometric feature: two straight lines, each stretching 50 miles or more, one north and the other south-southeast, meeting at an obtuse angle, as though a planetary engineer had sketched them with a triangle and pencil at a drafting table. East of the lines, light gray indicates an area so thinly vegetated that the almost bare soil dominates the amounts and wavelengths of light reflected from the surface. To the west, dark grays indicate dense vegetation in infrared wavelengths invisible to the naked human eye.

About the same time that the satellite orbited silently far overhead, I flew a few hundred feet above the ground in a single-engine aircraft, a Cessna 182, from whose noisy cockpit one of the boundaries was clearly visible (Figure 2.2). What appears as one line in the satellite view was revealed from this nearer vantage point as four: a railroad, a highway, and two firebreaks stretching into the blue haze of the horizon toward Nairobi. East of the line lay one of two parts of Tsavo, one of Kenya's largest national parks, covering approximately 5,000 square miles. It was the parkland that appeared brown from the satellite and could be seen now as desert-like dusty soil with a thin scattering of live and dead shrubs and trees, among which almost no game was visible. The land outside the park, to the west, which is red in the original Landsat image (Landsat's technology made green plants appear red), contained dense thickets of dark-green trees and shrubs. The scene was strange from this elevation, just as it was from the satellite, appearing as a photographic negative of one's expectation of a park. Rather than being an island of green in a wasted landscape, Tsavo appeared as a wasted island amid a green land.

The character of Tsavo was the result of 30 years of interplay between people and nature; of vagaries of climate, including a major drought that persisted through 1969 and 1970; and of a controversy over management that involved issues as old as Western civilization—the character of nature undisturbed by human influence, and the proper role of human beings in nature.[3] Perhaps nowhere else was the impact of well-meaning management of wild nature so visible from a planetary perspective. Something had failed at Tsavo, despite the best intentions, and the failed management of living resources was an example of a breakdown not merely in management but also in myths, beliefs, and fundamental paradigms that modern technological civilization held about nature.

The Elephants and Lessons of Tsavo

Tsavo became a park in 1948. Its first warden, David Sheldrick (1919–1977), looked at its dry, flat landscape, thickly vegetated by *Commiphora* trees and wild sisal, and knew that the park could attract tourists only to see wild game, not other scenic beauty. But much of the big game had been shot around the turn of the century by European settlers, and the remaining wildlife, mainly elephants and black rhinoceros, were under intense pressure from poachers.

Sheldrick devoted years to building roads, providing year-round water for the wildlife, and eradicating poaching by catching and driving off poachers through a kind of anti-guerrilla effort using Land Rovers, aircraft, and World War II repeating rifles. A thousand miles of roads were built to increase tourist access. The Galena River was dammed, and artesian wells were dug. These measures resulted in a rapid buildup of elephants, whose numbers climbed to 36,000. They took their toll on the vegetation, knocking down and killing trees and shrubs. The land seemed on its way to becoming desert. By 1959, areas where the vegetation had once been so thick that elephants were visible only if they actually crossed a road in front of tourists began to resemble a "lunar landscape," Sheldrick's wife, Daphne, wrote years later.[4]

Sheldrick became concerned. He wrote in a report that "during the past few years, the destruction of vegetation by elephant has reached serious proportions. If present trends continue, it is doubtful if the Park can continue to support the existing population much longer. What effect this will have on other species remains to be seen, but I think it is important that we should seek scientific advice regarding this problem as soon as possible."[5]

As with so many of these cases, the effects of human actions and the vagaries of the natural environment, including recurrent droughts, were compounded, and it was difficult to decide what to do. By 1966, most people believed that the park could be saved only by removing many elephants, and for a while Sheldrick agreed.[6]

But the rains came again, and the park seemed to recover; instead of the thicket of trees and shrubs, grasses sprouted and seemed to promise better food for the wildlife. The elephant population continued to grow, and the resource-management controversy worsened. The Ford Foundation agreed to sponsor a scientific study, and Richard Laws (1926–), one of the world's foremost experts on elephants and other large mammals, who had been studying elephants in Uganda, agreed to head the project. He and other scientists and some conservationists soon concluded that about 3,000 elephants should be shot to keep the population within its food supply and protect the game from the dangerous effects of its intrinsic capacity for growth.

Sheldrick at first favored such action, but in the end reversed himself and returned to the very old idea that nature can take care of itself and that human interference is undesirable, even though he had already interfered by building roads, damming the major river, and digging wells. He decided that "the conservation policy for Tsavo should be directed towards the attainment of a natural ecological climax, and that our participation towards this aim should be restricted to such measures as the control of fires, poaching, and other forms of human interference."[7]

At that time, the phrase *natural ecological climax* was taken to mean nature in a mature condition, the result of a long series of stages that occurs after a catastrophic clearing of the landscape and, once attained, persists indefinitely without change. Many, like Sheldrick, accepted the "climax" condition as the truly natural and most desirable state of wilderness. When the trustees of the park sided with Sheldrick and concluded that more studies, especially of the vegetation, were needed, Laws resigned and returned to England.

Sheldrick had come to believe that the die-offs of elephants during the droughts could be regarded as a natural culling, bringing the number of elephants "in line with the carrying capacity of their particular dry weather feeding grounds," producing a selective death of the weaker, ensuring a healthier population, and allowing the regeneration of vegetation.[8] In short, he believed that nature used the droughts to maintain or restore its proper balance. But the drought of 1969 and 1970 was much worse than previous ones of the century, and as an estimated 6,000 elephants starved to death, they destroyed the vegetation, producing the scene still visible a decade later from the air and from space. Elephants and human beings together had drafted the lines on the Landsat image.

Here was a controversy very different from the better-publicized environmental issues of the 1960s. Back then, as environmentalism was achieving worldwide recognition, most environmental issues seemed to produce two extreme camps that disagreed on goals: The environmentalists argued that the salvation of civilization and the human spirit lay in the preservation of nature, and their opponents responded that environmentalism threatened industry, economics, progress, and perhaps civilization. The disagreement at Tsavo, on the other hand, was among conservationists who shared basic goals and a fundamental love of wild nature. They wished to conserve in perpetuity fine examples of wild nature for its own sake and for people to view, but they disagreed on methods.

Daphne Sheldrick defined her husband's position strongly in her book *The Tsavo Story*. "Hasn't man always had a regrettable tendency to manipulate the natural order of things to suit himself?" she wrote.

> With amazing arrogance we presume omniscience and an understanding of the complexities of Nature, and with amazing impertinence we firmly believe that we can better it. . . . We have forgotten that we, ourselves, are just a part of nature, an animal which seems to have taken the wrong turning, bent on total destruction.[9]

The story of Tsavo illustrates the issues I have been discussing. There were two dominant opinions about what had happened at Tsavo. One was that undisturbed nature always achieves a balance, a constancy, a stability—the "natural ecological climax" that Daphne Sheldrick referred to in her book—and that people only interfere with and destroy that balance; therefore, the proper role for people is hands off. The other opinion, which I will call the Janus hypothesis, was that nature varies greatly and that human actions are required to create a balance. Janus is the god of the classical world associated with change, history, and the ability to look back and forward in time—hence his typical representation as two faces, one looking forward and one looking back. In calling the second approach "the Janus hypothesis," I mean the hypothesis that nature never has been in a true steady state, but is a non-steady-state, dynamic system. In regard to Tsavo, as with most other environmental issues of our century, the first opinion won out, and the devastation of Tsavo, at least in part, was a product of this policy.

There is a third possibility: that even Tsavo, large as it is, is too small to sustain an elephant population "naturally," and that before the European colonization of Africa and the establishment of modern African nations, the elephants, when subjected to one of the recurrent droughts, would have migrated from Tsavo to another part of Africa. This possibility accepts change as intrinsic at a scale as large as Tsavo. It is not consistent with the old idea of a balance of nature, which should take place locally as well as over vast distances, but is consistent with the new perspective that I am proposing and have been for more than 20 years. Even so, one who insisted that a balance of nature must exist somewhere, somehow, could still argue that at the scale of the entire African continent, the elephant population must have achieved a constancy over time before the imposition of a political geography over an ecological geography.

Forty years after that devastating Tsavo drought, elephants in Africa remained in great danger. The Tsavo population never did recover, in the sense of returning to its former maximum abundance of 36,000, Today, without Sheldrick's strong hand, illegal poaching for their tusks once again has become a major cause of elephant deaths, along with natural causes of mortality. By the 1980s their population in the park had been reduced to about 6,000.[10] By 2008, Tsavo's elephant population had climbed back to just over 12,000.

Clearly, direct control of poaching, as Sheldrick did so well, is necessary if elephants are to survive. But the lesson of Tsavo is that successful management cannot stop simply with the eradication of poaching. Successful conservation and management in the future might require harvesting excess elephant populations when or if they occur again in parks and preserves, but would also control killings that impose a rate of decline that exceeds the reproductive capacities of elephant populations. The rate of change is the problem, not the harvesting *per se*. In the last analysis, the survival of elephants will depend on our perception of them within the context of their total, variable environment.

Failing Fisheries

Tsavo was just one of many examples of unsuccessful attempts to manage the environment in the twentieth century. Around the time of the great elephant die-off at Tsavo, it was becoming clear that the management of living resources was in trouble in many parts of the world. For instance, far from Tsavo but linked by a common set of ideas about nature, management of another kind of wildlife, the major commercial fisheries of the world, seemed to be foundering. The allowed catch of these fisheries was determined by international agreements, and various mechanisms were tried to enforce the agreements. But even so, major fisheries failed. A classic case was the Peruvian anchovy fishery, once the world's largest commercial fishery. In 1970, 8 million tons of anchovies were caught, but two years later the catch had dropped by 75%, to 2 million tons. The fishery continued to decline, and although in 2010 some analysts said that a recovery was finally beginning, it has not yet rebounded to its pre-1970 abundance. Yet this fishery was intentionally and actively managed with the goal of achieving a "maximum sustainable yield."

The story was repeated elsewhere. In the 1950s, Pacific sardines, once a major species off the California coast, suffered a catastrophic decline that continued through the 1970s. The Atlantic menhaden catch, which reached a peak of 785,000 tons in 1956, dropped to 178,000 tons by 1969. Atlantic herring and Norwegian cod showed the same kind of decline. The North Atlantic haddock catch, which had averaged 50,000 tons for many years, increased to 155,000 in 1965, then crashed, declining to a mere 12,000 tons, less than 10% of the peak, by the early 1970s. The International Commission for the Northwest Atlantic Fisheries established a quota of 6,000 tons. Apparently, haddock had been able to sustain a 50,000-ton catch; but when the catch tripled, the population decreased so greatly that only a much smaller catch could be sustained.[11] The decline in Tsavo's elephants was curiously echoed by the decline in fish harvests and fish populations in faraway oceans.

The problems persist and if anything are much worse as I write this in 2012. Species that have declined greatly since 1990 include codfish, flatfishes, tuna, swordfish, sharks, skates, and rays. The North Atlantic, whose Georges Bank and Grand Banks have for centuries provided some of the world's largest fish harvests, is suffering. The Atlantic codfish catch was 3.7 million metric tonnes (MT) in 1957, peaked at 7.1 million MT in 1974, declined to 4.3 million MT in 2000, and climbed slightly to 4.7 in 2001.[12] European scientists called for a total ban on cod fishing in the North Atlantic, and the European Union came close to accepting this, stopping just short of a total ban and instead establishing a 65% cut in the allowed catch for North Sea cod for 2004 and 2005. Problems continue. In July 2011, the European Union proposed a major overhaul of fisheries policy "because scientific models suggested that only eight of 136 fish stocks in European Union waters would be at sustainable levels in 2022 if no action were taken."[13] Curiously, since the scientific models are a major part of the problem, as I will explain later, it is odd from the perspective I suggest to continue to use those methods as the primary basis for setting policy.

One widely used fishing method, long-line fishing, provides good data and shows that fish populations decreased quickly—about an 80% decline in 15 years. This continues in spite of worldwide attempts to manage the catch.

It is one thing for a population of African elephants to decline, of interest to conservationists and to those who trade (illegally) in elephant ivory; it is another when the decline has a major effect on world food supplies. Fish provide about 16% of the world's protein and are especially important protein sources in developing countries. Fish provide 6.6% of food in North America (where people are less interested in fish than are people in most other areas), 8% in Latin America, 9.7% in Western Europe, 21% in Africa, 22% in central Asia, and 28% in the Far East.[14] Along with timber, fish have been actively managed since the beginning of the twentieth century. How could this management continue to fail to achieve its goals? For one thing, it's difficult to control fishing in international waters, and violation of international fisheries treaties is not uncommon. For another, fishermen have to be motivated to follow the limits. This requires that they believe expert estimates and analyses are true, and that ways of regulating fishermen's activities work.

These requirements take a manager beyond physical and biological sciences into the field that has become known as environmental economics. These experts use what they call "policy instruments," meaning social, political, and economic mechanisms whose purpose is to create a situation in which, when a harvester acts in his own best interest, he is also acting in a way that will lead to the desired total harvest.

Fishermen are well known to be skeptical of those without direct fishing experience, so the task isn't simple. But most fisheries experts agree that the management itself is a major part of the problem. And the management practitioners are well aware of the continual failures. The answer, as with Tsavo's elephants, lies with the fundamental assumptions and scientific theory used in this management.

From Pencil-and-Paper Elegance to the Turbulent Real World

Throughout the twentieth century and persisting today, fisheries management has set allowable catches—those that were believed to be sustainable—from a well-known equation developed in the study of populations, the S-shaped logistic growth curve. First proposed in 1838 by Pierre-François Verhulst,[15] a Belgian scientist, the curve is a description of how a population grows from a small number to its final limiting and sustaining abundance, known as the *carrying capacity*. The major original contributions to the application of these mathematical models to populations occurred in the first decades of the twentieth century and were carried out by Alfred Lotka in the United States, Vladimir Vernadsky in Russia, and Vito Volterra in Italy, with some elegant experimentation conducted in the same period by the Russian Georgii Frantsevich Gause. The equations were based on physics and chemistry, not on direct biological knowledge, and the derivation of their ideas from physics and chemistry was made quite explicit by these scientists.

In the twentieth century, several laboratory experiments demonstrated that populations of bacteria and of certain insects grow according to Verhulst's equation if they are maintained in laboratory vessels under constant environmental conditions and provided with a constant supply of food. The results of one such experiment are shown in Figure 2.3. The mathematics that generates this curve includes the idea that the population is stable in a classic sense: It will achieve an abundance that will remain constant forever unless disturbed; and if disturbed (increased above or decreased below its carrying capacity), the population will return to the same abundance.

Alfred Lotka, one of the first of what we could call mathematical ecologists, explained the logistic growth curve this way (Fig. 2.3). Imagine a container, something like an aquarium but with screening rather than glass for walls. Into this container a scientist first puts a small number of fruit flies and then, each day, puts in exactly the same amount of food—let's say, a single banana. At first, when there are few flies, there is much more food per fly than any can eat. The flies do well and reproduce rapidly. With an increasing population, the flies begin to compete directly with one another for the food. At the least, they would interfere with each other in just trying to reach the banana, but eventually the competition might become more aggressive. The amount of food per fly decreases as the fly population grows, and the death rate climbs as competition increases.

Eventually, there is just enough food to sustain each fly and the death rate has risen to the point where it exactly equals the birth rate (which has meanwhile declined; the flies are too busy seeking food to have time for sex and are too unhealthy as well). At this point the population is constant as long as the food is constant. If the number of flies were to grow past this point, there would be less food per fly than required for survival; deaths would exceed births, and the population would decline to exactly the number that allowed just enough food per fly. This, Lotka explained, was the logistic carrying capacity. It was an exact and beautiful (if one can use that word for a container full of flies) description of the balance of nature.[16] So I'm not discussing a minor point here, but a central one in ecology and its application to the world food supply.

This hypothetical population regulates itself, not consciously but because of its assumed inherent characteristics. Ecologists refer to this as a population with "density-dependent" population regulation. For example, a paper published in 2010 discussed whether African elephant populations have density-dependent regulation and, if so, what physiological or behavioral mechanisms bring it about. The authors of this paper wrote that "density-dependent regulation is an accepted population phenomenon."[17]

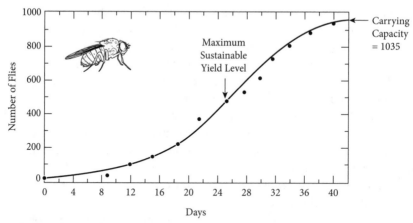

FIGURE 2.3 The growth of a population of fruit flies in a laboratory container. When kept in a constant environment (constant temperature, humidity, and so forth) and provided with a constant supply of food, a population of fruit flies increased following the classic S-shaped curve shown here, called the "logistic" and first proposed in 1838 by the Belgian scientist Pierre-François Verhulst. (Revised from Figure 4.6b, D. B. Botkin and E. A. Keller, *Environmental Studies: The Earth as a Living Planet*. Original data from R. Pearl [1932]. The Influence of Density of Population on Egg Production in *Drosophila melanogaster. Journal of Experimental Zoology, 63*, 57–84. Originally revised and reprinted by permission of Yale University Press from *An Introduction to Population Ecology* by G. E. Hutchinson [1978], Fig. 12, p. 25.)

A lot of ecological literature is about whether real populations do have this kind of self-regulation, or whether it is only the tough environment out there that causes population declines. The latter population is said to have "density-independent" regulation. For instance, dune grasses growing on the shores of a Caribbean island subjected frequently to hurricanes could grow rapidly between major storms, but could be knocked back greatly with each major storm, to the point where any biological regulation would be overwhelmed by the waves and winds. The extent to which density-independent regulation happens compared to density-dependent regulation is an ongoing discussion.

Implicit in the logistic equation are several assumptions about the characterization of a population. The logistic equation expresses in the simplest form our intuitive notions about the limitation of the growth of a population with finite resources, and this simplicity has an appeal to scientists. The equation views all individuals as equal: The population is described only by its total number, and an individual has no other attribute than its presence or absence. Every individual is thus assumed to contribute equally to reproduction, mortality, and growth, and to reduce the available resources by exactly the same amount. Moreover, a "logistic" individual decreases the availability of resources for its fellows regardless of how many others there are. In logistic terms, the elephant population at Tsavo should have grown smoothly to an equilibrium.

A logistic population therefore appears metaphorically and mathematically like a collection of identical colliding balls, the collisions resulting in a certain rate of destruction and the balls capable of identical rates of division. A logistic elephant is never a calf or senile; the equation cannot distinguish between a population composed primarily of young, nonreproducing individuals and a population composed primarily of reproductively active adults.

Strangely, although the logistic equation is supposed to be an ecological formula, the environment of a population does not appear in it in an explicit way. Strictly speaking, the logistic can accurately describe only a population to which all required resources are available at a constant rate, and whose members are exposed to all toxins (except those generated by themselves) at a constant rate. A logistic elephant responds instantaneously to changes in the size of the population; there are no time lags, no seasons, no history; a logistic elephant has no fat.

The borrowing of ideas and mathematical methods is explicit in the writings of Lotka. His classic book *The Elements of Physical Biology* is a thoughtful and careful discussion of many aspects of ecology and is worth reading for Lotka's intelligence alone, even if he pursued a line of reasoning that is out of date now. One can only say that at the time it was written it was an important and fundamental work, necessary in an initial scientific exploration of nature. Lotka, like most other mathematical ecologists, starts with an analysis borrowed from physics. His fifth chapter, "The Program of Physical Biology," is divided into sections on statics and kinetics, much like a textbook on Newtonian mechanics. His discussion of kinetics investigates the solutions of simultaneous differential equations, which he applies to population growth. This is an excellent beginning of an approach to an analysis of populations, but it isn't where we should end up today or should have been 50 years ago, as I hope to show in the rest of this book.

So the point here is that the problem hasn't been with Lotka's original work, but with the way it continues to be followed. In his book, Lotka wrote, "If we pause to hesitate in defining life, [let us] pass from legend to the world of scientific fact" and "borrow the method of the physicist."[18] The oddity is that he did *not* pass from legend to scientific fact, but relied on some theoretical mathematics that fit legend, as you saw earlier in his explanation about the growth of a population of flies in a laboratory container—something out of his imagination, not from actual observation.

Scientists extrapolated from these experiments to the world outside the laboratory windows and assumed that the logistic curve describes the growth of populations in the wilderness. When I wrote *Discordant Harmonies*, I searched the scientific literature as widely as I could to find any examples in nature of a population actually growing over a reasonably long period according to the logistic. I found none.[19] In spite of this, ecology textbooks still teach the logistic as a basic part of the growth of populations, and therefore the confusion continues about the difference between an idealized balance of nature and the reality of how populations grow.[20]

The logistic equation leads to a simple calculation of the population size that has the maximum growth, which turns out to be a population exactly half as large as its carrying capacity. This population size is known as the *maximum-sustainable-yield* population. If the mathematics were a true description of nature, the population could be allowed to grow each year above the maximum-sustainable-yield level and then be harvested down to it. Like clockwork, the population would grow back exactly the same amount each year, and a precisely sustainable harvest could be obtained year after year. And this is exactly what formal management practice has assumed.

In short, for the maximum-sustainable-yield concept to be true, the population must have an exact and single carrying capacity, and its growth must exactly follow the logistic curve. And for this concept to be employed successfully in management and conservation, a measurement problem must be solved—the manager has to know with complete precision and accuracy the total number of fish in the population to be harvested. It must be possible to know precisely both the carrying capacity and the present population size. It must be possible to obtain complete cooperation from all harvesters so that exactly the right number are harvested each year. But these levels of precision are rarely achieved in managing wild populations. In fact, in most cases we don't even come close to knowing how many fish of a species are out there. Fishing is tough, and counting fish isn't easy either. Just watch the currently popular TV program *Deadliest Catch* and imagine trying to count how many crabs are in the Arctic waters.

Soon after the publication of *Discordant Harmonies*, and in part a result of it, the state of Oregon passed a special bill to fund a scientific study of the relative effects of forest practices on salmon, and I was asked to direct that study. With this and other requests, I spent much of the next 10 years studying salmon and their management and conservation. The logistic growth equation was then and remains today the basis for scientific estimates used to set salmon harvest quotas. Although some government agencies claim they have moved away from this approach, careful examination of their practices has continually revealed the same logistic basis for calculating the harvest as was used throughout the twentieth century.[21]

The ideas underlying the maximum-sustainable-yield concept are the same ideas that were behind the management of Tsavo: the belief that nature undisturbed by people achieves a constancy that remains indefinitely; and that if disturbed, nature recovers its former status. The formal management of marine fisheries was based on the belief that nature undisturbed is constant and stable, just as the ancient Greek and Roman philosophers believed it should be.

One might ask, of course, whether the fisheries crashed not so much because of a theory but because real fishermen in a real world did not follow that theory. In the Philippines, for example, fishermen have used dynamite and cyanide to catch fish, which killed huge numbers of fish in an area and also destroyed their habitat. I was a Peace Corps volunteer in the Philippines and returned there in 1986 as part of a project with the East-West Center in Honolulu, Hawaii. By then familiar with their culture, on that second visit I spoke with some local fishermen about their use of dynamite and cyanide. They said they knew that these destroyed the fisheries but they "have to eat today to live tomorrow." It wasn't ignorance of the consequences that led to their unsustainable practices.

Such practices will destroy a fishery with or without a theory. International agreements can be compromised by fishermen circumventing them and harvesting more than the allowed catch. Regulation is mainly by inspectors on boats and inspection of boats on their return to harbor. To believe that major, managed world fisheries declined only because fishermen did not follow the maximum-sustainable-yield rule is to believe that the theory is correct. I am arguing that it is fundamentally wrong. If fishermen's compliance with established harvests were the only issue, management policies could have taken excess catches into account by estimating the amount by which fishermen would exceed the actual catch and then adjusting the official allowed catch downward so that the actual take would be at the maximum-sustainable-yield level.

There were attempts to get away from the tightly constraining maximum-sustainable-yield way of thinking. For example, the Marine Mammal Protection Act passed by Congress in 1972 grew out of concern about whaling and the declining numbers of most marine mammals to the point where they were becoming endangered. The act states that the primary objective for management of marine mammals should be "to maintain the health and stability of the marine ecosystems" and that "whenever consistent with this primary objective, it should be the goal to obtain an optimum sustainable population keeping in mind the optimum carrying capacity of the habitat."[22] What was this new idea—an *optimum sustainable population*? The members of the Marine Mammal Commission didn't know what it meant, and their science advisers weren't sure either. So the Commission asked me to analyze the law and tell them what could be ecologically and legally sound meanings of the term. I asked Professor Matthew Sobel, an expert on the application of mathematics to economic and environmental problems, to join with me in this effort. We wrote a report for the Commission.[23]

The act defines it as "the number of animals which will result in the maximum productivity of the population of the species, keeping in mind the optimum carrying capacity of the habitat and the health of the ecosystem of which they form a constituent element."

An advisory panel of scientists, headed by Douglas Chapman (1949–1988) of the University of Washington, an expert on fisheries management, was formed to recommend a scientific basis for management of marine mammals under the new law. But in their recommendations, the scientists returned to the same logistic growth curve that they had used in fisheries management, defining an optimum sustainable population in terms of this curve. The Act defined (and still defines) *optimum sustainable population* to mean "with respect to any population stock, the number of animals which will result in the maximum productivity of the population or the species, keeping in mind the carrying capacity of the habitat and the health of the ecosystem of which they form a constituent element." There is the rub: The defining sentence includes two terms from the logistic growth curve, *maximum productivity* and *carrying capacity*.

Therefore, one interpretation the law allows is that the optimum population is close to, but somewhat larger than, the maximum-sustainable-yield population of the logistic. Another interpretation possible under that law is that an optimum sustainable population is at the carrying capacity, modified slightly to consider the idea of an optimum. In practice, however, an operational definition of *optimum sustainable yield* was chosen that retreats to the logistic growth curve instead of moving away from a steady-state view. This definition focuses instead on a single requirement of the maximum-sustainable-yield concept: that the carrying capacity and the current size of the population must be known exactly.

The Marine Mammal Commission's request that I provide a scientific interpretation of *optimum sustainable populations*, as the legal terminology in the Act would allow, led to the beginning of my formal inquiry into the role that the ancient belief in the balance of nature played in the environmental policies. In truth, the Commission's scientific advisory panel was greatly disappointed with the findings that I and my colleague, Matt Sobel, had arrived at, even though these findings were all that the wording of the Act allowed. So not only was there a belief in the balance of nature, not only was the balance of nature the underlying principle of this Act, but also the scientists advising the Commission were unhappy that someone would point this out. I believe they were hoping for a single number, such as the maximum-sustainable-yield population size or the carrying capacity, that would make for easy management actions, and they were left with an interpretation that allowed a huge and therefore impractical range of goals.

Recognizing that it is generally impossible to satisfy that requirement, the officials who determined harvest policies settled on the idea that the allowed harvest should not bring the population all the way down to the supposed one-half carrying-capacity level, but should instead maintain the population at 10% above that level. Their definition of an optimum sustainable population implicitly retained the assumption that the population follows the classic logistic growth curve and that it does in fact have a fixed carrying capacity and absolute maximum-sustainable-yield level, but that these are difficult, if not impossible, to determine exactly.

In 2007 I participated in a meeting of the International Whaling Commission's scientific advisory committee. Setting an allowable catch of bowhead whales by Eskimos was a focus of this meeting, and my colleague, anthropologist John R. Bockstoce, and I had done studies of the historical records from the logbooks of ships hunting this species between 1849 and 1914. The data from these studies had become important in the negotiations about the allowed bowhead harvest.

In preparation for that meeting, I was sent reports and papers about the forecasting methods in use by the scientific committee for setting these levels. I found that the logistic equation continued to be the basis for these methods. I wrote to the committee that perhaps they could begin to consider a greater variety of forecasting methods, hoping that they would begin to entertain some of the ideas I am talking about in this book. The committee wrote back that the models in use for their forecasts had been "vetted" and it was agreed that the ones they wrote me about were the ones to use. I had no idea then, nor do I now, what it means to "vet" a computer forecasting model, but one thing was clear: The old ideas and approaches are alive and well.[24]

Thus even today, in both law and in practice, the scientific conservation of endangered marine species continues to be based on the idea that nature undisturbed is constant and stable, although some marine mammal scientists are beginning to argue that the Act's definition is impractical to apply in the real world.[25]

Myths of Nature and Management of Resources

The elephants of Tsavo, the management of fisheries, and the assumptions behind the Marine Mammal Protection Act raised a number of questions for me about the management of living resources during most of the twentieth century, even into the early 1970s. These are the questions I asked myself when I completed the report to the Marine Mammal Commission. The use of this concept for marine mammals seems especially peculiar because the Marine Mammal Protection Act was passed at a time when fisheries experts well knew that the logistic curve had not succeeded as the basis for managing fisheries. Why, then, was it used when an opportunity to go beyond it had presented itself? This management approach had by and large failed to achieve its goals, even in those cases where good intentions prevailed and there was agreement among all parties about the goals. Let us reconsider what accounted for these failures.

An important factor was that the management was based on beliefs so firmly rooted in our culture that they served as the basis for action and policy even when contradicted by facts. These beliefs had two roots. One was in the science of ecology, which had begun in the late nineteenth century. The other was in pre-scientific beliefs about nature. In his management of Tsavo, Sheldrick seemed motivated primarily by the latter, although he was influenced to some extent by the science of ecology. The management of fisheries and the Marine Mammal Protection Act were based on ecological theory first and on pre-scientific antecedents second.

It is worth repeating that the discussion of the roots of these beliefs has an importance that goes beyond the particular examples of Tsavo and marine fisheries. The failure of management of living resources is a symptom of what is wrong with the myths, beliefs, and fundamental paradigms that modern technological civilization holds about nature. Some of these myths concern the character of nature undisturbed by human influences—the character of wilderness. The qualities of a wilderness *without human beings* are crucial for us to understand so that we can know what is needed to preserve our surroundings, preserve ourselves, and understand the effects of our actions on nature. Later I will challenge the reality of the idea of wilderness as it came to be interpreted in the twentieth century, the idea that between the time of the evolution of human beings and the beginning of written history, most of the Earth was uninfluenced by human action.

But at a more personal and deeper level, nature is our mirror. The way we view ourselves, as individuals and as members of societies, is in part a reflection of how we see ourselves in relation to nature. Since the beginning of Western civilization, philosophers have held up the mirror of nature and found in it four questions about the biological realm of existence. What is the character of nature undisturbed by human influence? What are the effects of nature on human beings—on individuals as well as civilization and culture? What is the role or purpose of people in nature? And what are the effects of human beings—as individuals and as societies—on the living nonhuman world?[25]

In pre-scientific Western culture, wilderness was perceived in several ways. At one extreme, wilderness was an idealization of divine order, a place of perfect order and harmony to be worshiped or to serve as a source of spiritual awakening. At the opposite extreme, wilderness was a place that was chaotic, dangerously powerful, and to be feared or challenged.[26]

In addition, since ancient times, discussions of the character of nature have focused on the amazing adaptations of organisms to their needs, on the diversity of species and their connections with one another, and on the appearance of a balance of nature. *Balance of nature* has meant not only the constancy and stability referred to earlier but also the idea known classically as the Great Chain of Being, in nature a place for every creature, and every creature in its place; in modern parlance, that every creature and every species has its place (that is, its role and its location—its habitat) in the harmonious workings of nature and is well adapted to that habitat, and to that role, which ecologists call today a species' niche.

In the 1970s, flying just above the ground over Tsavo, my ears filled with the roar of the Cessna 182's engine, I focused on the desolate scene below, a scene that clearly illustrated that our attempts to manage wildlife and other renewable natural resources had failed. This seemed so obvious in any snapshot of Tsavo that I wondered why we were still applying outmoded ideas to elephants and anchovies. My conclusion, apparent then and worth repeating now, is that our actions were not based on facts alone—the view of Tsavo alone—but on beliefs hidden from view. We had taken Pierre-François Verhulst's imaginative suggestion of a century before, the logistic growth equation, an advance in scientific thinking at that time, and fixed it as though it were a permanent and final explanation.

During its 150 years—from Verhulst's 1838 formulation of the logistic to modern times—the science of ecology, along with its application in the management of living resources, has in the large viewed nature like a snapshot of Tsavo, as though it were fixed in time and space, as constant as the environment of a modern scientific laboratory. By the 1970s, a curious situation had arisen in which the accepted theories were failing to provide a successful basis for managing living resources. It seemed to me that it is one thing to play games in a laboratory and pretend that nature is like an artificial environment in a laboratory container full of fruit flies, but quite another to fool ourselves that such a game should be played out with the remaining treasures of wildlife and wild habitats in the realities of our complex and discordant world.

An irony is that it seems that everybody talks about how complex nature is— we see and hear about this complexity in movies like 2005's *March of the Penguins* and in television programs like *Nature* and many on the Animal Planet channel— but we are content to formalize nature in about as simple and simplistic a way as possible. It is all the more perplexing when we realize that since Verhulst first proposed the logistic curve in 1838, the continental railway was completed in the United States; California and many other states became part of the United States; the automobile and airplane were invented and developed into dominant means of transportation; submarines, heat-seeking missiles, dogfighting airplanes, machine guns, TNT, and dynamite were invented; the theories of thermodynamics, of relativity, and of quantum mechanics revolutionized physics and astrophysics and greatly revised our view of the universe and the relative importance in a solely physical sense of our sun and solar system; radioactivity was discovered, and theory developed for it led to the atomic bomb and nuclear power plants; Robert Goddard (1882–1945) invented the modern rocket, leading to the space age, men landing on the moon, satellites carrying millions of telephone conversations, e-mails, and GPS, which guides all of us everywhere; medicine went through several revolutions and can be said to be a completely different science, with completely new theory; DNA was discovered and genetic engineering has become commonplace. Many believe that ecology is still a "young" science, but in comparison to most modern sciences, it is not young but simply retarded.

We must change what we have been doing, but we can't if we stick to the surface level of things, merely observing the view through the plastic window of a single-engine airplane—and since 1990 we have pretty much stayed at that surface. To change the ways we manage our living resources and the framework within which my colleagues and I, as scientists, make our measurements, we must revise our ideas at a deeper level. We must accept the contradiction between fact and theory, and understand that to resolve this contradiction we must move to a deeper level of thought and confront the assumptions that have dominated perceptions of nature for a very long time. This will allow us to find the true harmony of nature, which, as Plotinus wrote so long ago, is by its very essence discordant, created from the simultaneous movements of many tones, the combination of many processes flowing at the same time along various scales, leading not to a simple melody but to a symphony at some times harsh and at some times pleasing.[27]

I said this in 1990, hoping that it would lead to progress; I state it again today, in 2012, with the same hope, perhaps a little closer to realization than 20 years ago, but not anywhere near what is necessary for true harmony between ourselves and nature.

FIGURE 3.1 Moose grazing in the shallows of Isle Royale National Park. (Photograph by the author.)

3

Moose in the Wilderness

THE INSTABILITY OF POPULATIONS

> The ecologist and the physical scientist tend to be machinery oriented . . . the machinery person tends to see similarities among phenomena as opposed to differences.
>
> —ROBERT H. MACARTHUR (1972)[1]

Nature Undisturbed

One year around 1900, a small number of moose left the shore of Lake Superior near the border of Minnesota and Ontario and crossed 15 miles over the lake's waters to a forested island called, since the arrival two centuries earlier of the first French explorers, Isle Royale. With their poor eyesight, the moose may not have realized that they had reached an island rather than just another promontory on the lake's wooded shore, for Isle Royale is large, covering 210 square miles, an area ten times that of Bermuda and almost nine times that of Manhattan. The moose had migrated from a region heavily influenced by civilization to one that was then and still is one of the best examples in the world of a wilderness undisturbed by human influence, what many would call the forest primeval.

People had visited Isle Royale during the past several thousand years, but had never settled for long. American Indians occasionally had crossed the icy waters of the greatest of all lakes to collect "native copper" (pure copper), which could be found at or near the surface in some rock outcrops. Archaeological studies suggest that they were visiting the island most actively between AD 800 to AD 1600.[2] Europeans, too, had tried copper mining, but found it uneconomical. The island had seen a few small, short-lived farms, occasional resorts, and vacation cabins. Moose had never been there before, but around the turn of the twentieth century the commercial fishermen who sought whitefish and trout in the deep lake waters and spent the summers in shoreline cabins on the island knew that changes were taking place. For reasons the fishermen could not explain, caribou, which had been common, were fast disappearing, apparently moving northward from the island into Canada. Now the moose had arrived, just as inexplicably.

From the air, Isle Royale appears as a series of parallel ridges and valleys lying southwest to northeast. They are made of sedimentary rocks laid down as long as 300 million years ago, interlaced with volcanic flows that covered sedimentary layers and then were themselves overlaid by newer, but still ancient for us, sandstones and conglomerates. Today, the harder volcanic basalts form ridges that are forested with species of trees and shrubs that moose prefer. Along the cold shores where the moose arrived were forests of white birch and fir, which provided spring, autumn, and winter food for the moose, and spruce, which the moose did not and do not eat. In the interior, where the ridges were warmer because they were protected from the cold winds that blew off Lake Superior, grew yellow birch, maple, aspen, wild cherry, and other trees and shrubs that provided summer forage of the moose.

In the island's valleys were 45 large lakes, beautiful stretches of open, shallow water that on a clear day appeared a deep blue against the dark green of the hills. Some of them graded into large marshes with patches of open water, floating mats of moss and sedge, and dense thickets of bog cranberry, Labrador tea, and graceful northern cedars. Many small streams had been dammed by beaver ponds. Near the coast, the valleys ended in long harbors whose shallow waters teemed with water lilies, rushes, and many other aquatic plants that moose fed on in midsummer.

As the twentieth century opened, the moose had arrived at an ideal habitat for them, lacking at that time their principal predator, the North American timber wolf. They had arrived at what we call and believe to be a primeval wilderness, the reality of an ideal that has played a central and important role in the thoughts of human beings since the origin of Western civilization. Here was nature undisturbed—indeed less directly disturbed by human influence than the African plains and savannas, where *Homo sapiens* and their ancestors have been residents for several million years, and less directly disturbed by our influences than much of the North American Arctic, where people have hunted, fished, and lived for thousands of years.

Although Isle Royale had been undisturbed by all but a few direct human actions, it had been observed, the subject of a natural history survey in 1846, and had been visited repeatedly by botanists who studied its forests and zoologists who studied its wildlife. Like many naturalists to follow them, including myself, they were fascinated by the forest primeval, by nature undisturbed by human action. Isle Royale exerted a kind of attraction, like a magnet to which we naturalists, like small iron filings on a piece of paper, were drawn with or without our direct acknowledgment, fulfilling within ourselves some hidden need, some necessity, some drive to discover that true nature, so much the cause of ourselves, the driver that through millions of years of evolution had somehow produced in us curiosity about how that nature worked.

When I did research on the island, I felt that pull. After even a week's stay there, I found myself going through a kind of cultural readjustment when I returned to the mainland. I would feel isolated and alone, separated from some kind of deep emotional pull of the wilderness, which I could feel but could not completely explain. Arriving on the mainland, I would often want to just turn around and go back; it wasn't an alienation from modern civilization and its accouterments, but the lure of the wild. I had felt something similar years before working in the forests of New Hampshire as a surveyor's helper, a feeling strong enough to lead me to become a scientist who studied nature and wanted to understand, for some kind of deep feelings within me, how that nature worked. I was sure that every leaf on the ground, every log, every living tree, every handful of soil, on Isle Royale had a story to tell. It was not put there by a person but arrived there and persisted because of some rules of nature, which I was driven to try to understand, not so much a choice and certainly not a casual choice, but an inner necessity. And I was sure that the island would be one of the keys to that understanding.

After the moose were first observed on the island in the early years of the twentieth century by fishermen and naturalists, their number increased rapidly. In less than a decade, the moose began to change the vegetation greatly, as reported in a series of scientific papers about the ecology of the island by William S. Cooper (1884–1978). Although there was not an accurate count of the animals, the increase in numbers was apparent from their effect on the vegetation. Two of their favorite foods—water lilies, which had almost completely covered some of the ponds, and yew, an evergreen shrub that had been the dominant ground cover in much of the island's forest—began to disappear. At the time of Cooper's study, done in the first decade of the twentieth century and published in 1913, both plant species seemed threatened with extinction on the island because of the moose.[3]

By the late 1920s and early 1930s, the moose's impact on the vegetation was so great that Adolph Murie (1899–1974), a well-known naturalist, warned that they were about to run out of food and that a major die-off was imminent. This die-off did occur in the mid-1930s. Murie estimated that the number of moose dropped from more than 3,000 to fewer than 500.[4]

Soon after the die-off, a fire burned more than one-third of the island. The forest that regenerated after the burn contained many low, young stems of species that moose favor, such as white birch. The moose population began a second increase, and after the mid-1940s, when the island, which had been a state park, became a national park, they again became numerous enough to greatly affect the vegetation. Concerned that a second major die-off might occur, the park personnel wondered what should be done.

The park was supposed to be a *natural* area undisturbed by human beings. Was it natural to have the moose dying in large numbers anywhere, but especially in such a park? And even if it was, should it really be allowed to happen when the moose were one of the park's main attractions? Indeed, one could ask whether the presence of the moose was natural at all, since they had not been on Isle Royale, as far as was known, prior to European settlement of North America. The issues raised remind us of the controversies over Tsavo National Park. (As we shall see, in later years some biologists and anthropologists, most notably and powerfully Paul S. Martin (1928–2010), would argue that nature wasn't nature since the arrival of human beings, and that in North America it was therefore our moral obligation to return the flora and fauna to what it was at the end of the last ice age—mammoths, mastodons, and all.)[5]

In the 1930s it seemed clear to the park personnel that the moose population was in an "unnatural" situation in one sense: Its major predator, the timber wolf, was missing from the list of species on the island. Another ancient belief was that a predator is necessary to maintain the balance of the population of its prey. Following this belief, the National Park Service decided it would be beneficial to introduce wolves onto Isle Royale.

Six wolves obtained from zoos were brought to the island in 1946. The zoo wolves were not accustomed to life in the wild, though, and instead of hunting moose, they were said to have hung around the Park Service headquarters looking for handouts. Although this intentional introduction failed, it was followed a few years later by a natural introduction—the arrival of a pack of wild wolves that probably crossed the lake during a particularly cold winter when the water froze across to the mainland. This pack, living mainly on the moose, increased from about a dozen in the 1940s to more than 20 in the early 1960s. Meanwhile, the moose population appeared to remain around 1,000 adults—a high but comparatively constant size, suggesting that the wolves were indeed holding the moose population in check and that Isle Royale, nature undisturbed, was indeed also an illustration of a constancy or balance of nature in the wilderness.

But was the balance of nature really established there in a constant number of moose and wolves and in a constant abundance of the plants the moose ate? Had Isle Royale, in fact, reached a steady state?

It was in the late 1960s that I began a study of the moose and their wilderness ecosystem at Isle Royale with my colleague Peter Jordan, an expert on wildlife who had been studying the moose on the island for years. I came to the island believing what I had been taught by my ecologist mentors, that there was a balance of nature. Here, I thought, was the opportunity to find out how that balance of nature worked— what were the mechanisms by which nature in its entirety could sustain itself in constancy indefinitely? That is, I arrived with the idea that here was a place where we could come to understand how a balance of nature might be achieved by the interaction of many species with one another and with their local, nonliving environment, the assumption being that the entire island was in such a balanced state. Our specific, beginning purpose was to determine the factors that controlled the population of the moose, in part with the belief that this natural control led to a balance of Isle Royale's nature.

We had an idea that some aspect of mineral nutrition would put an upper limit on the moose population, and we hoped to determine just which chemical element would be so limiting to the moose. Others believed that the wolves were the key to the apparent constancy of the moose population. The interaction between wolf and moose is complex. Packs of wolves have been observed to ambush moose. A wolf pack was once seen to divide into two groups, one of which hid along a heavily wooded and narrow pass in a moose trail, while the other drove a moose toward that point.[6]

After an initial scouting trip one spring, Peter and I and several students who made up a field crew arrived at Washington Harbor on the western end of the island and set up a research camp. We were accompanied by a large pile of equipment; in addition to the usual camping gear, we had carted to Isle Royale a large electromechanical Friden calculator, which must have weighed about 15 pounds and which we could plug into a single outlet that the Park Service provided to us at our campsite. I had brought the calculator along to help make certain statistical computations. The calculator worked slowly, with a great deal of whirring and clicking. It was especially slow at long division, and while it whirred and clicked over those calculations, I had time to pause and listen to the musical minor thirds being trilled by many white-throated sparrows or to the haunting call of the loons across the harbor.

That calculator was unwieldy and seemed out of place in the wilderness, but the machinery within it was an apt metaphor for the concepts that dominated the science of ecology at that time. It was a machine made up of many parts, each with a place and role and each within that place and role. As long as the motor, gears, wheels, belts, bearings, and so forth functioned together, each within their place and role, the machine worked as a thing, an individual system that performed its intended function, given to it by its inventors and designers, and most of the time it stayed in steady state.

As an aside, we learned that we could completely confuse the poor machine by asking it to divide a very small number by a very large one. The calculator had a fixed number of decimal places, and if the result was going to be smaller than the smallest number in the machine, it would get confused and go into an infinite search for that first and unavailable number. It seemed to me as if the machine were searching for a kind of Socratic idealized number beyond its universe and could only go on forever seeking to find it within its shadow world. Its gears, wheels, and dials would get whirring and it would lock into that position, as if it had drifted off into a kind of machine-meditation that sometimes seemed impossible to stop, even if we pulled the plug and stuck it back into the outlet. Although we avoided making this happen, we were amused when it did, so simple-minded did the machine seem. It's with that kind of mechanical device that it is easy to see its limits and therefore its inability to represent nature's complexity. Modern computers make that realization more difficult, as they appear to be much smarter.

With that calculator, a pile of forestry field devices, some plastic laboratory flasks and bottles and graduated cylinders, and a peculiar orange plastic, 18-foot, double-ended, twentieth-century version of a whaling dory to haul our heavy gear around the island, we felt prepared for our scientific study.

In the second year of our study we upgraded our equipment and brought with us one of the very first pocket electronic calculators, a Hewlett-Packard little miracle machine that could add, subtract, multiply, divide, and store one number, in absolute silence. This new calculator worked off a small battery, weighed only a few ounces, and had cost us only $450, a bargain at the time for a device that we could carry with us into the woods where we took measurements and samples. In one year our equipment had made the transition from a device based on gears and wheels to a device based on solid-state computer logic. Our ideas evolved more slowly.

We learned a lot about the moose, vegetation, and soil in the several years we worked on the island. By collecting samples of leaves and twigs of the plants that the moose ate and samples of mud, water, and moose tissue, we found that sodium was the most likely factor limiting the population of moose.[7] But this was a factor that could set an upper limit, not one that could neatly fix the moose abundance at a single level.

In the autumn, I began to set down some of what we had learned on the island in a computer program that I developed to provide a primitive model of the moose population. We needed a way to calculate what the total size and weight of the moose population would be for different birth and death rates. In the program, the moose were divided by age, and each age had a fixed rate of death and a fixed rate of reproduction. Moose lived to 17 years, so there were 17 ages of moose in the computer program, and determining how the number of moose changed from year to year took many calculations, more than one would care to do by hand, or with the old-fashioned Friden calculator, or even with the newfangled HP solid-state calculator that could store only one number at a time and couldn't do even the simplest statistical calculation.

The concepts that underlay this primitive computer program forced the computer-moose population to achieve, inevitably, a constant condition in which the number of moose and their total weight remained the same year after computer-generated year. Because a balance of nature was required by the assumptions I wrote into the computer code, it was necessarily achieved within the computer, consistent with the ideas then prevalent about natural populations and with what we thought was correct for the moose on Isle Royale. It didn't seem strange to us that the computer program, by design, forced this constancy on the population, and we did learn quite a lot from that primitive program. We could see which age classes had the highest mortality rate and therefore, by inference, had been most easily killed by the wolves. Also, the exercise of creating the computer software helped me soon after with much more imaginative and complex computer simulations of ecological systems.

Despite the laborious and troublesome character of computer programming at that time and skepticism with which it was viewed by most of my colleagues, I felt that there was a potential link between computer programs and field studies of wilderness, between contemplation of the numbers and artificial "languages" of the computer and contemplation of the character of nature as we observed it at Isle Royale. But the computer was as yet a minor factor in our work studying the character of wilderness, dealing with ancient questions about the character of nature undisturbed by human influence, about the idea of a balance of nature.

Not long after our study began, the moose and wolf populations began to change again. First the moose population began to increase; then the wolf population doubled to more than 40; later, in the 1970s, the moose population began to decline. Variations in the two populations have continued ever since. In 2012, the wolf population had dropped to 9, bringing up the same questions about whether human actions, on the island and elsewhere, were to blame, and even if not, should we reintroduce wolves—what should our role be in this natural of all natures?[8]

These complex patterns, which show great variations over time and in which the individuality of the wolves and moose seemed to play an important role, did not conform to the theories that had predominated in the twentieth century about populations or the interactions between predator and prey, theories similar to the ideas that underlay David Sheldrick's management of the elephants at Tsavo.

Modern Ideas About the Balance of Nature

The modern science that deals most directly with the ancient questions about human beings and nature is ecology, the study of the relationship between living things and their environment. That was the science I pursued at Isle Royale, and the term *ecology* is used in this book to mean this science. The word has acquired other connotations connected to *environmentalism*, which encompasses the social, political, and ethical movements relating to our use of the environment, and the original meaning of *ecology* is sometimes lost. Also, in the past two decades other major scientific fields—such as climatology, geology, physical and chemical oceanography—have made major contributions to the study of environment and to our understanding of what nature is like and what our roles are and could be within nature.

The word *ecology* was coined in 1866 by the German biologist Ernst Haeckel (1834–1919), less than ten years after the publication of Charles Darwin's *Origin of Species*, which, Haeckel said, provided the basis for his new science, and only two years after George Perkins Marsh published *Man and Nature*, the first major modern book suggesting that human activities were leading to negative effects on the environment.[9] (Haeckel was one of the early promoters and supporters of Darwin's theory of biological evolution.)

Ecology as a modern science developed in the nineteenth and twentieth centuries—the "machine age," historians of science and technology call it. Like the Friden calculator, it was a child of that age, which began with the Industrial Revolution but flowered between the two world wars and for several decades afterward. From the mid-nineteenth century through the end of World War II, mechanical metaphors dominated people's ideas of beauty, progress, architecture, and home furnishings. Machines even became objects of art, such as Charles Sheeler's painting *Suspended Power*, which shows a huge hydroelectric generator about to be put into place, painted to show its beautiful symmetry. Mechanical metaphors and machine models dominated the science of ecology through the 1990s. During that decade, research that some of us were doing, involving computers and other digital devices, seemed to be leading to radical changes. At least that is what I thought when I wrote *Discordant Harmonies*. At that time I wrote that a new perspective was developing "whose effects are only now beginning to be felt and understood."[10] But more about that possible departure from the machine age later.

Ecology began to develop in the latter part of the nineteenth century with three background elements: new observations of natural history, including the development and acceptance of the theory of biological evolution; pre-scientific beliefs about nature, including the age-old desire to find order and stability in nature; and a dependence on the physical sciences and engineering for theory, mathematical approaches, concepts, models, and metaphors. These led to the growth of increasingly sophisticated mathematical theory (formal models) that required and in turn led to exact equilibria, and to a worldview of nature as the great machine. These foundations resulted in an untenable situation: The predominant, accepted ecological theories asserted that natural, undisturbed populations and ecological communities (sets of interacting populations) would achieve constancy in abundance, an assertion that became inconsistent with new observations.

The mathematical theories that developed in ecology during the nineteenth and most of the twentieth centuries about individual populations asserted the same ideas we have met before: that a natural population left undisturbed would achieve a constant number, called the "carrying capacity"; that the population would remain at that abundance until disturbed; and that once the disturbance was removed, the population would return to that same abundance. Theories about vegetation ecology were similar: that if undisturbed, a set of interacting species, such as trees and shrubs in a forest, would also achieve a constancy, not only in the numbers of each of the species but also in the total number of species and in the total amount of organic matter stored in the forest.[11]

In sum, ecological theory of the period was reminiscent—actually a restatement—of the old idea of a balance of nature. Although that phrase was not commonly used by ecologists, their desire to find constancy and stability in observations was apparent and revealing. This perception of order was stated eloquently in 1925 by Stephen A. Forbes (1844–1930) in *The Lake as a Microcosm*. He wrote that "no phenomenon of life . . . is more remarkable than the steady balance of organic nature, which holds each species within the limits of a uniform average number, year after year, although each one is always doing its best to break across boundaries on every side." This balance is all the more remarkable because on the one hand, he wrote, "the reproductive rate is usually enormous," while on the other "the struggle for existence is correspondingly severe"; each individual has its enemies, and "nature seems to have taxed her skill and ingenuity to the utmost to furnish these enemies with contrivances for the destruction of their prey." In spite of these opposing forces, life continues and, Forbes believed, does not "even oscillate to any considerable degree, but on the contrary," a small lake is "as prosperous as if its state were one of profound and perpetual peace."[12] It was a lovely and appealing perspective on nature.

In the simplest natural-history terms, there are two threats to such a balance of nature, one from within and one from without. The external threat is from the physical and chemical forces of the environment that erode and degrade: on the land, wind, storms, rain, fire, and slow chemical leaching of waters; in oceans, lakes, and ponds, the irrepressible force of gravity pulling nutrients down, away from life at the surface; and winds, tides, and currents, which disrupt, degrade, and destroy these aquatic systems, as do wind and its colleagues on the land.

All of these external forces were quite apparent at Isle Royale when I was doing field studies. There were days when the wind blew so strongly that the firs and cedars near the harbor shore bent low like old men. Large areas of the island were difficult to cross; in some places we spent much time climbing over the remains of trees downed in storms. Fires occurred after summer thunderstorms and were watched for with great care and caution by the park personnel. How fragile nature's balance might be against these forces depends in part on the strength of biological forces of growth, their ability to resist the forces of erosion— nature's finger in the dike—or to replace what has been pulled away. These are the power of the trees and shrubs to grow and fix roots in the soil.

The internal danger to a balance of nature is the same power of growth, the potential for exponential growth, with which the moose population at Isle Royale may have threatened the island at the beginning of the twentieth century. The capacity of a population for such rapid growth is probably well illustrated in our century by two species whose numbers had been greatly reduced by human actions. One is the northern elephant seal of the Pacific Ocean, found on islands far from Isle Royale. The other is the sandhill crane.

One of the places where the northern elephant seals (*Mirounga angustirostris*) bred then and still breed is the Channel Islands, just off the coast of Southern California and Mexico. Hunted in the nineteenth century mainly for their blubber, which was converted to oil, the seals were said to be reduced to as few as a dozen individuals or perhaps between 20 and 100 around the turn of the twentieth century and were believed to be doomed to extinction.[13] (Indeed, the story is told that the British Museum sent out a team to find and hunt down the remaining few elephant seals and bring them back to London, on the grounds that it was better for them to be stuffed and exhibited in a museum than to die an ignominious death in the American wilderness.)

Instead, the population rebounded. The population had reached more than 60,000 in the mid-1970s.[14] Counts showed that by the end of the 1980s elephant seals had increased in number quite steadily at about 9% per year.

By 2012, the elephant seals had increased greatly, with numbers estimated today to exceed 120,000. According to the National Oceanographic and Atmospheric

Administration in 2010, "Though a complete population count of elephant seals is not possible because all age classes are not ashore at the same time, the most recent estimate of the California breeding stock was approximately 124,000 individuals."[15] (Obtaining accurate counts or even good estimates of the numbers of a wild species is one of the problems I will return to later.) The species is so abundant that it is not even listed as a "species of special concern" by the U.S. Marine Mammal Commission, the U.S. federal agency charged with the conservation of these animals.[16] The seals have reoccupied much of their previous habitat, and are easily seen along the Pacific shore of California at such famous tourist destinations as the Hearst Castle.

If this rate had continued, there would have been about 500,000 elephant seals by the year 2000, and almost 500 million by about 2070. Indeed, at this rate of increase, in a little more than 400 years the mass of elephant seals would equal the mass of the Earth.

Making such exponential calculations was quite fashionable in the 1960s, when environmentalism became popular and discussions about the limits to growth of the human population were common. Obviously, such growth cannot continue for long. Even 500,000 elephant seals would have difficulty finding a place to sleep on the sand and would likely destroy their breeding grounds.

On the land, a similar story can be told about the sandhill crane (Fig. 3.2A).[17] They were the first bird ever protected by a U.S. international treaty, the U.S.-Canadian Migratory Bird Treaty of 1918. This protection lasted until 1961. By that time, they had become so numerous that a single flyway in the western Great Plains states was estimated to have more than 130,000 birds that year, all flying as a single flock. They would land in farmers' fields and feed on the grain, a kind of vertebrate plague of locusts. The farmers complained to the federal government, and the Department of the Interior established hunting seasons for them in Texas and New Mexico.[18] Since then, the population has increased beyond 130,000, but has varied considerably over the years, rising to about 500,000 and averaging 400,000 between 1982 and 2007 (Fig. 3.2A and B).

Interestingly for our purposes, "The Cranes Status Survey and Conservation Action Plan Sandhill Crane (*Grus canadensis*)" asserts that this population is "probably stable."[19] Given that the sandhill crane population varied by as much as 40% from its average over this period and bounced from –30% to +30% from the average within a few years, the concept of "stable" used here has to be different from the classical idea of something at a fixed equilibrium, if it has any meaning at all. The simplest meaning to attach to this assertion, and for our purposes probably the best, is that the population, despite bouncing around considerably in numbers, appears quite unlikely to go extinct anytime soon.

Because *stability* has come to mean so many things, it would be helpful to distinguish this idea of stability from the rest, especially because in legal situations the stability of nature has often been taken to mean constancy in numbers. The main point for our purposes is that well into the twenty-first century there has been confusion over what *stability*—the modern term that can substitute for

SANDHILL CRANE POPULATION

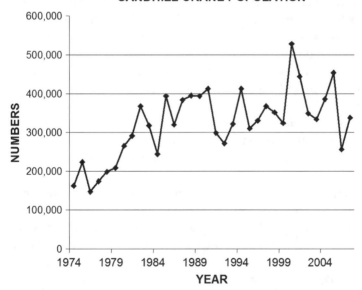

FIGURE 3.2A Sandhill cranes' abundance. This species was protected from hunting until 1961, when its numbers exceeded 100,000. Since then, the population increased to a maximum of more than 500,000 birds, but underwent considerable variation over time and has decreased in recent years.

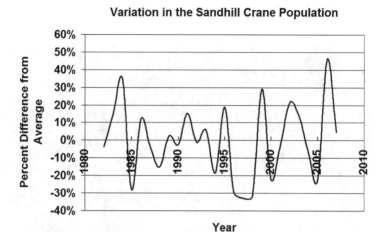

FIGURE 3.2B Variation in the sandhill crane population. This is for the central flyway, the main, largest population of the species. (Data from USGS.)

balance of nature—means. I will come back to the sandhill cranes and the notion of stability later.

Although many ancient philosophers and theologians believed in an absolute balance of nature, the incredible potential for the growth of biological populations has long been recognized, and it was also understood that if it did occur, it could disrupt any balance or harmony. Aristotle, whose writings contain the oldest discussions of longevity and other fundamental population phenomena, recounts a story in which a pregnant mouse was shut up in a jar filled with millet seed, and "after a short while" when the jar was opened, "120 mice came to light."[20]

Of course, it has long been recognized that this growth potential—the potential for exponential, geometric, or Malthusian growth—could never be maintained for long by any natural population, for it would soon be limited by some life requirement, as I just mentioned in regard to the potential growth of the elephant seal population. Exponential growth assumes that the population increases by a constant percentage in each time period, just as money does in a fixed-interest savings account. (It's interesting that the human population grew faster than that during some periods in the twentieth century. This is because, following the beginnings of modern medicine, the rate of increase itself increased, whereas exponential growth, by definition, is a constant rate of increase.) Thus we arrive at the famous Malthusian statement that organisms have the potential for geometric (his term for exponential) increase, while their resources can increase only arithmetically (his term for what today we call a linear increase, meaning that the quantity added in each time period is the same and therefore becomes a decreasing percentage of the total as time goes by; hence it greatly lags exponential growth).

Machine-Age Moose and Wolves; Pencil-and-Paper Predator and Prey

Much of the theory in ecology, from the origin of this science until the 1970s, relied on two formal models of population growth: the logistic for the growth of a single population, which occupied us first in Chapter 2, and the Lotka-Volterra equations for predator–prey interactions (obviously named after two of the first scientists to apply them to explain this kind of population growth). You will remember that in the logistic curve a population has the capacity for exponential growth, but it grows smoothly and continuously along an S-shaped line, eventually reaching a fixed, maximum size called the carrying capacity. The Lotka-Volterra equation for a combined predator and prey is formulated with the same assumptions, the same worldview, and the same mathematics, allowing only in addition the effects of the two species on each other. Two of these hypothetical populations, a predator and a prey, oscillate regularly, either continuing forever or dampening to a constant abundance, like two swinging pendulums. The two populations swing out of phase—the prey at a peak when its predator is at a minimum, the predator at a peak when its prey drops to a minimum.

Mathematician and physicist Vito Volterra (1860–1940), born in Ancona, Italy, became interested in his son-in-law's reports about Mediterranean fisheries. A decline in the catch during World War I led to an increased abundance of predatory fish. This suggested the idea that predator and prey fish would undergo opposing changes in abundance, with the prey decreasing as the predators increased, and vice versa.[21] Volterra recognized that such an interaction could be described by two simple mathematical equations, which also describe the interactions between two chemicals in a liquid medium. These two equations, one for the predator and one for the prey, have become famous as the Lotka-Volterra equations for predator-prey interactions. It is impossible to overestimate the influence of these equations in twentieth-century population biology. Like the logistic, they occur in every ecology and population-biology text, underlie hundreds of papers, and have been the subject of repeated, extensive mathematical analyses in long monographs and treatises.[22] As I write today in 2012, it continues to be true that one school of ecological theory still approaches the population dynamics of species as if it were a subclass of this kind of model.[23]

A set of Lotka-Volterra populations is much like a logistic population, except there are two colors of colliding balls (Fig. 3.3). A collision between balls of different colors results in the disappearance of one, the prey, and an increase in the number of the other, the predator. In the absence of "predator" balls, the "prey" balls increase either exponentially or, in later formulations, following the logistic. The predator balls simply die away at an exponential rate in the absence of prey.

Two kinds of stability are possible for the Lotka-Volterra equations: unending constant oscillations (like the vibrations of two strings exactly out of phase) or dampened oscillations that lead to a fixed single equilibrium (as when two strings are plucked once and then their vibrations, and therefore their sounds, slowly die away). The swings of predator and prey are out of phase, with one reaching a peak

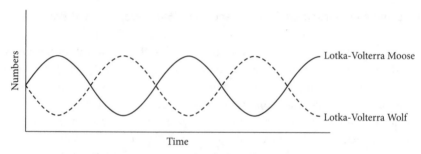

FIGURE 3.3 Lotka-Volterra theoretical moose and wolf population changes over time.

when the other reaches a minimum, as illustrated in Figure 3.3. A Lotka-Volterra Isle Royale wolf population would increase and decrease, chasing over time the similarly oscillating moose population, each population reaching its peak while the other descended to its trough. Interestingly, in the Lotka-Volterra formulation, predator and prey populations oscillate *because* of their interactions.

As with the logistic, the Lotka-Volterra equations do not distinguish individuals within either population. All Lotka-Volterra prey would be like a logistic moose; similarly, a Lotka-Volterra predator would be equally identical to its fellows. A wolf pack would not be divided into lead male and female; there would be no wolf pups playing at the adults' heels. The populations are viewed as though from afar, through the wrong end of a telescope, reduced to their simplest single character, each animal indistinguishable from others of the same species. These equations reduce the biological world to a mechanistic system.

Nature in a Test Tube

The growth of a single population and the effect of predators on prey were studied in the laboratory by the Russian scientist Georgii Frantsevich Gause (1910–1986), who used two species of single-celled microbes of the biological group Protista: *Paramecium caudatum* as the prey, and *Didinium nasutum* as the predator. Gause conducted scientific research at its best, combining formal theory with laboratory experiments, which he described in his 1934 classic and beautifully written book *The Struggle for Existence*. The theory he chose for population growth was the logistic equation and the Lotka-Volterra equations.

In Gause's experiments, the microbes were grown in laboratory flasks under constant conditions, with a uniform environment and a steady supply of food. In one experiment, he grew paramecia alone and found that as long as the food supply was constant and the environment constant in every other way as well, growth did follow a logistic (Fig. 3.4). The artificial conditions of his laboratory experiments and the unicellular paramecia match the assumptions of the logistic about as closely as is possible. Gause's successful single-species experiments were influential, but although other laboratory experiments with insects such as fruit flies kept

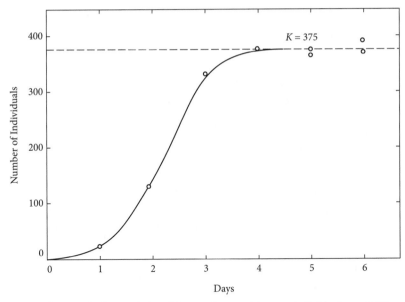

FIGURE 3.4 The growth of a population of *Paramecium caudatum* as studied by Gause in his famous experiments. (From G. F. Gause [1934]. *The Struggle for Existence* [Baltimore: Williams and Wilkins].)

in closed containers in a constant environment with a sustained food supply also yielded logistic growth curves, logistic growth has never been observed in nature.

In another set of experiments, Gause introduced five *Paramecium* prey into each of several tubes. Two days later, three *Didinium* predators were added. The paramecia increased in abundance, reaching 120 individuals by the second day, and then declined rapidly after the predators were introduced. The predators increased to about 20 individuals. By the fifth day, the paramecia were completely eliminated by the didinia, which eventually died of starvation (Fig. 3.5).

In other experiments, Gause repeatedly tried to obtain the Lotka-Volterra oscillations, but could not. At best, he was able to sustain several cycles before one or the other species went extinct. Furthermore, the swings were not properly out of phase, as predicted by the Lotka-Volterra equations (see Fig. 3.3). Gause concluded that the periodic oscillations were not a property of the interaction itself, as predicted by the equations, but seemed to result from "interferences."

Gause's analyses remain among the most scientifically complete in the history of ecology in that he considered concepts, formal theory, and experimental tests. Even though his tests disproved the theory 80 years ago, the Lotka-Volterra equations were used widely throughout the decades that followed and are still widely used today.[24] In a concluding statement in his book, Gause wrote:

> We expected at the beginning . . . to find "classical" oscillations in numbers arising in consequence of the continuous interaction between predator and prey as was assumed by Lotka and by Volterra. But it immediately became

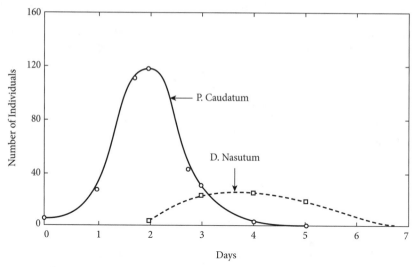

FIGURE 3.5 Gause's experiment with microbial predator and prey.

apparent that such fluctuations are impossible in the population studied, and that this holds true for more than our special case. . . . It is to be hoped that further experimental researches will enable us to penetrate deeper into the nature of the processes of the struggle for existence. But in this direction many and varied difficulties will undoubtedly be encountered.[25]

However, the mathematical analysis of these two models, the logistic and the Lotka-Volterra, continued. Several decades later, Raymond Beverton (1922–1995) and Sidney Holt (1926–) proposed a slight modification of the logistic, in which a fish population grows according to the logistic except that there is an additional mortality rate because of fishing. This equation was analyzed extensively by these authors in the now-classic 1957 book *On the Dynamics of Exploited Fish Populations*.[26] I got to know Sidney Holt when I began work on marine mammals in the 1970s. He is a delightful person with a great love of life and sense of humor, and when we talked about the Beverton-Holt equation, Sidney would laugh at how much it was still in use, given its simplicity and limitations. Fisheries biologists and ecologists unfamiliar with integral and differential equation calculus sometimes don't understand that the Beverton-Holt equation is a variant of the logistic. Sometimes when I talk to them about how the logistic still dominates fisheries management, they'll say, "No, we don't use that, we use Beverton and Holt," not realizing that it is just a modification of the logistic. When this happens, the discussion begins to take on an aspect of folktales and how they are spread—a story that someone thinks up and tells to someone else, who then repeats it with variations, and it becomes a part of a society's culture.

The moose and wolves of Isle Royale also do not fit the logistic and Lotka-Volterra theories. Accurate and methodologically consistent counts of these populations have continued from 1959 to the present (the earlier estimates that I discovered before were based on less formal and consistent methods). The populations have not undergone regular population cycles of fixed duration nor of fixed maxima and minima, as the Lotka-Volterra equations predict (Fig. 3.6). Moose peaked at 2,500 individuals in the early 1990s, close to twice as high as any other peak—two other peaks reached less than 1,500 moose. Similarly, the wolf population reached a single peak of 50 individuals around 1980, close to twice the other maxima for wolves.

The population history of moose on the island also contradicts the assumptions of the logistic growth curve, which led to another long-held tenet about such large herbivorous animals: that their populations will be self-regulating, just as a theoretical logistic population is, and for the same reasons—food limitations would lead to a stabilization in such a population. Rolf Peterson, a wildlife ecologist who has been one of the long-term researchers on the wolves and moose of Isle Royale, wrote in 1999 that the long-term studies of these populations on the island revealed "the end of natural regulation" in regard to self-regulation by an herbivore, because the moose population did rise to a very high level, from which it "crashed," rather than stabilizing as the long-standing theory would forecast. Consistent with the thesis of *Discordant Harmonies*, Peterson wrote that "the long-standing NPS [National Park Service] management tradition of nonintervention may not be compatible with the current policy that stresses maintenance of natural ecological processes, such as a predator-prey system."[27]

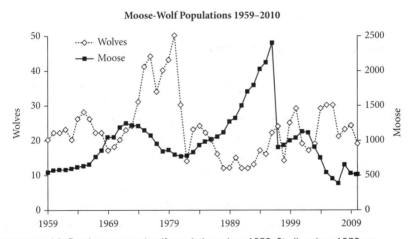

Moose-Wolf Populations 1959–2010

FIGURE 3.6 Isle Royale moose and wolf populations since 1959. Studies since 1959 are consistent in methods and more accurate than estimates of the moose and wolf populations of the island before that year.

(*Source:* J. A. Vucetich and Rolf O. Peterson. [2010]. *Ecological Studies of Wolves on Isle Royale, Annual Report 2009–10* [Houghton, MI: School of Forest Resources and Environmental Science, Michigan Technological University, Houghton, MI].)

Gause's predator-prey experiments, the elephants of Tsavo, the crashes of marine fisheries, and the conservation of marine mammals are a few illustrations, among many, of the deficiencies of the earlier theories in ecology that asserted a constancy of nature. At the heart of the issues are ideas of stability, constancy, and balance, ideas intimately entwined in theories about nature. Perhaps one reason that the deficiencies of the theories were not examined or tested adequately by observations in the field—out in nature—was that ecologists were typically uncomfortable with theory and theoreticians. Doing science and creating theory were commonly distinguished as separate activities. Theory was commonly considered unnecessary or unimportant to the practicing ecologist. Formal models were considered unnecessary, and conceptual models (idea models, constructs in your head, the kind scratched on the back of an envelope) were rarely if ever discussed explicitly. Yet, as I tried to make clear, theory played a dominant role in shaping the very character of inquiry and conclusions about populations and ecosystems—that is, about nature. As Kenneth Watt, an ecologist familiar with theory, wrote in 1962, ecologists had tended to believe that their science lacked theory, while in fact it had "too much" theory, in the sense that theory had been used and was influential even though it was not carefully connected to observations.[28]

"Field ecologists," those making measurements and observations in the forest and field, generally did not understand the mathematics of the logistic and of the Lotka-Volterra equations. But because physicists and mathematicians had the highest status among scientists, and because what physicists and mathematicians generally said was generally right, field ecologists tended to regard the logistic and the Lotka-Volterra equations as true. Lacking the understanding to analyze and thereby criticize these equations, they accepted them on the basis of authority.

Mechanical Stability Applied to Nature

In the scientific literature of ecology, the idea of ecological stability is often vague and implicit, but where it is stated explicitly, it is almost invariably equivalent to, and usually consciously borrowed from, the physical concept of the stability of a mechanical system, phrased in terms used in engineering, physics, and chemistry. Three concepts of stability were important to ecology and its applications in the first half of the twentieth century (and continue to be important as I write this in 2012): static equilibrium, quasi-steady state, and classical static stability. *Static equilibrium* means absolute constancy of abundance of all species over time (like a clock pendulum at rest). *Quasi-steady state* refers to variations that are persistent but small enough to be ignored (like a shaking clock pendulum). *Classical static stability* has two attributes: constancy unless disturbed, and the ability and tendency to return to the state of constancy following a disturbance (like a pendulum in motion).[29]

An equilibrium is a fixed rest point, a condition of constancy. The idea of stability is often confused with the equilibrium or constancy itself, but an equilibrium can be stable or unstable. A metal rod (or something with a similar shape, a thin rectangular solid, like a cell phone, an iPad, a typical candy bar out of its wrapper) balanced on end is in an unstable equilibrium. As long as it is not pushed, it remains in a constant vertical state. A pendulum, in contrast, has a stable equilibrium in its vertical position.

These ideas suggest, at least metaphorically, the possibility of fragile and resilient balances in nature. It is possible that there can be balance without stability. Conversely, the ability to return to an original condition might be quite strong, whether or not the original condition had the appearance of perfect order or relative chaos. There are many physical or mechanical examples of fragile and resilient stability. Imagine a traveler on a train who is sitting in a club car with a drink and building a house of cards. The house of cards has considerable order, but it will collapse at the slightest vibration; one could not expect the house of cards to last very long on the journey. The liquid in the glass may jostle about considerably, but it won't spill unless the train hits a particularly violent bump. Is Isle Royale like the house of cards or the liquid in a glass?[30]

It is comparatively easy to consider the characteristics of constancy and stability of physical entities: a ball, a pendulum, a tower, an airplane. It is much harder to attach clear meanings to the ideas of constancy and stability for populations, communities, and ecosystems, and the difficulty increases in that order. But with the development of the machine metaphor for the balance of nature expressed in formal mathematical models, the ideas of constancy and stability became the basis for the management of fisheries, wildlife, and endangered species. You can find an excellent exposition on the application of the concept of classic static stability to logistic and Lotka-Volterra classes of models in papers and books by Lord Robert May, a physicist-turned-ecologist who has had a distinguished career that included serving as the Queen of England's environmental adviser.[31]

In the social and political movement known as environmentalism, ideas of stability may have been less formal, but the same underlying beliefs in a balance of nature dominated. It should be clear by now that although modern environmentalism seemed to be a radical movement in the 1960s and 1970s when it rose to great popularity in the United States, Canada, and Western Europe, the ideas on which it was based represented a resurgence of pre-scientific myths about nature blended with early-twentieth-century studies that provided short-term and static images of nature undisturbed.

Rabbits, Cats, and Trappers

I realy [sic] do not know what we should have done for Victuals, as not one partridge has been served out this winter nor Rabbits, nor Fish to be got.

—George Atkinson *(January 9, 1785)*

There are in some seasons plenty of rabbits, this year in particular, some years very few, and what is rather remarkable, the rabbits are the most numerous when the cats appear . . . the cats are only plentiful at certain periods of about every 8 to 10 years.

—Peter Fidler, report to the Hudson's Bay Company *(1820)*

Might I turn your attention to the remarkable circle of increase and decrease that each decades [sic] exhibits. In nearly all the Furbearing animals this is observable, but particularly so in the Martens. The highest years in the decade 1845–55 being the extremes and the lowest 1849.

—Bernard Ross, *letter to Spencer Baird of the Smithsonian Institution (November 26, 1859)*

(All above quoted in L. B. Keith, Wildlife's Ten-year Cycle [Madison: University of Wisconsin press, 1963], p. 6.)

Analysis of fur returns and pelt collections using autocorrelations and spectral techniques demonstrates that some populations exhibit unequivocally regular oscillations while others exhibit fluctuations which are either non-cyclic or questionably cyclic. In Canada, lynx in the Mackenzie River district and snowshoe hare oscillate with a period in the neighborhood of ten years.

—James P. Finerty, Ph.D. thesis, *Yale University (1971)*

Another approach to determining whether nature really is constant over time is to examine long-term histories of populations, ecological communities, and ecosystems. However, ecology is not yet 200 years old, and a century is too short a time in which to resolve by direct observation whether populations are constant or stable. Among the small number of scientific observations, there are few in which the methods of observation have been consistent for more than a decade. Few scientists have maintained an interest in one population for more than a decade, and in cases where they might have, rarely has any government or private organization been willing to fund such lengthy projects.

Counting the number of animals of one species for a long time is a rather idiosyncratic activity. This is illustrated by two Australian biologists, Louis C. Bich (1918–2009) and Herbert Andrewartha (1907–1992), who conducted one of the longest direct scientific counts of a single population of insects—thrips, an insect about 1 millimeter long that lives in the petals of flowers, where it feeds on pollen. The two biologists went for a walk every morning between 1932 and 1946, except on Sundays and "certain holidays." They collected roses at precisely 9:00 a.m. and counted the number of thrips per rose. Over the 14-year period, there was great variation in the population size, with the peak number of thrips per rose varying from as few as 100 in 1934 to as many as 500 in 1939. Variation, not constancy, was the rule (Fig. 3.7), with the number of thrips changing both annually and seasonally; in the latter case, the largest number were found in midsummer.[32]

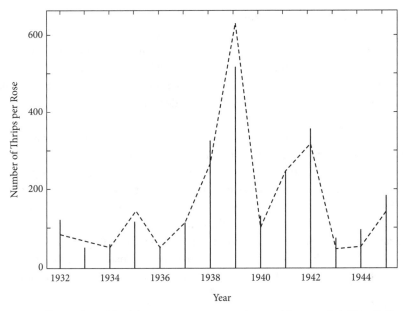

FIGURE 3.7 The number of the thrips per rose, as measured over a 14-year period. (From H. G. Andrewartha and L. C. Birch [1954]. *The Distribution and Abundance of Animals* [Chicago: Univ. of Chicago Press].)

Insect populations are in general highly variable, as suggested by outbreaks of locusts and by short-term observations of other insects that attack crops and trees. However, the dearth of long-term scientific studies prevented any definitive conclusions about the constancy and stability of insect populations. For many years, censuses were taken of insect pests in German coniferous forests. According to one record, the *Lasio-campid* moth maintained a comparatively low and constant population in all but two years during an 80-year period, but in those two years the population increased to epidemic numbers. At the peak, the insect was 10,000 times as numerous as it had been at the minimum. There are a few other studies, including 150 counts of the moth *Lymantria monacha*, of which there were six outbreaks in Germany. Highly erratic fluctuations in the chinch bug population were observed in the Mississippi Valley between 1823 and 1940, and similarly erratic variations have been observed in wheat-blossom midges in England and tsetse flies in Africa.[33]

Amateur bird-watchers and the participants in annual American Audubon Society Christmas counts have made shorter-term direct observations of some species of birds. The longer records of breeding bird populations are from Europe; these are on the order of decades and also show great fluctuation rather than constancy.

Perhaps surprising to those with scientifically sophisticated views, trappers in northern Canada provide us with the longest continual records of animal populations. Moravian missionaries began trading for furs with the Eskimos in Labrador at the beginning of the eighteenth century and the trade grew. The Hudson's Bay Company entered the fur trade around 1830, and records have been kept since. Charles Elton (1900–1991), one of the most important ecologists of the first half of the twentieth century, famous for his books (including *Voles, Mice and Lemmings: Problems in Population Dynamics*), realized the importance of these records and examined them in the 1940s.[34] His analysis made them among the best-known and most studied population histories, mentioned in every ecology textbook. In the 1980s, while working on *Discordant Harmonies*, I obtained additional data accumulated since Elton did his study from the Hudson's Bay Company. The data from both Elton's original publication and those obtained by myself are combined (Fig. 3.8).

As many have done, we can examine the Hudson's Bay Company records to see whether the animal abundances were constant over time. We can also, as some scientists have tried to do, analyze these historical records to find out whether predator and prey oscillated as would a Lotka-Volterra pair of populations.[35]

A 220-year history of one of the animals, the Canadian lynx, is shown in Figure 3.8. The number of lynx pelts obtained varied tremendously, from fewer

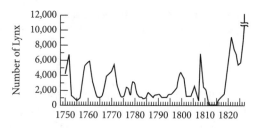

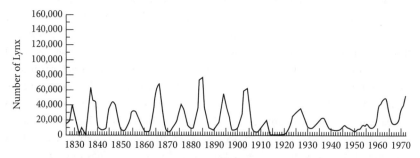

FIGURE 3.8 The 220-year history of the Canadian lynx based on fur trading records from the Hudson's Bay Company. (Data until 1940 from W. S. Allee, A. E. Emerson, O. Park, T. Park, and K. P. Schmidt [1949]. *Principles of Animal Ecology* [Philadelphia: W.B. Saunders]. Data since 1940 kindly made available to the author by the Hudson's Bay Company. To my knowledge, this and *Discordant Harmonies* are the first and still only publication of Hudson's Bay Company's records that extend beyond 1940. Ecology textbooks continue to reproduce the original records, part of a tradition, like the tradition of teaching the logistic and Lotka-Volterra equations, continuation without justification or reason.)

than 1,000 in the 1790s to more than 80,000 in 1885, but oscillated fairly regularly, with peaks about every decade. Other furbearing animals showed the same kind of variation, and fell into two groups: herbivores (such as the Arctic hare, a prey of the lynx), whose abundance cycled approximately every four years; and carnivores (such as the lynx), whose abundance cycled about every ten years. The obvious conclusions to draw from this history are that animal populations are not constant and that predator and prey do not oscillate as a Lotka-Volterra set would.

But the data might be faulty. The number of animals trapped may have varied even though the abundance of the animals was constant. For example, the number of trappers and traps and the interest in catching animals may have changed, perhaps with some kind of ten-year economic cycle. Elton considered this, but concluded from the records that there were always more trappers and more traps than animals to be caught, and on this basis the trapping would be an accurate reflection of the animal populations. While the observations of the trappers and other residents over the years support Elton's conclusion, it is not certain that this was the case. It would be much better to have records that were independent of trapping intensity, and in the 1980s it came to light that the data are faulty in another way: The major trapping of Arctic lynx took place in a different area of Canada from that of hare, and so the Hudson's Bay Company's records are not in fact a record of the populations of a predator and its actual prey, but only of a predator and an animal that it could have preyed on if the two were in the same habitat.[36]

In the cases I have discussed, predators do not seem to control the abundance of their prey in an exact sense. However, predators and parasites do have large effects on prey and host populations. The evidence is especially strong for parasites, which are used routinely today in the biological control of pests, and some do so efficiently. A plant spray containing *Bacillus thuringiensis*, a bacterium that causes disease in garden insect pests, is widely available. Parasitic wasps are used to control caterpillars. As for predators, mosquito fish are placed in rice paddies to control the growth of mosquitoes. Herbivores can devastate the vegetation that is their prey, as is clear for the moose at Isle Royale. But although predators and parasites can greatly reduce the abundance of their prey and hosts, there is little evidence that the result is a constancy of nature, a perfect balance in the classical sense or in the Lotka-Volterra sense of a fine-tuned oscillation.

Fitting Facts to Theory

By the 1950s and 1960s, evidence indicated that variation in population size is the rule, that animal populations are not in a static equilibrium. But some ecologists continued to argue against it, claiming that the data show merely minor variations and are therefore proof of a quasi-steady state. This was the view expressed in the 1950s by the famous British ecologist and ornithologist David Lack (1910–1973) in an influential book, *The Natural Regulation of Animal Numbers*.[37] Lack reviewed the evidence available then, just as we are doing now, and concluded that "most wild animals fluctuate irregularly in numbers between limits that are extremely restricted compared with what their rates of increase would allow."[38] This means each population is capable of exponential growth, and the fact that exponential growth did not occur over the entire period of observation was taken to be sufficient evidence that the variations are not different from constancy: The pendulum may be shaking, but it is fundamentally stable.

This is an unfair argument, made on shifting grounds. The word *regulate* has two meanings: to maintain an exact constancy; to remain within some general range. One starts the argument with the first meaning, by stating that nature undisturbed is constant and therefore is regulated. Turning to the evidence of lynx pelts, for instance, one finds variations in number from 800 to 80,000. Confronted with this huge range, no contemporary ecologist I know would claim that was an example of population regulation as it has been meant within the field. But, Lack would say, if that lynx population started at 800 and grew exponentially at 5% per year, in 220 years the lynx population would exceed 47 million, 60,000 times as many as the lynx numbered at their peak. *Ipso facto* in comparison, the variations in the lynx population are "small," and therefore some mechanism must be at work to keep the population in check, when "in check" means somewhere between hundreds and tens of thousands. Then the second meaning of *regulate* is invoked in its extreme—anything less than an exponential rise to infinity is "regulated." The discussion ends with the conclusion that the population is regulated, suggesting that the first meaning of the term is the intended one, when it cannot be.[39]

The fallacy can be illustrated by analogy. This argument would not be acceptable for a stability analysis of an airplane in flight as viewed by a passenger. Suppose a pilot guaranteed that an airplane was stable and very constant in flight, but its path through the air traced out the curve of the lynx's population history, varying from 800 to 80,000 feet in altitude. Who would call that airplane stable? Lack's interpretation violates the assumptions of the logistic and the more qualitative assumptions about animal populations generally held by ecologists at the time that Lack wrote his book. In addition, his interpretation violates ancient and modern ideas of a balance of nature that assume and require strict constancy and tight regulation. And perhaps most important, it violates any idea that appears in ecological scientific writing about what it means for a population to be regulated. The facts were interpreted to fit the theory, not vice versa.

When I first went to Isle Royale, it was generally accepted that animal populations in undisturbed wilderness are unchanging over time. The dominant ecological theory of the machine age implied and predicted that populations achieve a constant abundance or undergo exacting periodic oscillations and that the growth of populations is stable—that is, populations return to their original constant abundances or exact predator–prey cycles if disturbed. However, the available evidence, discussed in this chapter, led to a different conclusion. When I worked on *Discordant Harmonies*, I read everything I could find in the scientific literature that provided histories of populations. It was clear that variation, rather than constancy, in the abundance of animal populations was evident in observations and experiments. In only the simplest possible case—the growth of a single population of Paramecia under constant laboratory conditions—is the theory supported by experiment. Even two-species laboratory populations fail to support the theory, and long-term records of predator and prey populations in the wild show cycles that are not similar to those predicted by theory, with huge variations in amplitude, and periodicities of different lengths. Yet despite the absence of agreement between theory and fact, the theory was used in science and as a basis for resource management, and continues to be used in both today.

But you could say that so far I have considered rather short-term histories of populations—even 220 years of lynx is not long enough for us to decide about a balance of nature. Perhaps a longer history of nature, involving larger groups of species than predator and prey, will provide additional insight about the character of undisturbed nature and its stability or lack of stability, and that is what we turn to next.

FIGURE 4.1 A NASA helicopter, part of research to study rates of change in the northern forests of Minnesota, approaches to land in the U.S. Superior National Forest, adjacent to the Boundary Waters Canoe Area wilderness. This research, directed by the author, showed that even these forests of a comparatively cold climate burned frequently and were never in a single condition for very long. (Photograph by the author).

4

Oaks in New Jersey

MACHINE-AGE FORESTS

The individual trees of those woods grow up, have their youth, their old age, and a period to their life, and die as we men do. You will see many a Sapling growing up, many an old Tree tottering to its Fall, and many fallen and rotting away, while they are succeeded by others of their kind, just as the Race of Man is. By this Succession of Vegetation the Wilderness is kept cloathed with Woods just as the human Species keeps the Earth peopled by its continuing Succession of Generations.

—THOMAS POWNALL (1784)[1]

[*The new settlers of Pennsylvania take*] little account of Natural History, that science being here (as in other parts of the world) looked upon as a mere trifle, and the pastime of fools.

—PETER KALM (1750)[2]

A Virgin Forest Begins to Change

In 1701, Cornelius Van Liew, a Dutch settler, purchased land near what became New Brunswick, New Jersey, clearing all of it but one small area of magnificent old trees, which he kept as a woodlot. Half a century later, the forests of this locale were viewed by Peter Kalm, a Swedish botanist sent to the New World by Carolus Linnaeus to collect plants. Traveling from Philadelphia north to New England in 1749, Kalm described the forests as composed of large oaks, hickories, and chestnuts, so free of underbrush that one could drive a horse and carriage through the woods.[3]

By the mid-1950s, the chestnut blight had eliminated the chestnut as a major tree of the eastern United States, and all the original oak-and-hickory forests of New Jersey had been cut except for Van Liew's small woodlot, barely 65 acres. Recognized for its uniqueness as the sole remaining virgin forest of the region, it was purchased in the 1950s by Rutgers University and became known as the Hutcheson Memorial Forest. The establishment of this unique preserve became a minor media event. An article in *Audubon* in 1954 described this wood as "a climax forest . . . a cross-section of nature in equilibrium in which the forest trees have developed over a long period of time. The present oaks and other hardwood trees have succeeded other types of trees that went before them. Now these trees, after reaching old age, die and return their substance to the soil and help their replacements to sturdy growth and ripe old age in turn."

The replacements for the current mature trees were thought to be "largely white, red and black oaks," with some hickories—that is, the same species that dominated the forests in the 1960s.[4]

An advertisement in a major national magazine by Sinclair Oil, which had helped in the purchase of the forest, referred to the woods as a place where "nature has been working for thousands of years to perfect this 'climax' community in which trees, plants, animals, and all the creatures of the forest have reached a state of harmonious balance with their environment. Left undisturbed, this stabilized society will continue to perpetuate itself century after century." In the advertisement was a picture of Rutgers University's leading ecologist, Murray Buell, examining a cross-section of one of the huge logs that were the remains of an old tree.[5] (Buell [1906–1975] was my major professor for my Ph.D. and one of my favorite people ever. He was one of the best naturalists I ever met and a kind and gracious New England gentleman if there ever was one.)

A major article in the November 8, 1954, issue of *Life* magazine repeated the theme. This woodland "stands as a 'climax' forest community, which means that it has approached a state of equilibrium with its environment, perpetuating itself year after year essentially without change, secure against the invasions of all other forest types that might seek to displace it."[6] The article went on to describe scientists' opinion about "how long nature might take to re-create the woods were they now despoiled and converted to cropland."

Buell and his contemporaries regarded populations of trees as complex assemblages made up of individuals of different sizes and different rates of growth, reproduction, and mortality. Beginning with bare ground, a forest was believed to develop through a series of stages, each following the preceding one in clockwork fashion until a final permanent form, the "climax stage," was achieved. This process was called *succession*, a term perhaps first used by Thomas Pownall (1722–1805) (quoted at the beginning of this chapter), who had been governor of the Massachusetts Bay Colony and of South Carolina, and lieutenant governor of New Jersey. He used it in his book *Topographical Description of the Dominions of the United States of America*, published in 1776.[7] The term was reintroduced in 1860 by Henry David Thoreau (1817–1862) to describe the development of pine woodlands following the logging of hardwood stands in New England, and used again in 1864 by George Perkins Marsh. This succession, which led inevitably to the same climax forest, was described in *Life*. It was assumed to be so exact that even the duration of each stage was regular and predictable, with the entire process taking "no less than four full centuries."[8]

Such a climax forest would make a pleasant still-life painting, composed of trees of many sizes. Dominating the landscape would be the old, 100-foot trees, beneath which would grow younger trees, then saplings, and, close to the forest floor, seedlings. As the old trees died, they would be replaced by vigorously growing younger trees, and so on, so that the view of the forest at any one time would be identical to the view at any future time.

But Hutcheson Forest did not stay constant. By the late 1960s it was clear that something unexpected was happening: The forest was changing. There were few young oaks beneath the dying giants. Instead, saplings of sugar maple were everywhere. The future forest would be maple, not oak and hickory. What had happened to the forest of permanent form and structure? The answer lay within the trees themselves.

Trees that had fallen during a hurricane in 1950 were cut, and their growth rings were counted and studied. These studies showed that fire had once been common in Hutcheson Forest. When a fire burns through a forest, trees not killed may be wounded, with patches of bark burned away, leaving scars, over which new wood grows. Annual growth rings allowed Rutgers scientists to date each fire and determine the average time between fires. Buell found that before European settlement in 1701, fires had occurred about every decade, but none had burned after European settlement, when they had been intentionally suppressed.[9]

Oaks and hickories, it turned out, dominated the forest because they are more resistant to fire than are sugar maples. Burned oak trees can sprout new stems from their remaining roots, and acorns and hickory nuts can survive light fires better than sugar maple seeds can.

Historical evidence suggests that the American Indians occasionally may have intentionally set fires. When fires are frequent, little fuel builds up in the time between any two, so the fires have only light effects. The effects of fire were visible in another way: The open woods that Peter Kalm had seen were a result of frequent, light fires. The 1960s and continuing to the twenty-first century Hutcheson Memorial Forest was a dense thicket of shrubs and saplings crowded against the tall trees (Fig. 4.2). The forest had taken more than 200 years to respond to the suppression of fire in a way that people could recognize. The forest primeval was revealed as, in part, a human product.[10]

Since the realization that the appearance of Hutcheson Memorial Forest was a result of frequent light fires, others have observed the same kind of openness of forests in the eastern United States. In 2000, forest scientist Neil Sampson summarized what had been learned for forests of the eastern shore of Maryland. He writes that pre-settlement forests burned every seven to ten years, in some places even more often. And like Peter Kalm, travelers within the eastern shore of Maryland found the forest open. Captain John Smith said in the early seventeenth century, "A man may gallop a horse amongst these woods any waie, but where the creekes or Rivers shall hinder."[11]

FIGURE 4.2 Hutcheson Memorial Forest in 1967. One of the few remaining old-growth oaks is surrounded by a dense thicket of saplings, mostly of other species, and the ground cover includes sugar maple seedlings. (Photograph by the author.)

In the 1960s, Hutcheson Memorial Forest was changing in other ways. Exotic species from other continents—such species as Japanese honeysuckle, Norway maple, and the Chinese tree of heaven—were abundant even in the 1960s as young plants. These exotics were brought to North America intentionally for various purposes, usually decorative. Norway maple had become a fashionable urban tree; Japanese honeysuckle was planted in many urban gardens. Once these species were established in the New World, their seeds were transported by natural means—wind, birds, or mammals—and spread into Hutcheson Forest. It was clear in the 1960s that the future Hutcheson Memorial Forest would be unlike any ever seen, dominated by native and introduced maples and dense with exotic species.

A few years ago I revisited Hutcheson Memorial Forest, which is still maintained by Rutgers University as a nature preserve. It had been more than 30 years since I'd last been there. Peter J. Morin, professor in the Department of Ecology, Evolution & Natural Resources and director of Hutcheson Memorial Forest, kindly arranged for a group, including himself, to take me around. As we walked through it on that lovely warm day, it was clear that the forest had continued to change, to move away from the twentieth-century ideal of a climax forest in steady state. Many of the grand old

FIGURE 4.3 Hutcheson Memorial Forest in 2004. One of the grand old trees has fallen into its neighbors, as if symbolizing the end of what had been called the forest primeval. The young trees nearby are mostly sugar maples, which, according to twentieth-century beliefs, were not supposed to grow there. (Photograph by the author.)

trees that were supposed to represent the climax forest, and still grew in the 1960s, had disappeared; a few had decayed and fallen over, leaving only stumps (Fig. 4.3).

The story of Hutcheson Memorial Forest shows that people thought about forests the same way they thought about animal populations, described in Chapters 2 and 3. Scientists, as well as the public, believed in the constancy of nature and in nature's ability to recover its exact former structure following a disturbance. In the 1950s these ideas were presented as results of modern science, but they were in fact an exact repetition of nineteenth-century ideas, which were in turn repetitions of much earlier beliefs.[12]

Static Theories

The classic statement of belief in the constancy and stability of nature was made in 1864 by George Perkins Marsh in *Man and Nature*:

> Nature, left undisturbed, so fashions her territory as to give it almost unchanging permanence of form, outline, and proportion, except when shattered by geologic convulsions; and in these comparatively rare cases of derangement, she sets herself at once to repair the superficial damage, and to restore, as nearly as practicable, the former aspect of her dominion.

He went on to write that "in countries untrodden by man," all factors balance one another, so that "the geographical conditions may be regarded as constant and immutable."[13]

Botanists had divided the landscape into what they called vegetation "associations," groups of species that occur together over wide geographic areas. The groups were named for those species that were believed to be the dominant ones in undisturbed conditions. Through the first half of the twentieth century—before the invention of radioactive dating methods and pollen analyses of ancient vegetation—scientists believed that these groups had persisted for a long time anywhere that was not influenced by man, and that if they did move in response to climate change, they moved as a group, as a unit. This meant that North America before European colonization had kinds of vegetation associations that had persisted with little if any change—that is, the vegetation was in static equilibrium.

Once again, we can turn to Isle Royale, our American wilderness, for an example. In 1909, William S. Cooper, a biologist from the University of Chicago, went to Isle Royale to study its forests and their development after natural disturbances by windstorms and fires. The result was one of the classic forest ecology studies of the early twentieth century.[14]

Cooper studied primarily the northeastern part of the island, which he found was largely a forest of balsam fir, white spruce, and white birch, with some bogs and areas of maple. The forest, he wrote, "is the final and permanent vegetational stage, toward the establishment of which all the other plant societies are successive steps," and "both observational and experimental studies have shown that the balsam-birch-white spruce forest, *in spite of appearances to the contrary* [my italics], is, taken as a whole, in equilibrium; that no changes of a successional nature are taking place within it," even though "superficial observation would be likely to lead to exactly the opposite conclusion." (The key here is the phrase "in spite of appearances to the contrary." This seems to typify much ecological thinking in the twentieth century. Cooper had to have been a very good naturalist; the rest of his writings make this clear. Yet he could stand there and say to himself, "It might not look this way, but it has to be.")

Cooper saw this forest as stable in three ways. First, it had classic stability: When disturbed, any patch recovered to its original condition, which persisted until disturbed again. Second, given a long enough time without disturbance, any bare surface—be it a barren, dry, rocky ridgetop; an open-water bog; or a fertile, well-drained soil—would develop into the same final balsam-spruce-and-birch forest: "In other words, those phases of the vegetation that are not uniform in character with the main forest mass are plainly tending toward uniformity." The collection of species that formed this balsam-spruce-and-birch forest had to be the end result on any kind of surface—pond water that filled into bog and then solid ground, sandy beaches along lakes, or well-drained and fertile soils. Third, although there were natural disturbances, such as fire and windstorms, they were local and they balanced out in the larger picture: "The climax forest is a complex of windfall areas of differing ages, the youngest made up of dense clumps of small trees, and the oldest containing a few mature trees with little or no young growth beneath," but "the changes in various parts" balance each other so that "the forest as a whole" remains the same. Every disturbed area returned to the same final condition, which remained constant until disturbed again.

Moreover, the forest was believed subject to a uniform rate of small disturbances so that its average composition was always the same. Any small area was thus characterized by a stability like that of a pendulum: If upset or pushed, it returned to its former condition. The forest as a whole, on a large scale, was composed of a constant number of patches of any one stage at any time, and was therefore uniform. Nature without human influence was constant in both the small and the large. Cooper's study and conclusions were being repeated in other areas, and this concept of succession was expounded by major botanists in the first decades of the twentieth century.

A century later, formal scientific statements about succession had changed little. For example, Robert Whittaker (1920–1980), one of the major plant ecologists of the twentieth century and one of Murray Buell's closest plant ecology colleagues, wrote in 1970 that Hutcheson Memorial Forest was behaving as it should according to the old ideas that "communities go through progressive development of parallel and interacting changes in environments and communities, a succession," and that "the end point of succession is a climax community of relatively stable species composition and steady-state function, adapted to its habitat and essentially permanent in its habitat if undisturbed."[15] By this time, Buell was well aware that the forest was not behaving in this way, and at some level Whittaker should have been, too, because he was a careful, scholarly, and influential ecologist, one of the most thorough and intellectual plant ecologists of the time.

In a 1969 paper that summarized the then-current major beliefs about ecology, Eugene Odum (1913–2002), whom many consider to be the father of modern ecosystem ecology, wrote that succession is "an orderly process of community development that is reasonably directional and, therefore, predictable" and that succession "culminates in a stabilized ecosystem."[16] Odum was therefore reasserting the mechanistic beliefs about ecology even as the machine age waned.

These same ideas about the constancy and classic stability of ecosystems persist today. In his 2009 book *Heatstroke: Nature in an Age of Global Warming*, paleoecologist Anthony D. Barnosky writes that "complex systems are pretty robust and resilient. Unless something unusual happens, they are slow to change, and even when they are buffeted by events from outside the system they tend to return to the same stable state. Ecosystems are no exceptions."[17]

As we have seen, Hutcheson Forest did not conform to these ideas. The character of nature, or the interactions of human beings and nature, or both, led to different dynamics. Was this merely a peculiar exception or the general case? Trees would seem more likely than animals to remain in constant abundance over long periods, not only because individuals cannot migrate but also because most trees can live much longer than animals.

The view of nature that we have been talking about is rather local, what you can see in a casual day's walk around Hutcheson Forest or on a hike of several days at Isle Royale. When we fail to find constancy at that scale, the natural thing to do is to take a broader perspective, hoping that conditions will average out or that there are features of a larger area that are constant even if the local scene is not. And so we need to take a longer look back in time and a broader scan beyond the horizon.

Trees and other flowering and coniferous plants provide much better evidence about long-term patterns than do animal populations. With vegetation, three kinds of data are available. First are written historical accounts. In North America, for example, there are the accounts written by early explorers, like Peter Kalm; notes of land surveyors; and studies by natural historians. Because forests are important economic resources and forestlands were often cleared for farming, ranching, and human settlement, there is a considerable written history of forest conditions.

Second, many species of trees produce annual growth rings, from which we can determine their age and how well they grew during each year of their lifetimes. By extension, the age and development history of an entire forest can be determined by direct measurement of the growth rings of sample trees, as was done in Hutcheson Forest, enabling us to reconstruct the history of a forest for hundreds and sometimes thousands of years.

Third, tiny grains of pollen provide the longest record of the history of forests. Pollen is widely dispersed, and its outer casings persist for very long periods. When pollen grains are deposited in lakes, they drift to the bottom and become part of the lake sediments. As a lake gradually fills in, the sediments form a history of the local and regional vegetation, with the oldest sediments at the bottom and the most recent at the top. Scientists study these records by driving long corers into the sediments, preserving the cores by freezing, and using radioactive isotopes to date the samples. The patterns on pollen grains are distinct, and the grains are inert; thus, grains millions of years old can be used to identify the species that were present and to estimate their relative abundance at each time.

A Dynamic, Climate-Changing, Icy Past

Continental glaciation was not known to modern science until the nineteenth century and was not influential in the study of natural history until the twentieth century.[18] But by the time Odum and Whittaker and others were writing about the natural stability and constancy of ecosystems, it was well known that Earth's surface had gone through major periods of glaciation during the past 2 million years, and that those glaciations had greatly disrupted the distribution of life on the land.[19]

The farthest extent of continental ice is demarcated by terminal moraines, hills of mixed deposits of soil and rocks that extend for hundreds and thousands of miles. There are a number of distinct terminal moraines in Europe and North America. Four predominant moraines suggested to nineteenth- and early-twentieth-century scientists that there had been four major eras of glaciation. (More-recent evidence suggests that there may have been as many as 16 major periods of glaciation.)

The last great Pleistocene continental glaciation—beginning between 70,000 and 50,000 years ago when the continental glaciers underwent their major expansion, and ending about 12,500 years ago—had an impact on Earth that is difficult for us to imagine. As now reconstructed and generally accepted among scientists, the continental ice sheet covered 6 million square miles in North America, 2.3 million in Europe, and 6 million in Asia at the time of its greatest extension.[20] The North American ice sheet was bigger than that of modern Antarctica, and the ice was as much as 4,000 feet thick. Its southernmost extension in North America was more or less along latitude 37°30′N, meaning that it ran from Cape Cod through Long Island, Lake Erie, central Ohio, Indiana, Illinois, and approximately along the Missouri River valley to the Rockies. Cape Cod and Long Island are part of the terminal moraine, the deposits left by the glacier at its greatest extent. In machine-age terms, think of the glacier as a gigantic bulldozer, digging up soil and cutting up the bedrock and pushing the debris ahead. Moraines are the deposits of these materials along the sides and at the end of the ice. That's why they are above sea level and are large areas with little exposed bedrock.

The effects of the glaciers extended south beyond the limits of the ice sheets themselves. Decreases in temperature and increases in rainfall created large lakes south of the glaciers. One that modern geologists named "Lake Bonneville" covered 20,000 square miles 12,000 years ago, an area ten times bigger than its current remnant, Utah's Great Salt Lake. (The Great Salt Lake is salty because most of the original glacially deposited freshwater evaporated, leaving dissolved salts behind.) Because of intense winds along and to the south of the southern edge of the continental glacier, large parts of the central North American Midwest were covered with loess, windblown sands and silts as much as 20 feet thick. These formed the very fertile soils of the Great Plains, extending from Louisiana to eastern Colorado, and including 20,000 square miles of sandhills in Nebraska and South Dakota, and other sandhills in Texas and New Mexico.

Scientific understanding of continental glaciers developed in the early nineteenth century. Jean Louis (Rodolphe) Agassiz (born in Switzerland in 1807, died in the United States in 1873) discovered continental glaciation when he roamed the Swiss hillsides and spoke with farmers. At first he was skeptical that there could have been such major changes in the land, but the farmers showed him rocks and strangely shaped hills of soil debris that were far from any mountain glaciers; they had to have been transported somehow to where they now stood. These looked just like deposits from mountain glaciers plowing their way over the land. Something had to move those deposits long distances. Glaciers had moved them down mountains. Agassiz therefore recognized that certain landforms created by mountain glaciers also occurred widely on flatter lands and could only have been produced by giant glaciers.

He announced this in 1837, the year before Verhulst proposed the logistic growth curve for populations, and the date we can call the beginning of ecology. It was becoming clear that Earth's surface had undergone vast changes due to gargantuan geological forces. The landscape was not static and did not merely age, but pulsed back and forth over long periods.[21] (It is also interesting that farmers, unaware of current scientific theories to the contrary, learned directly from nature, and what they learned became the foundation for an important part of an environmental science, geology, and its subdiscipline, geomorphology. This foreshadows other impressive observations by foresters, fishermen, and farmers who, unacquainted with formal ecological theories, depended directly on what they saw, rather than what they had been taught to believe, and were able to provide wonderful insight. Such people did not observe nature and then preface their conclusions with "In spite of appearances to the contrary . . .")

For those who believed that vegetation was static over long periods, the evidence of continental glaciation raised several major questions: How can a balance of nature incorporate the history of the periods of glaciation? What can be said about the constancy and stability of nature since the last glaciation or during interglacials, the warmer periods between times of glacial expansion? In light of these major climate disturbances, is there any period for which one could talk about a constancy of nature?

These questions are of particular interest to us with respect to the Pleistocene epoch, usually estimated to have begun approximately 2.5 million years ago, corresponding closely with contemporary scientific estimates of the evolutionary origin of our genus, *Homo*. The earliest of our genus, *Homo habilis*, appears to have evolved sometime between 2.4 and 1.5 million years ago, and *Homo sapiens*, about 200,000 years ago[22] Thus our species appears to have emerged during the Pleistocene, and man the toolmaker appears to have arisen near the beginning of the modern glacial ages.

In this context, our original question can be rephrased: Is there evidence of any kind of constancy or stability of nature since the origin of *Homo sapiens* or since the appearance of hominid species who were toolmakers? Until very recently, it was generally accepted that periods of glaciation represented atypical climate conditions, while the interglacials represented the typical or average climate of the Pleistocene. In the first half of the twentieth century, scientists thought the normal climate of the past 2 million years was much like today's. Indeed, the same belief underlies, in an odd and self-contradictory way, current global climate models and assertions by some of the major climatologists and those writing about potential environmental effects of global warming.

First, all the publications I know of that discuss possible global warming refer to climate "anomalies," meaning differences between a standard climate and that of some other time. And the standard for all the climate work I know of is either the climate from 1960 to 1990 or the climate from 1960 to 1980. These are taken as standard and normal. One could argue that this is simply a mathematical convenience, but when you read what is written about how the world may change with global warming, it means much more to the writers. But I am getting ahead of myself; I will deal explicitly with global warming in Chapters 12 and 13.

The interglacial periods were believed to be much longer than the glacial periods—the glaciations were viewed as comparatively short anomalies or disturbances of an otherwise comparatively uniform climate. Furthermore, the impact of glaciation was thought to have been limited to the higher latitudes, leaving the subtropics and tropics untouched. Thus, until the second half of the twentieth century, about the time I first began working on *Discordant Harmonies*, ecologists and botanists believed that the tropics, having been spared climate change, were even more constant than the constancy attributed to ecosystems and populations of the higher latitudes.

Now we know that climate changed at all latitudes; and that throughout the tenure of our species on this planet, the climate has varied greatly. With this in mind, we have to rephrase the question: What has been the natural state of the vegetation in temperate and higher latitudes under continual climate change since our genus, and our species, has been on Earth? And we are forced to realize that the answer, once believed to be so simple, isn't. The state of the climate wasn't just one thing. And living things didn't just sit there like a train on a siding, like an image in a permanent still life.

Wilderness History

A little more than 100 miles west of Isle Royale, beginning just on the western shore of Lake Superior, lie a million acres of lakes, streams, marshes, and forests set aside in 1964 as a legally designated wilderness known as the Boundary Waters Canoe Area, extending from Minnesota to an adjacent parkland in Ontario, Quetico Provincial Park. Moose meandered there as they did on Isle Royale, and every summer people came, and still come, to canoe and hike and to discover wilderness. The Boundary Waters had been a favorite hiking, fishing, and hunting area for decades before, and its designation as a legally protected wilderness was the result of a long battle between those who believed the area should be open to motorboats, airplanes, and other kinds of modern recreation, as well as logging and perhaps even mining, and those who wanted it preserved in as "natural" a state as possible.

But what is the true character of this wilderness that people seek in the Boundary Waters? This was a question that intrigued Miron (Bud) Heinselman (1920–1993), a career U.S. Forest Service scientist who was a native of northern Minnesota and one of the major proponents of the wilderness designation for the Boundary Waters. I got to know and admire Bud during the 1980s when I was doing research in and near the Boundary Waters. He was another of those fine naturalists like Murray Buell, a great observer who paid attention to what he saw and never, to my knowledge, reached conclusions "in spite of appearances to the contrary." He studied the history of the vegetation of the Boundary Waters using three kinds of evidence: written history, existing forests, and lake sediments, and here's what he found out.

Pollen deposits from the Lake of the Clouds within the Boundary Waters Canoe Area indicate that the last glaciation was followed by a tundra period during which the ground was covered by low shrubs now characteristic of the Far North, as well as reindeer moss and other lichens and what botanists informally call "lower plants"—mosses, club mosses, and their relatives. As the climate warmed a bit, the tundra was replaced by a forest of spruce, species now found in the boreal forest of the North and dominant in many areas of Alaska and Ontario. The climate continued to warm, not at a constant rate but with a good bit of variation. About 9,200 years ago the spruce forest was replaced by a forest of jack pine and red pine, trees characteristic of warmer and drier areas. Paper birch and alder immigrated into this forest about 8,300 years ago; white pine arrived about 7,000 years ago; and then there was a return to spruce, jack pine, and white pine, suggesting a cooling of the climate.[23]

Thus, every thousand years a substantial change occurred in the vegetation of the forest, reflecting in part changes in the climate and in part the arrival of species driven south during the ice age and slowly returning. (Species that could live in the same climate did not all arrive back in that climate at the same time, because seeds of each species move at different rates from one another.)

The question is, which of these forests represented *the* natural state? Or, putting it a slightly different way: If our goal were to return the Boundary Waters Canoe Area to its natural condition, which of these forests would we choose? Each appears equally "natural" in the sense that each dominated the landscape for approximately 1,000 years and each occupied the area when human influence was nonexistent or slight. The range of choice is great—from ice to tundra to boreal forests to hardwood forests to prairie—representing kinds of vegetation communities now distributed thousands of miles apart and some that no longer exist.

In short, the goal of keeping the Boundary Waters Canoe Area in a constant state leads us to a blind alley. We know that allowing this area to become tundra is not what we had in mind. Nor do we want it to become an open prairie. It is my impression that what most people really want from a visit to the Boundary Waters Canoe Area is a sense of openness, contact with nature, a chance to get away from all modern pressures, to experience what Henry David Thoreau called "wildness," a feeling you get from being out there—some will call it a spiritual feeling, others an aesthetic one. And those who are naturalists or outdoorsmen and have read and thought about nature and its history in North America would want the sense of wilderness as it was experienced by the voyageurs 200 to 300 years ago, not only the feeling of wilderness as a place untouched by people, as it has been expressed in recent decades, but also the feeling of the north woods. That goal is a much less stringent one than a constant nature. It requires that we manage the forest to allow for a natural range of variations in space and time. It would not allow us to strip-mine the area or suppress fires to the point where the character of the forest changes.

Let us consider in some detail the goal of managing a particular spot in the Boundary Waters Canoe Area to retain its constant, single "natural" state. This creates a quandary not only in the short term but also in the longer term, because as the glaciers receded and the climate changed, the species composition of the vegetation communities also changed—the abundances of each forest-tree species over time at a particular location were not constant. In light of this history, what meaning can we now attach to a stability of nature? Or, as managers of a wilderness, what would we recommend as the "natural forest" for Minnesota or Isle Royale? And if we manage an area to maintain some desired state, isn't it no longer nature undisturbed?

One answer that ecologists proposed during the first half of the twentieth century was the idea of the "climatic climax" forest. This idea is that even if the forests moved around during the ice ages, they did so as units that remained intact, just as when you move snapshots around in the family album, the location of a picture changes but the people in each picture remain the same. Proponents of this idea granted that the glaciers caused a great dislocation of vegetation communities, but assumed that these communities migrated intact southward in front of the advancing ice, and then migrated northward again as the ice retreated. For example, for many years a standard text about ice-age geology pictured vegetation communities during the height of the glaciation as having been pushed south, unchanged.[24] It showed a tundra border just below the edge of the ice, spruce forests south of the tundra, and temperate mixed-species forests south of the spruce.

If the "climatic climax" idea was true, then the communities had a continuous existence throughout the Pleistocene, and the "natural" condition of a particular spot would be the biological community that was characteristic of the present climate, whatever that happened to be. (Note that, ironically, this would mean that if global warming happens as climatologists forecast, the "natural" vegetation a century from now might not be very much like what is there today. Instead of sounding an alarm that global warming was going to eliminate the boreal forest from the Boundary Waters, a believer in climatic climax would simply conclude that the change was natural and ought to be.)

But groundbreaking research by a small number of paleoecologists who studied fossilized pollen had also proved the simple idea of a climatic climax false by the time *Discordant Harmonies* was published. Margaret B. Davis, one of the most prominent of these scientists, studied pollen deposits from 26 sites scattered across the eastern and central United States and reconstructed the migratory paths of the major tree species returning north as the North American continental ice sheet melted during the last 13,000 years.[25] The trees migrated at different rates, depending on the size and mobility of their seeds. Light seeds, like those of poplar, are readily blown over long distances by the wind and moved northward most rapidly. Heavy seeds, like those of beech, are moved by squirrels and other small animals and as a result migrated much more slowly. Many studies since Davis's have confirmed her findings for vegetation around the world.

The different species moved northward not only at different rates but also from different directions. The hickories returned to Hutcheson Forest by moving northeastward from a refuge in the southern Midwest or West, while chestnuts moved westward from a refuge east of the Carolinas, in what is now the Atlantic Ocean but was dry land during the glaciation. (The glaciers contained so much of the Earth's waters that the ocean level was several hundred feet below its present average.) Thus, this research demonstrated that the species composition of forests changed markedly over the millennia, and the species that seem most important in the modern forest had been scattered in different directions, forming forest communities that no longer exist.

In our search for constancy in nature, we have had to abandon the ideas of local constancy and of a climatic climax, but we still have a few more possibilities. Perhaps the vegetation recovered very quickly to its present condition after an ice age. If this were the case, then one could talk about an interglacial forest that would be constant throughout its entire range for most of the interglacial period. But the evidence is against this possibility also. A map of tree-migration routes indicates markedly different rates of return for different species. Hemlock reached Massachusetts 9,000 years ago, approximately 2,000 years before beech, although now beech and hemlock grow in the same region.[26]

The migration process seems not to have reached its limit even today. Hemlock reached the Upper Peninsula of Michigan 5,000 years ago and moved westward slowly, reaching the western shore of Lake Superior 1,000 years ago near where the moose of Isle Royale must have entered the waters of that Great Lake at the turn of the century. Beech, however, seems to be still migrating westward, with the present western boundary of its range in the middle of the Upper Peninsula. Given that information, one could make a last-ditch attempt to hold on to a belief in the constancy of nature, arguing that recovery is still taking place and a true equilibrium in the distribution of the forest trees has yet to be reached. But this interpretation forces us to conclude that the true and constant state of nature will be achieved in the future and has never been seen by human beings. This is a peculiar and undesirable idea of a natural condition. The point is academic anyway, because the migration rates of trees during the previous interglacials show that such an equilibrium never was reached at the middle and high latitudes even by the end of an interglacial.

A cynic could argue that since ice has been the most common feature during the past 2 million years, it must be the identifying feature of nature undisturbed north of the moraines. Obviously, the line of reasoning we have attempted to follow, searching for a constancy of structure, is in error.

There is still one last possibility: If the vegetation is not constant, perhaps it is the pattern of change that repeats itself so that vegetation still responds like a clock pendulum to climate change. If this were true, then the pattern of reestablishment would have been repeated in every interglacial. For example, beech would always retreat to the same refuge during an ice age and return from it following the same routes at the same times. If so, then one could claim that the natural state of the forest would be the appropriate stage in the sequence that occurred following the end of each glaciation.

But this idea doesn't seem to work either. Pollen records from England, where the history of the trees has been studied for six interglacial periods, show that the pattern of species migration does not repeat itself from one ice age to another. For example, as the glaciers melted at the end of the most recent period, hazel rapidly became very abundant in East Anglia, Great Britain, and then essentially disappeared. But in the previous interglacial, hazel in Great Britain became abundant slightly later, and slowly became less abundant throughout the period. In the oldest interglacial, hazel in Great Britain was never very abundant but did persist throughout the period. The apparent disappearance of hazel in Great Britain during our own time could be attributed to human activities. But the great differences in its patterns of reestablishment in the earlier interglacials argues against regularity in the pattern of recovery during interglacials. The evidence, as far as the picture is complete, is consistent for the forested regions of North America and Europe.

To further complicate the picture, forests also change slowly in response to changes in soils, which develop continually throughout interglacial periods, suggesting perhaps that interglacials are too short to produce soil in a steady state, and providing another reason that vegetation might fail to achieve a constant condition even over a long period of time between glaciers.

Space Age Perceptions

In the mid-1980s, NASA posed three questions: whether there was such a thing as a global ecology; if so, whether the scientists from all the different disciplines involved in studying Earth's environment from space could communicate with each other, and if so, whether there would be a role in this for NASA. Scientists who had started at Johnson Space Center began to experiment in the 1970s and early 1980s with the detection of vegetation from space. The first NASA project came about when the then Soviet Union bought a large quantity of wheat from the U.S., claiming that there was a terrible harvest that year. Soon after, the USSR was selling that U.S.-grown wheat on the open market, at a much higher price than was paid for it. And soon after that, Congress funded NASA to develop the technology to measure Soviet wheat growth and to forecast Soviet wheat harvest. Landsat satellites, which could detect four color bands in visible and infrared light, provided the successful observations.

The leader of this research was Bob McDonald. He and I began to work together in the early 1980s to see whether these techniques could be applied to natural ecosystems. Funding appeared in the spring and we were supposed to begin research that summer. So I chose a place I was familiar with, the Boundary Waters Canoe Area of Minnesota and the adjacent Superior National Forest (Fig. 4.1).

Soon we had students making measurements on the ground of small areas of the forest, just large enough to more than fill the smallest image the Landsat satellite could see. Helicopters hovered above these with the same device in Landsat, so we could calibrate the satellite observations with what was on the ground.

We discovered that Landsat could distinguish five different stages in forest succession, from a cleared area to the oldest kinds of woodlands in the BWCA. We took two Landsat images ten years apart, one in 1973 and one in 1983, overlapped them and compared them. As I have discussed, according to the old beliefs and theories, the old-age forest stands should have persisted, certainly for a mere ten years. But on the contrary, Landsat showed us that only slightly more than half of the oldest stands stayed in the same state in 1983 that they had been in ten years earlier. The forest was incredibly dynamic even over very short times, not at all what anyone expected even then. [27]

Wherever we seek to find constancy, we discover change. Having looked at the old woodlands in Hutcheson Forest, at Isle Royale, and in the wilderness of the Boundary Waters, in the land of the moose and the wolf, and having uncovered the histories hidden within the trees and within the muds, we find that nature undisturbed is not constant in form, structure, or proportion, but changes at every scale of time and space. The old idea of a static landscape, like a single musical chord sounded forever, must be abandoned, for such a landscape never existed except in our imagination. Nature undisturbed by human influence seems more like a symphony whose harmonies arise from variation and change over every interval of time. We see a landscape that is always in flux, changing over many scales of time and space, changing with individual births and deaths, local disruptions and recoveries, larger-scale responses to climate from one glacial age to another, and to the slower alterations of soils and yet larger variations between glacial ages.

Oddly enough, since the publication of *Discordant Harmonies*, the tables have turned with respect to what is thought of as "normal," but not the basic concept that used to be called the "climatic climax" vegetation community, a term whose use has declined while its meaning has persisted. Today "normal" is put forward almost all the time as close to what we were experiencing at the end of the twentieth century. This, as I pointed out previously, is reinforced by modern climatologists in their writing about global warming, and by some of the leading ecologists doing research today about possible ecological effects of global warming.[28] We have replaced one imaginary constant nature with another, but not changed our fundamental beliefs about what nature must be like. More on this in Chapters 12 and 13.

Tropical Wilderness

There is so much concern today about the fate of natural areas in the tropics, especially the rain forests and the great plains and savannas of eastern and southern Africa, that it is important to turn our attention to the character of nature in these latitudes. It was commonly believed among ecologists before the 1960s that the tropics, even more than the middle and high latitudes, had been characterized by climate constancy and that the patterns of vegetation in the tropics had varied little in past millennia. This concept of the tropics changed in the 1970s, when sufficient evidence from pollen studies became available.

What, for example, has been the history of the great dry savannas at Tsavo, in the Serengeti, and down to the Cape of Good Hope, land that had felt the elephants' massive but gentle stride for centuries and had been considered by some the last sanctuary on Earth for wild nature? Although the great plains and savannas of Africa are often portrayed as the last natural wilderness, they have been influenced by human beings for millions of years through hunting and through fire. Contrary to popular myth, the populations of wildebeest and other large herbivores have varied greatly in the twentieth century. They were first decimated by rinderpest, a viral disease introduced from Europe at the turn of the century and usually fatal to cattle. They began recovering only in recent decades and have been undergoing rapid population increases ever since.[29]

While much less is known about the history of Africa than of regions in the northern latitudes during the Pleistocene, the picture that was emerging in 1990 was similar to the one for the North. Some of the evidence comes from the study of mountain glaciers in East Africa. In the land of the elephant, as in the land of the moose, the climate has changed greatly. Daniel Livingstone (1927–), an expert on African lakes and their history, studied pollen cores from Lake Victoria and wrote, "The African environment is capricious, not stable, and apparently has been so for at least several million years."[30] At times in the past, glaciers existed in the equatorial mountains, and the climate was much wetter and cooler than it is today, with rain forests growing where there are now dry grasslands.

Recent research, although still spotty and sparse compared to knowledge of changes in vegetation during the past 2 million years in North America and Europe, continues to confirm these general patterns—in particular, that climate changes that imposed cold and warm periods in the North led to wet and dry changes in the low latitudes.[31] If the vegetation changed, so must have the soils, the soil organisms, the large mammals, the insects, and the birds. Rain forests lack the great number of large mammals found in the tropical savannas and grasslands of Africa. Thus the changes in the sediments of Lake Victoria and the other African lakes, the variations in the extent and distribution of the African mountain glaciers, suggest that Africa, too, has been subject to biological change rather than constancy. Over the past thousands of years, deserts, grasslands, savannas, and woodlands have marched slowly across the landscape to the beat of the changing climate.

In other tropical areas, as in Africa, the ice-age history of the vegetation is confounded by the effects of human civilization, so only the earlier periods are useful in seeking to uncover the pattern of nature without human beings. In the tropics, as in the North, it is not easy to find wilderness undisturbed. One likely area that has been subject to some study is the Western Highland District of New Guinea, where mountains reach an elevation of 14,000 feet. Here cores of sediments from swamps and ponds have been collected along a 60-mile distance, from elevations of 5,200 to 8,300 feet, providing information about conditions as far back as 30,000 years ago.[32] Hiking down from these summits today, a traveler passes through alpine grasslands at the highest elevations, a subalpine forest between 13,000 and 11,000 feet, and a transitional forest called mixed montane, which changes to a Southern Hemisphere beech forest at 9,000 feet and to a Southern Hemisphere oak forest below 8,000 feet.

The pollen deposits in the sediments show that it was much colder on the mountains more than 22,000 years ago than it is today; alpine and subalpine plants, which occur now above 11,000 feet, grew between 6,000 and 7,000 feet, where today there are oak forests. Between 22,000 and 18,000 years ago, the climate warmed and the tundra was replaced by beech forests. At 8,000 feet, where beech forests now grow, alpine grasses grew 16,000 to 12,000 years ago; subsequently, beech and mixed montane forests, typical of the colder climates, alternated over thousands of years.

There is also evidence of an opening up of the forests at the lower elevations during the past 5,000 years, suggesting human influence, and human artifacts have been found at two sites dated at 2,300 years old. The scientists who reported these findings concluded that change not only is likely but is a demonstrable fact even in vegetation formerly thought to be more stable than most, and change has been continual.[33] (Of the bodies of water studied, five were below 9,000 feet, meaning they were within Southern Hemisphere oak or beech forests that are and have been affected by human agriculture, in particular by pigs, and by fire set to clear land.)

Almost 1,200 miles directly to the south of Mount Hagen lies Lynch's Crater at an elevation of almost 2,000 feet above sea level. Lynch's Crater is one of Australia's few natural lakes. Now filled with a swamp, this natural basin was formed by an ancient volcanic explosion. The crater lies in the Atherton Tableland of northern Queensland, which today is in the subtropics, wetter than much of Australia, with an average rainfall of approximately 100 inches per year.

In the swamp, little of the pre-British-settlement vegetation remains, most of the region having been cleared for cattle grazing. However, a tropical rain forest exists several miles to the east of the crater, suggesting the character of the pre-settlement landscape. A sediment core taken from the swamp tells a story of change similar to those of New Guinea and Lake Victoria.[34] The oldest sediments from Lynch's Crater, 60 feet below the present surface, indicate that forests occupied the surrounding areas between 60,000 and 38,000 years ago, but the species included some that no longer occur in northeastern Queensland, and the relative abundances of others are not like those found today. Similar forests grow today in areas that receive less than half the rain that falls in the region of Lynch's Crater. Between 38,000 and 27,000 years ago, the rain-forest species gradually diminished and were replaced by plants of dry woodlands. Thus, the area near Lynch's Crater appears to have undergone a long, slow warming and drying.

From 27,000 to 12,000 years ago, plants of dry woodlands were dominant. Then rain-forest species returned, but the collection of species has no known modern equivalent in northern Queensland. For example, small plants typical of the understory were much more abundant than they are in modern rain forests.

Other studies in Australia and New Guinea suggest a similar vegetation history.[35] Some allow an even longer history to be reconstructed. Taken together, the various studies conducted in the last quarter of the twentieth century suggest that the climate in the tropics between 60,000 and 40,000 years ago was drier than it is now, became colder during the next 10,000 years, and even colder in the subsequent 10,000, with the greatest extent of mountain glaciers in New Guinea occurring about 17,000 years ago. Like those in East Africa, these glaciers, it has been estimated, would have formed as a result of a 6° to 10°F decline in the average annual temperature. Temperatures began to rise rapidly around 15,000 years ago. By 10,000 years ago, lakes filled in southern Australia and most of the ice was gone from the New Guinea mountains. Between 8,000 and 5,000 years ago, the climate was somewhat warmer than today's.

The evidence from Australia and New Guinea is consistent with that from East Africa. The tropics, which were believed to have been constant throughout the past 2 million years, are now known to have varied greatly. This seemed obvious to Daniel Walker one of the scientists studying these tropical areas. He concluded that if the naturalness of change is not taken into account in planning the future, those plans are doomed to failure in one way or another: "Planning which is ecologically rational needs a measure of flexibility simply because ecosystems are dynamic, evolving all the time and at varying rates. . . . Once we have determined our aims," he continued, "we must look forward to the need for continuous management to achieve and maintain them. And a knowledge of the processes of vegetation change, on all time scales up to tens of thousands of years at best, is likely to prove important in that management."[36]

In sum, near the end of the machine age, the available evidence overwhelmingly contradicted the idea that populations or communities have been constant in abundances or in the relative importance of species. At the time I wrote *Discordant Harmonies*, many scientists believed that the low latitudes had escaped the catastrophic changes brought about by the Pleistocene glaciations and therefore offered the best hope of finding a constancy in nature. As the twenty-first century approached, a radical change took place in the interpretation of the Pleistocene history of the tropics. Everywhere that careful studies have been made, from Africa to New Guinea and Australia, a similar picture emerged: alternations of times of cool and wet climate with times of warm and dry; changes in the distribution of communities of plants and in the kinds of plants that made up these biological communities. The picture, although cruder, is like the one we found for the higher latitudes in the Northern Hemisphere. And thus those land areas that had seemed, and been interpreted by ecologists to be, the most likely home of nature's biological constancy were revealed not to be.

The classical areas of wilderness, those last outposts of nature undisturbed—the forest of Isle Royale, the grasslands and savannas of East Africa, the mountains of New Guinea, the marshes of Australia—show a temporal mutability that scientifically we cannot ignore, but unfortunately, during the past 20 years we have continued to ignore it in terms of laws, policies, and beliefs. The sediments in lakes are evidence of the great mutability of biological communities: changes in the combination of species; differences in the migration rates of species; transformations in biological communities with changes in climate; differences among the interglacial periods in the species that composed the forests and grasslands, and in the relative importance of the species that have existed throughout the Pleistocene. Clearly, in both temperate forests and tropical regions, nature is not constant, is not like a single tone indefinitely sustained, but is composed of patterns that themselves change, like a melody played against random background noises.

Since 1990, scientists have published a wealth of information that shows past climate change and the responses of vegetation to it, too numerous to mention all of them. The general moving-picture show is consistent with what I wrote 20 years ago: The climate changed, vegetation responded to it; the climate changed again, vegetation responded to it again. The ability of species to respond to environmental change is responsible for the persistence of life on Earth. One example relevant to today's concerns about global warming in the eastern United States is described in a scientific paper published in 2006 about a period of warm and dry climate in southern New England from 10,100 to 7,700 years before the present. Tree species that are called "intolerant" of such conditions, including beech and hemlock, declined, and the landscape became an open savanna, perhaps in silhouette resembling the many now-familiar movies and television programs about East African and South African parks. Once the climate cooled and precipitation increased, hemlock and beech returned, at individual rates and times. The scientific papers about this do not discuss it in terms of disasters, catastrophes, or tipping points, but do point out that the closest analogue since that time was the open landscape created by eighteenth-century New England farmers.[37] For our purposes, perhaps the most remarkable and telling insight about North America, from new research, is that in spite of 20 ice ages in the past 2.5 million years, "in the intensively studied flora of North America only one species of tree, *Picea critchfiedii*, is known to have gone extinct in the last ice age."[38]

When I worked on *Discordant Harmonies*, I had to dig far to find scientific papers that reconstructed the response of vegetation to environmental change in both the wet and dry tropics. But today, scientific knowledge of these histories has increased greatly as well. Mark B. Bush of the Florida Institute of Technology and Henry Hooghiemstra of the University of Amsterdam, after reviewing scientific findings up to 2005 about how tropical life responded in the past to climate change, wrote: "The most important lesson of the past 40 years of tropical climate research is that tropical systems do not have stable climates."[39] Their first emphasis is how extensive and how dramatic environmental change has been in the tropics, in contrast to the classic twentieth-century perception that the tropics had undergone little environmental change. They report that in the past 20 million years there were three critical geological events: "the rise of the Andes, the rise of the Tibetan Plateau, and the closure of the Isthmus of Panama."[40] In the Western Hemisphere, this of course allowed the migration of species of mammals from North to South America. The greatest of these occurred about 2 million years ago, and the invasive species from the North led to the extinction of "equivalent forms through intensified competition and predation."

Species of trees also migrated south, including alders and oaks. About 3 million years ago, the vegetation included groups we could recognize now—for example, in the temperate deciduous forests of higher elevations, and, in warmer country, in savannas and rain forests. There are two key conclusions. First, that "tropical biodiversity has withstood the coming and going of about 20 ice ages in the past 2.5 million years. . . . It is apparent that the coming and going of ice ages did not eliminate tropical biodiversity." But, second, that "large-scale changes in climate generated the patterns of modern biodiversity, induced migration of species, and spurred extinction and speciation."[40]

Yet, in the oddest of ways, we have continued to ignore this reality. It is as though we learned by the end of the twentieth century that nature was like molten glass, changing as if flowing, responding flexibly to external as well as internal forces, but that with the arrival of man, or a least industrial man, the glass hardened and, confronted with any change, will shatter. That's the tipping point we hear so much about at the end of the twenty-first century's first decade. We got the history but we missed the message. We heard a bacchanalian song but we saw it only as a view in Apollo's mirror.

It is fair to ask whether there might be some truth in the hardened-glass view of ecological systems. Bush and Hooghiemstra, while acknowledging the naturalness of change in nature and the resilience of species to it, phrase the response this way: "Whether that same resilience should be predicted within landscapes fragmented and altered by human activities is far less certain."

Indeed, this is where our effects on nature are most profound—in destroying and isolating habitats, not in disrupting life by allowing or promoting large-scale environmental change. But more of that later.

Sometimes Change Is Necessary

Local changes sometimes are necessary for life: because, first of all, the chemical clock sometimes runs down and chemicals become unavailable; and second, species evolve with change, and many are specifically adapted to it.

In Australia there are sand dunes that have existed for 100,000 years. This is highly unusual because dunes are typically subjected to intense storms and are eventually blown open by the wind. The Australian dunes form a sequence, with the youngest nearest the shore and the oldest the farthest inland. One can trace the history of a typical area by walking inland from the coast, from one dune to another. Because the dunes are older the farther they are from the ocean, this walk is a journey back through time and should therefore take one to older and older stages in ecological succession. At first, the vegetation seems to follow the classical model of succession: From dune to dune, the plants become larger and denser. Near the shore are a few scattered hardy plants, small and with shallow roots that hold the sand in place, which allows seeds of other plants to sprout. Farther inland grow larger woodland plants of greater diversity. But at the oldest dunes, the pattern deviates from the classical one: The vegetation becomes smaller and less diverse as woodlands give way to shrubland.[41]

Studies of this sequence of dunes have shown that chemical elements needed for life can build up in the sandy soil from salt spray from the ocean and from organic matter added by the plants as they die. As the dunes age, however, the chemical elements are leached downward by rain, and plants need deeper and deeper roots to survive. Eventually, most of the chemicals settle in a layer below the reach of most kinds of plants; a scrubland of relatively few species survives. Only an intense disturbance will turn the sand over and bring the chemical elements necessary for life back to the surface.

A similar story can be told for rain forests along the west coast of New Zealand's South Island, which I visited myself. There, one can follow ecological succession by walking from the edge of a glacier toward the shore, a brief walk. Near the glaciers, where plants are just beginning to become established, I saw what appeared to be the normal pattern of ecological succession: first lichens and mosses, then low flowering plants, then trees, then temperate rain forest with a dense growth of many species. I bought color photographic slides from the park office at this location. Consistent with the classic mythology, these showed only the development from bare rock to temperate rain forest. This part of New Zealand is famous for these forests, mainly because they are among Earth's few temperate rain forests, and because New Zealand's have a relatively large number of tree species, a high species diversity.

But wandering a bit beyond the usual tourist route and past what the professional photographs showed, I found that this was not the last stage; farther toward the shore, the rain forest disappeared and was replaced by scrubby grass and shrubs. Here, as on the Australian dunes, chemical elements necessary for the plants had been leached downward into the soil, below the reach of the trees; as a result, only a few plants adapted to very low-nutrient soils could grow there (Fig. 4.4). Another glaciation that would turn over the soil could bring these elements to the surface, and the succession could start again. Without such a "disaster," or climatic "tipping point" (to put it in early-twenty-first-century parlance), eventually all the forests would run down, their successional clock slowing and slowing, until, like one of those old windup alarm clocks we all had in the twentieth century, it stopped.

Yet another example of the necessity for change is the natural regrowth of forests along the Alaskan coast as the glaciers melted back in the twentieth century. Again, the succession of plants at first seems to fit the classical pattern. Alders are among the first trees to become established; they have nitrogen-fixing symbiotic bacteria in their roots, and together the alders and their bacteria enrich the soil. In the enriched soil grow other trees, including spruce, which eventually grow taller than the alders, shading that species and preventing its reproduction. Slowly, lacking the alder and its bacteria, the soil becomes less fertile. When spruce trees die, beds of sphagnum moss develop. They make the soil acid and soak up water. The area becomes uninhabitable for trees, and what was forest becomes bog, not at all like the "climax" forest of spruce that was supposed to be the final stage.[42] Once again, the successional clock unwinds.

FIGURE 4.4 The old stage observable in a New Zealand temperate rain forest succession. Under very heavy rainfall (on the order of 245 inches a year), the chemical elements essential for vegetation are leached below the reach of the roots, except for some deep-rooted small grasses and shrubs. The leached layer appears almost white; the layer into which the chemical elements are leached appears darker, below the whitish layer. Rather than ending as the grand climax of maximum biomass and diversity, this former forest has declined to a state that was never believed possible in an undisturbed forest. (Photograph by the author.)

Another ice age would start the Alaskan forests again. Even a slight cooling period, perhaps no colder than it was during the little ice age, from the mid-sixteenth to mid-nineteenth century, could rebuild the glaciers enough to cover Glacier Bay and locally renew forest progression, as it would for the New Zealand rain forests.

A smaller, shorter change in the environment might be enough to restart the progression of dunes along the Australian coast, perhaps an episode of severe storms. But in all these cases, an environmental change is necessary for the recurrence of what used to be thought of as the climax stage, which, proponents say, would remain indefinitely under undisturbed conditions. Of course, if we want to keep some of these stages fixed in their current conditions, we ourselves can change the soils by turning them over or adding fertilizers. That is, we can replicate nature's natural processes with our own energy, time, and resources. We can do this in some cases for a long time, and in all cases for a short time, *but we cannot freeze all of nature indefinitely in a single state.* To paraphrase Abraham Lincoln, you can hold all of nature in one state for a short time, and you can hold some of nature in one state for a long time, but you can't hold all of nature in a single state for a long time.

Conserving a Small Warbler: A New Direction in Managing and Conserving Living Resources

In the 1960s, the facts about the naturalness of change and the adaptations of living things to change began to force their way through the myths of constancy to affect a least a little of the management of living resources. One of the first and classic examples of the beginnings of this new direction was the attempt, still ongoing, to save a small friendly bird, Kirtland's warbler (Fig. 4.5A).

In 1951, Kirtland's warbler became the first songbird in the United States to be subject to a complete census. About 400 nesting males were found. Concern about the species grew in the 1960s and increased when only 201 nesting males were found in the third census, in 1971.[43] Conservationists and scientists began trying to understand what was causing the decline, which threatened the species with extinction.

Kirtland's warbler winters in the Bahamas and then flies north to Michigan, where it breeds in an exacting set of conditions in jack-pine woodlands (Fig. 4.5B). Although jack pine grows widely throughout the boreal forests, especially in Canada, Kirtland's warbler nests only in jack-pine woods on one soil type, called Grayling sands, which occurs in central Michigan at the very southern end of the jack-pine range. These warblers build their nests on the ground and, apparently because the nests must remain dry, prefer to build them on dead tree branches still attached to a tree at ground level, and only on coarse, sandy soil that drains away rainwater rapidly. Young jack pines provide the dead branches at the ground surface, and Kirtland's warblers are known to nest only in jack-pine woodlands that are between 6 and 21 years old, ages when the trees, 5 to 20 feet tall, retain dead branches at ground level. Males are territorial, and each defends an area as large as 80 acres in a uniform jack-pine stand.

These requirements leave the warbler with few nesting areas, all the fewer because jack pine is a "fire species"—it persists only where there are periodic forest fires. Jack-pine cones open only after they are heated by fire, and the trees are intolerant of shade, able to grow only when their leaves can reach into full sunlight. Even if seeds were to germinate under mature trees, the seedlings could not grow in the shade and would die. Jack pine produces an abundance of dead branches that promote fires, which is interpreted by some as an evolutionary adaptation to promote those conditions most conducive to the survival of the species.

Kirtland's warbler thus requires change at rather short intervals—forest fires approximately every 20 to 30 years, which was about the frequency of fires in jack-pine woods in pre-settlement times.[44] But where was the warbler during the past 10,000 years, when the continental ice sheets were retreating? With such specific nesting habits, the warbler must have followed the jack pine as it migrated north-ward, and must have nested in trees on outwash plains, where roaring rivers cre-ated by melting ice deposited coarse sands. At the time of the first European settlement of North America, jack pine may have covered a large area in what is now Michigan. Even as recently as the 1950s, jack pine was estimated to cover nearly 500,000 acres in the state. At best, the warbler could have reached that area no sooner than 6,000 to 8,000 years ago, when the jack pine returned there. But the males seem unwilling to set up new territories away from old ones—they nest only a short distance from where they were born—so the species may have migrated northward even more slowly than the forests.

European settlers first reached the warbler's habitat in 1854, when Tawas City was founded on Lake Huron, after which lumbering of red and white pine in the area began in earnest. Jack pine, a small, poorly formed tree, was then considered a trash species by commercial loggers and was left alone, but many big fires fol-lowed the logging operations when large amounts of slash (branches, twigs, and other economically uninteresting parts of the trees) were left in the woods. Else-where, fires were set in jack-pine areas to clear them and promote the growth of blueberries. Some experts think the population of Kirtland's warblers peaked in the late nineteenth century as a result of these fires. After 1927, fire suppression became the practice, and control of forest fires reduced the area burned and the size of fires. Where possible, people were encouraged to replace jack pine with economically more useful species. All of this shrank the areas in which the warblers preferred to nest.[45]

Although it may seem obvious today that Kirtland's warbler requires forest fire, this was not always understood. In 1926, one expert wrote that "fire might be the worst enemy of the bird."[46] Only with the introduction of controlled burning

A

B

FIGURE 4.5 Kirtland's warbler (A) nests only in young jack-pine forests (B). (A, courtesy of Dominic Sherony; B, photograph by the author.)

after vigorous advocacy by conservationists and ornithologists was habitat for the warbler maintained. The Kirtland's Warbler Recovery Plan, done in cooperation with the Audubon Society and Michigan's Department of Natural Resources, was published by the Department of the Interior and the Fish and Wildlife Service in 1976 and updated in 1985. It created 38,000 acres of new habitat for the warbler, land where "prescribed fire will be the primary tool used to regenerate non-merchantable jack-pine stands on poor sites."[47]

In this case, the facts had become unavoidable and could not be hidden in myth: Unlike trees that may seem to extend indefinitely to the horizon, or fish in the oceans whose numbers are difficult to count, the Kirtland's warbler population consisted of a small number of individuals that were subject to a complete census and whose habitat requirements were absolutely clear. Those who wanted to save this species acted from observation.

This episode, perhaps small in the grander scale of Earth's millions of species, nonetheless marked a turning point in the modern perception of the character of nature and the requirements for managing and maintaining nature. Within a decade after an article in *Audubon* described Hutcheson Forest in New Jersey as "a cross-section of nature in equilibrium," the same magazine ran an article entitled "A Bird Worth a Forest Fire," which explained the necessity of fire for the persistence of a bird and marked the beginning of the present transition in ideas about nature.[48] In Chapter 13, we will return to the problem of conserving the warbler in a globally warmed world. At this point in our journey, the key point is the following.

The old ideas are beginning to change, but even today the lesson of Kirtland's warbler is understood by some but not yet widely enough heard or widely enough accepted for other forests and their animals. In the years since the publication of *Discordant Harmonies*, when I have given talks around the United States, I have often asked whether anyone in the audience has heard the story about Kirtland's warbler. From naturalists in Michigan, the answer is often yes. Elsewhere, few have heard of it.

It is worth repeating that the breakdown in ideas and myths is discomforting, even frightening. And in this case, to some who want to make wise use of our natural resources, the conclusion that one must manage at least one species by promoting change may seem to open up a Pandora's box of terrible consequences. With the warbler, we must now confront the question I raised earlier: If one admits that some changes are acceptable, how can one reject any changes out of hand? Again, there are clear answers to this question, for the fact that some changes are natural and necessary does not imply that all changes, regardless of time, intensity, and rate, are desirable. There are both natural and unnatural changes, and there are natural and unnatural rates of change. To recognize that melodies and themes are made up of changing tones does not imply that any noise is music.

The key to a new but wise management of nature is to accept changes that are natural in kind and in frequency, to pick out the melodies from the noise.

Understanding these distinctions requires that we explore the origin and history of beliefs that have so dominated our treatment of nature. Only by doing so can we free ourselves from the old myths and create a new mythology—in Joseph Campbell's sense of an agreed-upon explanation of how the world came about and how it works—consistent with scientific facts and appropriate for our time.

PART TWO

Background to Crisis

FIGURE 5.1 *The Creation of the Animals*. Engraving by Charles Health in *The Holy Bible* (London: White, Cochrane, 1815. Courtesy of Library Special Collections, University of California, Santa Barbara.)

5

Mountain Lions and Mule Deer

NATURE AS DIVINE ORDER

Everything in the world is marvellously ordered by divine providence and
wisdom for the safety and protection of us all. . . . Who cannot wonder at
this harmony of things, at this symphony of nature which seems to will the
well-being of the world?

—CICERO, *THE NATURE OF THE GODS* (*44 BC*)

An Orderly Universe

One of the themes that run through twentieth-century ecology is the idea of a
highly ordered universe with several characteristics important to life. The orderliness is itself extremely well suited to support life, and life is part of the ordered
structure, with every species having its role in nature, its function as a necessary
part of the entity. The perception that the universe is remarkably structured and
ordered to support life was expressed in the early part of the century in the book
The Fitness of the Environment, by Lawrence Henderson.[1]

As mentioned briefly in Chapter 1, Henderson reflected on the many ways in
which Earth has just the right characteristics to support life. Among Earth's other
remarkable life-support properties, Henderson points out, are that it is large
enough to hold an atmosphere within which life can evolve and persist, is near
enough to the sun to be warm but not close enough to be too hot, and rotates on
its axis so that one side does not become very hot while the other remains very
cold. Henderson's book evokes an image of a universe of extraordinary order at
every level, from the biochemical to the astronomical, beyond the likelihood of
mere chance, and it leaves the reader with a puzzling contradiction: Science was
supposed to explain the world around us without recourse to purposefulness and
religion, and yet an analysis of nineteenth- and early-twentieth-century scientific
discoveries, those available to Henderson at the time he wrote his book, implied
that the character of our planet and of our universe seems too perfectly designed

for life to have happened by chance. The universe appeared to be an elegantly complex machine suited to life's needs.[2]

Order and Disorder on the Kaibab Plateau

DEAD OF ITS OWN TOO-MUCH

The perception of nature as highly ordered was common among scientists, naturalists, and conservationists in the twentieth century. Not only was the environment thought to be highly ordered, but also species were perceived to interact in a highly ordered way. This orderliness was exemplified by the role of predators in nature, as revealed by a famous case in predator control.

A rapid decline in mule deer on the Kaibab Plateau, whose edge forms part of the north rim of the Grand Canyon, was the focus of a widely known controversy in American conservation during the first decades of the twentieth century (Figs. 5.2 and 5.3). According to an account made famous by the great American conservationist Aldo Leopold (1887–1948), the decline was the result of an earlier "irruption" (the term wildlife experts used at the time) of the deer, during which these browsing animals had destroyed the trees and shrubs on which they fed and depended. Having destroyed much of their food, the deer starved and the population crashed. Leopold blamed the problem on "overcontrol" of the deer's major predator, the North American mountain lion, which he believed had kept the deer population in check so that the two species had existed in a natural balance.

This account of the Kaibab mule deer, first made famous by Leopold, was repeated in many standard ecology and wildlife-management textbooks and scientific papers, in which Leopold's account was accepted as true.[3] The trouble began, he wrote, around the turn of the century, which was a time of "predator control"—large predators were felt to pose a danger to domestic stock, and there was considerable hunting of mountain lions. From 1906 to 1931, hunters hired by the government killed an estimated 781 mountain lions, 30 wolves, 4,889 coyotes, and 554 bobcats on the Kaibab. One hunter, "Uncle Jim" Owens, claimed to have taken 600 lions himself between 1906 and 1918.[4] The small population of mule deer, estimated to have numbered 4,000 in 1904, was said to have increased rapidly after removal of the lion.[5] By 1930 a population peak was reported, with estimates as high as 100,000. Then 50% of the herd was said to have starved to death during the two following winters, and the population suffered a decline to only 10,000 animals, according to some reports.

From Leopold's perspective, the lion, along with the wolf and other major predators, played an important and necessary role in the workings of nature. "The cow man who cleans his range of wolves does not realize that he is taking over the wolf's job of trimming the herd to fit the range," Leopold wrote in his famous and influential book *A Sand County Almanac.* "When wolves are removed from mountains, the deer multiply," he continued, and "I have seen every edible bush and

FIGURE 5.2 The Kaibab Plateau begins as the north rim of the Grand Canyon in Arizona. In this picture, we are looking south from the Kaibab, past a dead pinyon pine, into the canyon. (Photograph by the author.)

FIGURE 5.3 A mule deer mother and young, the species that Aldo Leopold said died on the Kaibab Plateau in Arizona from "its own too-much"—that is, its overpopulation. (Photograph by the author.)

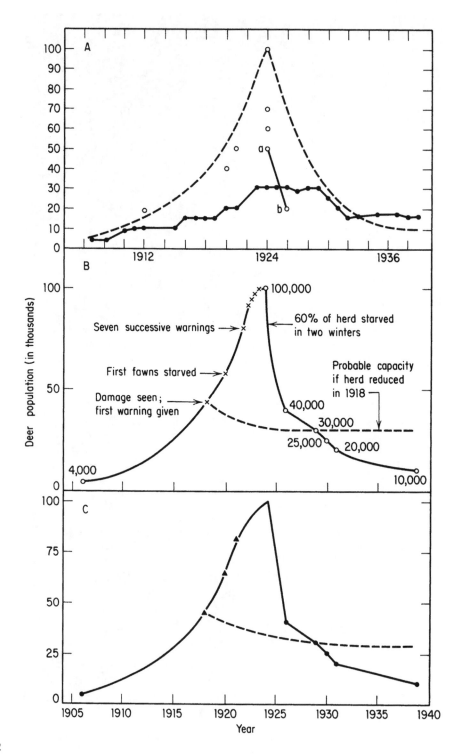

Deer population (in thousands)

A

100
90
80
70
60
50
40
30
20
10
0

1912   1924   1936

a
b

B

100,000

Seven successive warnings →

60% of herd starved
in two winters

First fawns starved →

Probable capacity
if herd reduced
in 1918 →

Damage seen;
first warning given →

40,000

30,000

50

25,000   20,000

4,000

10,000

C

100

75

50

25

0

1905   1910   1915   1920   1925   1930   1935   1940

Year

seedling browsed, first to anaemic desuetude, and then to death. . . . In the end, the bones of the hoped-for deer herd, dead of its own too-much, bleach with the bones of dead sage, or molder under the high-lined junipers."[6]

In saying that predators are required in nature to regulate the abundance of their prey, Leopold was stating a belief common throughout the century in wildlife management and among conservationists and ecologists. These experts in general agreed with Leopold that in their undisturbed condition the mountain lions and the deer lived in a natural balance, with the lions killing just enough deer to keep the population constant. Considerable policy had followed from this belief, including the National Park Service's attempt in the 1940s to introduce wolves onto Isle Royale in order to regulate the moose, which were also undergoing an "irruption."

The facts of the Kaibab Plateau story were pieced together by New Zealand ecologist Graeme Caughley (1937–1994), who had studied patterns in populations of wild ungulates—herbivorous, cud-chewing mammals, including deer and cattle.[7] Leopold had based his analysis on a paper written by D. I. Rasmussen, another wildlife naturalist, who had given not one but three sets of estimates of the size of the mule deer population: a forest supervisor's estimates made each year, but not by a quantitative method; observations by others who visited the Kaibab Plateau; and Rasmussen's own estimates (Fig. 5.4).[8] These three sets of figures differed markedly from one another. Whereas one source estimated the peak abundance in the 1920s to be 100,000, others estimated the peak abundance at 70,000, 60,000, 50,000, and 30,000. This last was in fact simply the number that most naturalists believed to be "sustainable," meaning that the vegetation growth on the plateau each year could sustain a population of 30,000 deer indefinitely. If the population had actually been 30,000, there would have been no irruption and no crash; the population would have been essentially constant.

Even if there had been an irruption, the role of predators was not clear. If it occurred, the growth of the population of Kaibab mule deer coincided with a reduction in the number of sheep and cattle on the plateau. As recently as 1889, Caughley reported, there had been 200,000 sheep and 20,000 cattle grazing on the plateau, but there were only 5,000 sheep and few cattle by 1908. Thus, the

FIGURE 5.4 The changing story of the Kaibab deer population's "irruption."

(A) Population estimate of the Kaibab deer herd from D. L. Rasmussen (1941). Biotic communities of Kaibab Plateau, Arizona. *Ecol. Monographs, 3,* 229–275. Linked solid circles are the forest supervisor's estimates; open circles give the estimates of other people; those labeled *a* and *b* are those of a Mr. B. Swapp, and the dashed line shows Rasmussen's own estimates of the trend. (B) A copy of Leopold's interpretation of the history. (A. Leopold [August 1943]. Deer Irruptions. Wisconsin Conservation Bulletin. Reprinted in Wisconsin Conservation Dept. Publ. 321, pp. 3–11.) (C) A copy of the history given by ecologists. This is based on a graph in D. E. Davis and F. E. Golley's 1963 book *Principles in Mammalogy* (New York: Reinhold), which in turn is based on a graph in W. C. Allee et al. (1949). *Principles of Animal Ecology* (Philadelphia: W. B. Saunders); redrawn from A. Leopold (August 1943). Deer Irruptions. Wisconsin Conservation Dept. Publ. 321, pp. 3–11; altered from D. L. Rasmussen (1941). Biotic communities of Kaibab Plateau, Arizona. *Ecol. Monographs, 3,* 229–275. (Source: Graeme Caughley, who put all these together, with their individual sources, in G. Caughley [1970]. Eruption of Ungulate Populations, with Emphasis on Himalayan Thar in New Zealand. *Ecology, 51,* 54–55. Reproduced with permission of the Ecological Society of America.)

increase in the deer population may have resulted from reduced competition rather than decreased predation. Other experts suggested that the increase in the mule deer population may have resulted from a greater supply of edible vegetation due to changes in the frequency of fire and other disturbances or changes in weather patterns.

Caughley analyzed all known cases in which large ungulates were introduced into new habitats, and found that a population irruption and crash had occurred every time, regardless of the presence or absence of predators. After a severe disruption of a habitat, such as could have occurred under the grazing pressure of cattle and sheep in the nineteenth century, the response of the deer could have been very similar to that of ungulates after their introduction into a new habitat.

Thus, the facts about counts of the animals leave us up in the air. The famous "irruption" of mule deer on the Kaibab Plateau may or may not have occurred—and if it did, it may have been entirely unrelated to the presence of predators. It's surprising that such careful and observant naturalists as Leopold, Rasmussen, and the others who examined the Kaibab history and to whom the study of nature was important would have accepted one explanation among many when the facts were so ambiguous. Many interpretations are possible, yet for many years, until Graeme Caughley analyzed them and wrote about them in 1970,[9] only one of the possible stories was accepted, a story that painted a clear picture of highly ordered nature within which even predators had an essential role.

Leopold's Kaibab deer story was commonly told in textbooks until late in the twentieth century, well past Caughley's analysis, and was highly influential, but it doesn't appear in any of the current ecology textbooks I know of. Still, the story hasn't gone away; it played such a central role in American ecological thinking about predators and prey that it continues to fascinate. In 2006, Dan Binkley of the Department of Forest, Rangeland, and Watershed Stewardship, Colorado State University, in Fort Collins, and three colleagues revisited the story in a scientific paper entitled "Was Aldo Leopold Right About the Kaibab Deer Herd?"[10]

They point out that Caughley's report of 200,000 sheep and 20,000 cattle grazing on the plateau was actually a report of the numbers estimated to be "in the surrounding desert country and the Kaibab Mountain," a much larger area than the Kaibab itself, and that there were not good estimates for the Kaibab alone at the end of the nineteenth century. "Cattle numbers increased from about 9,000 in 1906 to a peak of over 15,000 in 1913 . . . declining to 7,000 to 10,000 through the mid-1920s, then dropping below 5,000." Meanwhile, "estimates of sheep numbers dropped from an initial [rough estimate of] 20,000 to 5,000 by 1910, followed by a steady decline through the 1940s." They also point out that sheep were mainly grazed below the elevation of most aspen. They considered Caughley's revelations about the different reports of the deer population size, and wrote: "Estimates of the deer population in the 1920s appeared to differ substantially between the pattern preferred by Rasmussen (1941) and Leopold (1943) and those of the forest supervisors. The apparent contradiction had disappeared before Rasmussen and Leopold published their estimates, as Forest Supervisor Walter Mann revised the forest supervisor estimates to more than 100,000 in the mid 1920s."

I gather that a revisionist approach by the forest supervisor about the three different estimates is taken as a scientifically valid change and therefore the authors continue as if there actually had been a population explosion of the Kaibab mule deer. It's hard to tell with this kind of reconstruction of the past, but it appears that the ambiguity remains.

They do add an interesting and apparently legitimate scientific analysis of the abundance of aspen, an important mule deer food, which they said was high before the supposed "irruption" and low afterward, supporting Leopold's conclusion. However, aspen is only one food of the mule deer, which eats widely (the *Guide to Wildlife Food Habits*[11] lists more than 40 plants) and changes its diet with seasons and availability. They especially like grasses and in summer eat a variety of flowering plants and their fruits and seeds, including berries. Their winter food includes twigs of Douglas fir, cedar, yew, aspen, willow, dogwood, serviceberry, juniper, and sage, depending on where they are. So the abundance of aspen may or may not be a good indicator of the availability of food for the deer, and the abundance of aspen is therefore not a sufficient guide to what was happening on the plateau.

The mule deer population on the Kaibab Plateau never has increased to anywhere near the maximum that Leopold speculated it reached in the 1920s. One study showed that the population climbed to around 50,000 in the early 1950s, but since then, through 2006, according to the Arizona Game and Fish Department, the population has ranged from about 5,000 to about 20,000, well within the range that the forest supervisor gave for the population from around 1906 to the late 1930s, the period that Leopold discussed.[12]

So where does this leave us in the second decade of the twenty-first century? Basically with the ambiguity that Caughley described to us, made in some ways more ambiguous: We don't know much about the cattle and sheep on the plateau before the supposed irruption; the size of the deer population remains unclear; and whether the deer's food changed enough to affect the deer's abundances is also unclear. Indeed, the debate still goes on among ecologists.[12]

To question the truth of this story on the grounds of ambiguous and incomplete scientific information—to proclaim that we did not have enough information to know whether an irruption of mule deer was caused by removal of the mountain lion, or even whether the irruption occurred at all—was to speak against deep-seated beliefs about the need for predators as well as all other creatures on Earth. Once again we come across an ecological story believed to be true "in spite of appearances to the contrary," or at least in spite of appearances that offer no evidence, or even clues, about cause and effect.

The story of the mule deer on the Kaibab Plateau was only one of many from the first half of the twentieth century about the removal of predators, irruptions of prey, and people's role in these processes. For the role of predators in the balance of nature was regarded as only one example, albeit an outstanding one, of the incredible and wonderful order of nature, with each species having its place in the working of the whole system.

In the context of this book, the justification for protecting predators is revealing because these animals are otherwise among those most disliked by people and most often considered pests, whose existence has been viewed as undesirable. To argue that even they have a necessary role in nature, worthy of their conservation, is to argue from an extreme case about the perfection of nature, but it fits within the general context seen at its most universal by Henderson.

Echoes of the Idea of Order

> Hyt ys no nede eke for to axe
> Wher there were many grene greves,
> Or thikke of trees, so ful of leves;
> And every tree stood by hymselve
> Fro other wel ten foot or twelve.
> So grete trees, so huge of strengthe,
> Of fourty or fitty fatme lengthe,
> Clene withoute bowgh or stikke,
> With croppes brode, and eke as thikke—
> They were not an ynche sonder—
> That hit was shadewe overal under,
> And many an hert and many an hynde
> Was both before me and behynde.
> Of founes, sowres, bukkes, does
> Was ful the woode, and many roes,
> And many sqwirelles, that sete
> Ful high upon the trees and ete
> And in hir maner made festes.
> . . . hyt was so ful of bestes,

> (Translated into modern English, this reads:)
> There is no need for to ask
> If there were many great green groves
> So thick with trees, so full of leaves
> And every tree stood by itself
> From one another ten feet or twelve.
> So great these trees, so full of strength,
> At forty or fifty fathoms in length
> And without branches near the ground,
> With huge broad crowns and each so sound

Left not an inch upon the ground
That lacked the shadow from the crown
And many a rabbit and red deer
Were far way from me and near
Of fowls, sows, bucks, and does
The woods was full, and with many roes
And many a squirrel that set
Full high up on the trees and fed
And in their manner had their feast. . . .
The woods was full of all kinds of beasts.
—Geoffrey Chaucer, The Book of the Duchess[13]

Just as Geoffrey Chaucer's poetry about a forest echoes the modern belief in a balance of nature, so does the explanation for the existence of predators, if unsupported by twentieth-century facts, have curious echoes in the past. Two centuries before Leopold, for example, Thomas Jefferson, whose many interests included fossils and natural history, gave the same justification for the necessity of predators as did Leopold. Jefferson wrote of a "benevolent persuasion that no link in the chain of creation will ever be suffered to perish." (Here Jefferson refers to the great chain of being, one of the central ideas of the balance of nature, which, as I mentioned earlier, is the idea that there is a place and function for every creature and every species, all linked and functioning together.)

Jefferson believed that the reason for the comparative rarity of predators was what he called the "ordinary economy of nature. . . . If lions and tygers multiplied as rabbits do, or eagles as pigeons, all other animal nature would have been long ago destroyed, and themselves would have been ultimately extinguished after eating out their pasture."[14]

Jefferson's statements, in turn, echoed ideas common before him in the eighteenth century. Among the many writers one can quote, one of the most important was William Derham (1657–1735), who wrote *Physico-Theology: or, A Demonstration of the Being and Attributes of GOD, from His Works of Creation.*[15] Derham's book discussed the discoveries by European explorers and naturalists of new species of animals and plants, discoveries begun during the age of exploration and continuing in his own time. First was the discovery of new lands—the Americas, Australia, the Pacific islands—followed by the increasing exploration of the wildlife of Africa and Asia. Derham's purpose was to explain the discoveries of natural history within a Christian context. He struggled with several issues, including the question: If there are so many kinds of creatures on Earth and each kind has a great capacity for reproduction, what prevents the world from being overpopulated and falling into disorder?[16]

"The whole surface of our globe can afford room and support only to such a number of all sorts of creatures," he wrote. These creatures could, by their "doubling, trebling, or any other multiplication of their kind," increase to the point where "they must starve, or devour another." That this did not occur, Derham took as evidence of divine order and purpose. Maintaining the balance of nature, he wrote, "is manifestly a work of the divine wisdom and providence." Order in nature is maintained because God gave creatures the longevity and reproductive capacity that are "proportional to their use in the world." Long-lived animals have a small rate of increase, and "by that means they do not over-stock the world," while those creatures that reproduce rapidly have "great use," as they are "food to man, or other animals."[17]

Derham tried to deal with the more difficult problem of why there should be, on an Earth made by a perfect God, vicious predators, such as the then newly discovered Peruvian condor, which he called that "most pernicious of birds" and described as "a fowl of that magnitude, strength and appetite, as to seize not only on the sheep and the lesser cattle, but even the larger beasts, yea the very children, too." They were observed to be the rarest of animals, "being seldom seen, or only one, or a few in large countries; enough to keep up the species; but not to overcharge the world." He gave many other examples of predators, which in all cases were rare in comparison with their prey. Derham wrote that this was a "very remarkable act of the Divine providence, that useful creatures are produced in great plenty," while "creatures less useful, or by their voracity pernicious, have commonly fewer young, or do seldomer bring forth."

This, then, was the mechanism that maintained the "balance of the animal world," which is, "throughout all ages, kept even." Thanks to "a curious harmony and a just proportion between the increase of all animals and the length of their lives, the world is through all ages well, but not over-stored."[18] This indeed was a divine order.

Although he was one of the more influential authorities of the time, Derham was not alone in his explanations, which were extended to many kinds of animals. Jacques-Henri Bernardin de Saint-Pierre (1737–1814) wrote in the same century about the "noxious insects which prey upon our fruits, our corn, nay our persons." He blamed human activities for these unpleasant and undesirable events, for "if snails, maybugs, caterpillars, and locusts, ravage our plains, it is because we destroy the birds of our grove which live upon them," or, by introducing the trees of foreign countries into another, "we have transported with them the eggs of those insects which they nourish, without importing, likewise, the birds of the same climate which destroy them." He believed that there is a natural balance within each country, which is upset by human actions; every country has birds "peculiar to itself, for the preservation of its plants."[19] Birds that feed on insects have their role in the divine order, just as do condors or lions, which feed on large mammals. You can see hints of modern environmentalism in these assertions.

Like Jefferson and others who followed him, Derham's explanations, although presented in terms of the discoveries of his age, merely repeated the argument by earlier Christian writers that echoed the explanation by the philosophers of classical Greece and Rome. Derham restated what has become known as the "design argument." As Clarence J. Glacken (1909–1989) described it in 1967 in his classic book *Traces on the Rhodian Shore: Nature and Culture in Western Thought from Ancient Times to the End of the Eighteenth Century*, one of the most important books of the twentieth century about people and nature:

> Living nature has been one of the important proofs used to demonstrate the existence of a creator and of a purposeful creation; in the pursuit of this proof there has been an intensification, a quickening, and a concentration of interest in the processes of nature itself. Proof of the existence of divine purpose involved consideration of the assumed orderliness of nature, and if this orderliness were granted, the way was open for a conception of nature as a balance and harmony to which all life was adapted.[20]

For the basis of this argument, Derham and other Christian writers relied most heavily on the ideas of four classical philosophers: Cicero, Seneca, Plato, and Xenophon.[21] To the ancient question as to what is the character of nature undisturbed, these theologians and philosophers had answered: a world of divine order, with its great chain of being.

Again, there are many writers one can quote. Herodotus, for example, wrote in the first century BC that "timid animals which are a prey to others are all made to produce young abundantly, so that the species may not be entirely eaten up and lost," whereas the "savage and noxious" animals are "made very unfruitful."[22] It's the same argument that was given in each century following, but while Derham and his contemporaries sought an explanation of nature in the universal purposes of the Christian God, Herodotus and the philosophers of his era had sought the same explanation in the universal purposes of their many gods.

In the first half of the twentieth century, scientists, conservationists, and wildlife managers sought to explain the same observations using universal laws of science. As we saw in the first part of this chapter, they arrived at the same explanation as the ancient philosophers and pre-scientific theologians: that nature is highly ordered, as illustrated in many ways but perhaps most dramatically by the existence of predators, which are necessary to regulate their prey. This brings us back to Leopold on the Kaibab Plateau, where we started.

Facts of nature other than the presence of predators are similarly explained from the assumption of a divine order. The writings of Cicero (106 BC to 43 BC), who among the classical writers perhaps best summarizes many of these explanations, include the more general context. In *The Nature of the Gods*, Cicero wrote that there are "signs of purposive intelligence" in the anatomy and morphology of organisms.[23] The shape and form of animals and plants are miraculously well suited to their needs. Trees have bark to shield them from heat and cold. Animals have hides, fur, feathers, scales, and spines to protect and insulate them.

Similarly, the food habits of Earth's creatures seemed to suggest a purpose behind the order in nature. Nature gave animals not only an appetite to seek food but also senses to distinguish the right food, and special physical adaptations to obtain it: "Some animals catch their food by running, some by crawling, others by flying or swimming. Some seize it with teeth and jaws, some snatch and hold it in their claws." Each has adaptations suited to its size: "Some are so small that they can easily pick [their food] from the earth with their beaks," while the elephant "even has a trunk, as otherwise the size of his body would make it difficult for him to reach his food."

Evidence to support the ideas of order and purpose could be taken from the interactions among species, as in the way that one kind of organism captures another, or the way two species cooperate. Cicero describes a shellfish he called the "naker" that "has a kind of confederacy with the prawn for procuring food. It has two large shells open, into which when the little fishes swim, the naker, having notice given by the bite of the prawn, closes them immediately. Thus, these little animals, though of different kinds, seek their food in common; in which it is a matter of wonder whether they associate by any agreement, or are naturally joined together from their beginning."[24] Sea frogs were said to cover themselves with sand and creep along the water's edge as bait for their prey. These adaptations were considered "marvelous," and many examples were repeated by the classical philosophers.

These observations of natural history—some more accurate, some less—suggest that the biological world is one of marvelous order. "What power is it which preserves them all according to their kind?" Cicero asked.

Cicero's ideas, in turn, are echoes of Plato, who lived from approximately 427 BC to 347 BC. His dialogue *Timaeus* (believed to have been written around 360 BC) is often cited as a source of the design argument in Western civilization. Plato wrote that "nothing incomplete is beautiful" and that nature must be "the perfect image of the whole of which all animals—both individuals and species—are parts."[25] The arguments are justified sometimes on the basis of observations and sometimes on the basis of belief. *Thus—and this is important—an argument in defense of a perfect balance of nature can begin either with a few observations that suggest that balance, order, and perfection in the world, or from a belief that there must be such order, and that if we study nature correctly we will discover it.* In our modern terms, the choice would be from a beginning in science or from a beginning in ideology. And in our usual somewhat muddled way of going about things in our lives, we can mix up the two, unaware of the confusion ourselves—a kind of Tom Stoppard confusion in the use and meaning of words.

What I have been trying to point out is that in Western civilization there has been a religious need for a belief in a balance of nature, a need that appears in the modern world as what people call ideological when they prefer not to invoke religion directly. Glacken made this clear when he pointed out that there were three kinds of proof for the existence of a divine providence: (1) physiology and anatomy; for example, the eye is an amazing device to give human beings a way to view the world and thus to appreciate divine creation; (2) the cosmic order; and (3) Earth as a fit environment for life, and the way Earth's creatures appear to have been given exactly the characteristics they need to survive.

Thus, life and nature provided the basis for two of the three proofs of the existence of God. Aristotle, for example, argued that nature does nothing in vain. He made the comparison between a machine, which requires a person to make it, and Earth, a divinely ordered entity that requires a divine creator. While Plato believed that nature is ordered and beautiful, Aristotle added the idea that nature was designed to meet humanity's needs. Similar ideas can be found in the Bible; for example, Psalm 104 suggests that God gave every creature its place in nature and gave order to the world.[26]

A Perfect Symmetry

The Greek and Roman philosophers believed that the world not only was perfect across time but also had perfect spatial symmetry. Geographic symmetry was believed necessary, desirable, and beautiful. The seventeenth-century bishop of Gloucester, Godfrey Goodman, echoed ideas of the Greeks and Romans about the necessity that nature be symmetrical, writing that the height of the tallest mountain above sea level had to be equal to the depth of the ocean floor beneath the ocean's surface, because God created the natural world with a rule of proportion.

The belief in a perfect spatial symmetry in nature continued into the nineteenth century. As I wrote in *Beyond the Stony Mountains: Nature in the American West from Lewis and Clark to Today*, this belief nearly doomed the Lewis and Clark expedition.[27]

The Lewises and the Jeffersons lived on the James River in Virginia and were friends. When Jefferson was elected president, he invited Meriweather Lewis to be his personal secretary and live with him in the White House. There, Lewis extended his education, and he and Jefferson began to plan the expedition from St. Louis to the Pacific Ocean, to learn about the land of the new Louisiana Purchase.

Lewis and Clark spent the winter of 1803–1804 in St. Louis, planning the expedition. They sought the best available maps of the West beyond, which were, of course, sketchy and incomplete. Jefferson believed, as did most people of his time, that the western mountains of North America must be symmetrical with the eastern Appalachians and would therefore have the same width and height. This was an important aspect of Jefferson's plan to search for a water route to the Pacific. It suggested that the western mountains could be crossed in a day or so, without great difficulty, and therefore an inland passage across the western half of the continent could be provided by the Missouri and Columbia rivers. This was believed to be not only possible but necessarily true. In fact, according to a late-eighteenth-century treatise promoting settlement of the West, the western mountains were "passable by Horse, Foot or Wagon in less than half a day."

The map used by Lewis and Clark was drawn by Aaron Arrowsmith, who was one of the best mapmakers of the late eighteenth century. Arrowsmith wrote that the western mountains were "3520 Feet High above the Level of their Base." That Arrowsmith accepted the belief in symmetry enough to provide a precise, though totally wrong, number tells us that this was not a casual belief.

When Jefferson (born in 1743) was 17, he was educated by a Reverend Maury, who assured him that the western rivers that flowed into the Pacific should reach as far east as the Missouri reached west—again, based on the certainty of geographic symmetry—and that the two rivers would be separated by a short and easy communication. Therefore, as president, Jefferson instructed Lewis to follow the Missouri River to its source, then cross the mountains and follow the Columbia River to the ocean. He did not suggest a more open-ended goal of finding the most practical route to the Pacific, because nature's symmetry had already dictated what that would be.

If the asymmetry of the mountains of North America had been known to Jefferson, or considered a likely possibility, the planning for the Lewis and Clark expedition might have been much different. First, Jefferson might have posed the goal differently: Seek the best water route to the Pacific. This route, as later pioneers learned, involved taking the Platte River west from where it flowed into the Missouri, leading to an easier crossing of the Rocky Mountains. Second, he might have realized that a small party of men was unlikely to succeed in this endeavor, and he might have held off until the United States could, and was motivated to, fund a much larger military expedition.

Lewis was similarly educated and influenced by the predominant eighteenth-century belief in the balance of nature. But like Jefferson, he was also fascinated by natural-history observations and by the potential of the recently developed scientific process. He was open to observations that might revise his ideas; thus throughout the expedition he and Clark were wonderful observers of nature, naturalists in the best sense, with a curiosity and ability to analyze what they observed that I find strikingly modern and at times amazingly insightful.

The belief in symmetry caused the expedition great difficulty when they reached the Rocky Mountains in 1805 and realized that crossing these would not be a simple few days' endeavor (Fig. 5.5). What got them through, in addition to the Indians who guided and helped them in many ways, was their great abilities as leaders, and as diplomats in their dealings with the Indians, paired with their observational excellence, adaptability, inventiveness, sheer bravery, and stamina.

It's important to reflect on their experience in comparison to ours. In today's world, with our cell phones, computers, and all the rest, the idea of a perfect balance of nature, including even geographic symmetry, may seem merely fanciful, the stuff of movies about life far, far away or a long, long time ago. But the belief in a balance of nature or an acceptance of the true dynamics of nature, as observed directly, could make the difference between life and death for Lewis and Clark and other early travelers from European culture in the American West. For us, today, the parallel is that what we take for granted and assume must be true about our surroundings can make the difference between conservation and great damage to our environment. The more we are surrounded by computer-based gadgets that make our lives feel safe

FIGURE 5.5 Travelers' Rest, near Missoula, Montana, where Lewis and Clark stopped before beginning to ascend the Rocky Mountains. From here, they could see the foothills of the Bitterroot Mountains, part of the Rocky Mountains that they would have to cross. Their maps showed these as symmetrical with the Appalachian Mountains, relatively easy to cross, which of course they were not. (Photograph by the author.)

and comfortable, the less dependent we believe we are on such things as a metaphor, image, or myth. But on the contrary, the more artificial our lives—artificial in that we are separated from the need to seek and obtain the primary necessities of life—the more vulnerable we are, because we are less able to fend for ourselves, and the less we feel impelled to understand nature and our relationship with it.

Failure of the Divine Order

The assumption and conclusion that there is and must be a divine order in nature leave a major question unanswered: If there must be such an order, how do we explain its absence? Two contradictory answers have been given to this question, but both point the finger at ourselves. One answer claims that an imbalanced nature is the result of what we have done; the other claims that it is the result of what we have *not* done.

The first claims that the absence of a balance of nature is not natural or God's will but must be the result of human interference with the natural state of affairs. Although to environmentalists of the 1960s and 1970s this idea may have seemed new, it is quite ancient. Among the classical philosophers, for example, Pliny pointed to the beauty and bountifulness of Earth without human interference in contrast to the imperfections of people who abused the Earth. He speculated that there was a purpose for wild animals that are hostile to humans: They were intended to be guardians of Earth, protecting it from us.

The other major explanation of the failure of the divine order lay in a belief that human beings are the final cog in the machine to create the divine order. This is interpreted as our purpose on Earth, a purpose from which we have strayed when we either fail to do our job at all or do it incorrectly, altering and thus destroying the divine order. As Alexander Pope put it in his "Essay on Man," the great chain of being extends from God to human beings to "Beast, bird, fish, insect, what no eye can see," and, he said, "From Nature's chain whatever link you strike, Tenth, or ten thousandth, breaks the chain alike."[28]

Nature as the "External Throne of the Divine Magnificence," or as Just Horrible

In the history of Western thought, nature has repeatedly been viewed as a wilderness in the worst sense, full of dangers and evils, and lacking the symmetry, order, and therefore beauty of the domesticated landscape created by civilization. These ideas were given voice in the eighteenth century by Georges-Louis Leclerc, Comte de Buffon (1707–1788). In his *Natural History*, he wrote that although nature is the "external throne of the divine magnificence," people "among living beings" establish "order, subordination and harmony." Human beings are granted by God "dominion over every creature," and it is our role to add "embellishment, cultivation, extension and polish." It is man who "cuts down the thistle and the bramble, and he multiplies the vine and the rose."[29]

From this perspective, nature without the proper action of human beings is not divinely ordered. Buffon describes the unpleasantness and the horror of nature undisturbed—that is, unhusbanded by human beings. "View those melancholy deserts where man has never resided," he admonishes. They are "overrun with briars, thorns, and trees which are deformed, broken and corrupted." Seeds are "choked and buried in the midst of rubbish and sterility." In wildness, nature has the appearance of "old age and decrepitude." Instead of the "beautiful verdure" of managed landscape, there is "nothing but a disordered mass of gross herbage, and of trees loaded with parasitical plants." The wetlands are a particularly awful example of nature undisturbed; they are "occupied with putrid and stagnating waters" and are impassable, useless, and "covered with stinking aquatic plants, serve only to nourish venomous insects, and to harbour impure animals." The unmanaged forests are equally unpleasant; they are "decayed," and in them "noxious herbs rise and choke the useful kinds." In savannas, there are "nothing but rude vegetables, hard prickly plants, so interlaced together, that they seem to have less hold of the earth than of each other, and which, by successively drying and shooting, form a coarse mat of several feet in thickness."

Human beings, forced to enter or live in such barbarous landscapes, feel horror and fear. Pursuing wild animals, a hunter is "obliged to watch perpetually lest he should fall victim to their rage, terrified by their occasional roarings, and even struck with the awful silence of those profound solitudes." Nature uncultivated is "hideous and languishing," and human beings alone can make it "agreeable and vivacious." Thus we must drain the marshes and transform the stagnant waters into canals and brooks. We should set fire to "those superannuated forests, which are already half consumed" and finish the clearing "by destroying with iron what could not be dissipated by fire." We are admonished to carry out our role in nature, just as every creature is meant to carry out its role.

Buffon's detailed description of a disagreeable wilderness is of interest to a modern naturalist, for it seems to contain merely the plants that grow in wetlands or in abandoned fields recently under cultivation or in pasture, rather than in the true forest wilderness that existed in Europe before any influence of civilization. For example, Buffon describes wild nature as having the thistle and bramble, "overrun with briars, thorns," which are generally characteristic of new woodlands that grow on comparatively open but abandoned fields, rather than of the dense shaded understory of old and undisturbed forests. Similarly, his description of "nothing but rude vegetables, hard prickly plants, so interlaced together . . . by successively drying and shooting" that they "form a coarse mat of several feet in thickness," is reminiscent of abandoned agricultural fields or pastures within 10 or 20 years after their last use, or of marshes, which are usually the last areas to be cleared for human occupation. Buffon's "wilderness" appears to have been simply the abandoned and poorest lands in an otherwise heavily domesticated landscape, not nature untrammeled by human beings, which is the meaning of wilderness as defined by the U.S. 1964 Wilderness Act, still in force today.

Whatever the reality of Buffon's surroundings, we find in his writings a repetition of a view of nature that predominated from the time of the classical Greeks through the Romans and to the nineteenth century: Although there may be elements of order and balance in the biological world, attainment of order and its beauty requires human action to create symmetry and harmony. Cicero, for example, summarized this classical view of the human role in nature as "the protection of some animals and plants and indeed there are many which could not survive without human care."[30] Buffon's ideas can also be traced to the Bible. Psalm 8, for example, says that God gave man "to have dominion over" the works of His hands.

These two perceptions of our role in nature will be familiar to those interested today in the environment, because they continue to be debated: What is our role in nature? To stay out of it or to improve it (including active conservation of nature and care for endangered species and of individual plants and animals among those endangered species)?

Nature as Designed for People

Modern discussions about people and nature commonly concern whether we have an obligation to protect it and make it better, or whether it is nothing more than arrogance on our part to think we could do such things and, even worse, have a moral right to do them. This, too, has an ancient lineage.

As we have seen from Plato, Aristotle, and others, the view developed that there is order in nature and a purpose behind that order. "Someone may ask, 'But for whose sake has this mighty work of creation been undertaken?'" wrote Cicero. To assume it was for the trees and other plants would be "absurd, for though they are sustained by nature they are devoid of sense or feeling." Equally absurd would be to assume it was for nonhuman animals, for "it seems no more likely that the gods would have undertaken so great a labour for dumb creatures who have no understanding." Who then is left? "For whom then shall we say the world was made? Surely for those living creatures who are endowed with reason. . . . For reason is the highest attribute of all. We may therefore well believe that the world and everything in it has been created for the gods and for mankind."[31]

Not all the writers about nature believed this, of course. Perhaps the most famous, if not the first, of the classical philosophers to oppose the predominant view that the world was created for human beings (part of the design argument, as discussed by Glacken) was Lucretius (Titus Lucretius Carus, who lived from ca. 99 BC to ca. 55 BC and was therefore a contemporary of Cicero). He wrote in *De Rerum Natura* that "this world of ours was not prepared for us by a god. Too much is wrong with it." Two thirds of the world "that pair of thieves, fierce heat, insistent cold, have robbed men of," and the rest, where human beings might live, "nature, as violent as either one, would occupy and homestead with fence of briar and bramble." Rather than life's necessities being given too easily to people, they must "groan as they heave the mattock" and "break the soil shoving the plow along," for without the plow "nothing at all could, of its own initiative, leap forth." Moreover, the vagaries of weather make it clear that the world was not made for people: "How many a time the produce of great agonies of toil burgeons and flourishes, and then the sun is much too hot and burns it to a crisp; or sudden cloudbursts, zero frosts, or winds of hurricane force are, all of them, destroyers." Also, if the world were made for people, why are there "the dreadful race of predatory beasts, man's enemies on sea and land?"

Indeed, the animals are better off than people; they grow without such care as we require, "they don't need rattles, they don't need the babbling baby-talk of doting nurses," nor do animals need clothes suited to each season or the protection of "weapons and walls," because for them "earth and nature, generous artificer, supply their every lack."[32] From these ideas, Lucretius argued that if there is a purpose behind the order, and if there is an order, the goal of this purpose cannot be the well-being of human beings.

In every century, there have been those who shared with Lucretius a skepticism about the perfection of nature, the design of nature, and the purpose of that design. As I showed earlier, throughout the history of the West one of the dominant themes about nature has been the belief that the universe, the solar system, and Earth are incredibly well suited to the requirements of life, so well suited as to exceed the likelihood of mere chance. To this we can now add that throughout most of Western history, this belief led to another: that the design of the universe must be the result of some purpose and purposeful creation and is necessarily a realization of a divine order.

In the twentieth century, these ideas became more complex, and the simple central belief in a divine order became obscured. Scientific observation was understood to deal only with what is, with the what and how of the universe, not with the why of metaphysics and religion. Scientists sought an understanding of the way the universe functions, an understanding of the rules that govern the phenomena that we observe. But ironically, as we saw earlier, the pursuit of these explanations brought Joseph Henderson, in *The Fitness of the Environment*, back to an ancient sense of wonder at the remarkable features of Earth that are so peculiarly suited to the emergence, evolution, and continuation of life. Unable as a scientist, acting in his role as a scientist, to make the next step in the argument, about the "why" rather than just the "what," Henderson merely wondered, whereas Cicero had asserted that such orderliness could not be the result of chance.

But the sense of wonder at the orderliness of the universe has never disappeared from the ideas of Western civilization. It recurs within the modern environmental debate. In modern times, especially since the beginning of the twentieth century, the role and uses of science have been clearly differentiated from the roles and functions of religion and metaphysics. It is a distinction so well known that it is commonplace in physics, chemistry, and molecular biology. However, environmental sciences, especially ecology, have ironically reinforced and simultaneously presumed the ancient idea of a wonderfully ordered universe and therefore the theme of nature as divine order. The belief in divine order carried with it a long history and thus a kind of mental—intellectual and emotional—momentum. The powerful observations and theories of nineteenth-century science, which revealed amazing order in nature and the suitability of the environment for life in realms never before known—the universal chemistry and physics of water, the size of our planet and its distance from the sun—were impelled by this momentum and reinforced it.

Here, then, is where our cultural history, augmented by modern environmental sciences, has brought us: Reinforcing the powerful beliefs in a divinely created and ordered nature were the historical interpretations of biological nature—for example, the ancient explanation for the existence of predators in an otherwise perfect universe. These ancient ideas were repeated as though newly discovered in century after century, revived in the age of exploration and 200 years later in Derham's discussion of why a perfect God would have made that "most pernicious of birds," the Peruvian condor. Such explanations became codified in the twentieth century through eighteenth- and nineteenth-century mathematics, put forward first as scientific truths, which at another level of our existence seemed to explain and justify the ancient, pre-scientific beliefs. But in fact—and this is the deep irony of the whole thing—the mathematical equations of early-twentieth-century ecology had developed from those ancient beliefs in a specific kind of order.

It is also ironic that the strong intentional separation of science from religion tended to obscure the underlying connections between them in the explanations about the character of biological nature. The idea of a divinely ordered universe that is perfectly structured for life has persisted, often beneath the surface but still influencing the interpretations of environment, nature, and the role of human beings in nature in our own time. Ecologists, seeing themselves in their role as scientists separate from religion and ideology, with the "how" and not the "why," have believed themselves truly separated in their work from cultural beliefs that preceded science and remain apart from science. Those who did not understand the science were, in general, unable to perceive and appreciate the ironic connectedness between ancient beliefs and twentieth-century ecological assertions. The final and greatest irony is that this connectedness between ancient beliefs and ecology and other environmental sciences continues today in spite of warnings by me and a few other environmental scientists, such as Stuart Pimm.

In reviewing briefly the history of the idea of a divinely ordered universe and the scientific observations of order, it has not been my purpose to argue either for or against a religious interpretation of the character of nature, but merely to show the parallel and the historical connection between the ancient, religious, and metaphysical perspectives on nature and modern beliefs that have been accepted as scientific. This is a humanistic goal, necessary if we are to make wise use of nature and seek some kind of harmony with our surroundings in the future. What we learn from the mountain lion and the mule deer is about what we believed, not about what we know.

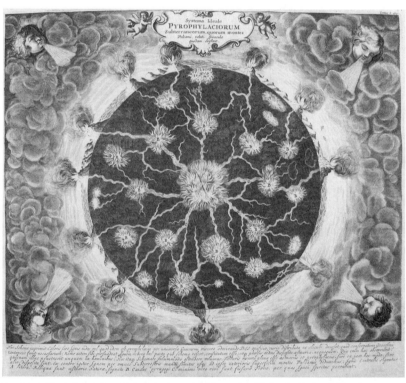

FIGURE 6.1 The interior of a volcano as drawn by Athanasius Kircher in his book *Mundus Subterraneus*, published in 1665. (Courtesy of the William Andrews Clark Memorial Library: University of California, Los Angeles.)

6

Earth as a Fellow Creature

ORGANIC VIEWS OF NATURE

Time does change the nature of the whole wide world; one state develops from another; not one thing is like itself forever, all things move, all things are nature's wanderers. . . .

—LUCRETIUS, DE RERUM NATURA *(FIRST CENTURY BC)*[1]

From Thales to Galileo, the world had been animate. It was permeated by mind, intelligent, alive. . . . The world and the universe lived as man lived; the world and even the universe, like man, was subject to decay and death. . . . With the death of an animate world and the breaking of the circles of the universe, old truths began to lose their force, gradually becoming the language of poets rather than of scientists.

—MARJORIE H. NICOLSON, MOUNTAIN GLOOM AND MOUNTAIN GLORY *(1959)*[2]

Nature Within the Volcano

One of the first people to descend into the crater of an active volcano and return to write about it was Athanasius Kircher, a seventeenth-century Jesuit priest who later wrote a book, *Mundus Subterraneus*, that described what he had seen. Father Kircher descended into Mount Vesuvius in 1638 and then into Mount Aetna. Climbing to the summit of Vesuvius the night before he descended into it, he saw "a horrible combustion" and heard "horrible bellowings and roarings of the Mountain." There was "a stench of Sulphur and burning Bitumen," with "smoaks mixt with darkish globes of Fires." Descending into the volcano, he found underground passages with bodies of water and areas of fire, which together suggested to him a kind of order. "The underground world is a well fram'd House, with distinct Rooms, Cellars, and Store-houses, by great Art and Wisdom fitted together," he wrote, "not, as many think, a confused and jumbled heap or Chaos of things."[3]

Rather than explain the caverns and their fires and waters in terms of force, energy, and pressures that might have lifted water and rock, perhaps by an analogy **121** with the internal belts and pulleys of a windmill or another machine—that is, in terms of the principles of physics, as a modern geologist would do—Kircher began his explanation of what he had seen with a quote from Virgil, who had written that the "belching rocks" of volcanoes were the torn entrails of the mountains. Kircher believed that water and ashes were the food of the fires. Along with rain, hail, and snow, the fires were fed by "veins of the Sea." Fire and water thus "sweetly conspire in mutual service, and with an inviolable friendship and wedlock" mix together their "several and distinct private lodgings" for the "good of the whole. . . . Out of ashes mixt with water, a new food and nourishment of everlasting fire is generated," he wrote.[4] The water mixed with dust and ashes produced a continual "conception and birth" of fires. The fires then grew and matured and, becoming ripe, erupted. To Kircher the volcanoes were like a rose, nurtured by water and minerals so that a seed could sprout, grow, mature, and burst into fiery bloom.[5]

In short, *Father Kircher described the action of the volcanoes and the inner workings of Earth in terms of organic, or biological, metaphors.* In his drawing of Earth (Figure 6.1) he pictured the planet as having a structure like that of a living creature, or of a cell within such a creature. In his analysis of these subterranean phenomena, Kircher was a spokesman for the second major idea about nature that runs through the history of Western civilization: the organic view, that Earth either is *like* a living creature or actually *is* a living creature.

It is interesting to compare Father Kircher's experiences inside a volcano with a trip made in 2006 by Ken Sims of the Geology and Geophysics Department of the Woods Hole Oceanographic Institute. According to an article in *Oceanus*, Sims was studying volcanoes to learn "how our planet is evolving and how volcanic gases cause climate changes that may have led to the extinction of dinosaurs," as well as "what effect gas emissions may have on human health." He and several friends, who previously had "climbed mountains, frozen waterfalls, and rock walls around the world for sport," descended into Nicaragua's active Masaya volcano in March 2006.[6]

Another member of the expedition, British volcanologist Tamsin Mather, explained: "We are trying to understand volcanoes as a mercury source. . . . How much comes out of volcanoes? How does it move around the Earth? What role does it play in the environment? If we can understand this at a couple of volcanoes, we can start thinking about it on a global scale and how such emissions are distributed."

The description of what Sims and his colleagues found it to be like inside the volcano is quite different from Father Kircher's. The *Oceanus* article continues:

> The crater floor of a volcano is an alien environment. Scorched brown and black rocks ranging from the size of peas to SUVs litter the ground where the crater's walls have collapsed. There is little wind, no vegetation, and zero shade. The sloping floor crunches and slips under the climbers' hiking boots. "It's like walking on ball bearings," John Catto, another member of the expedition, said. The rocks are sharp, too. "If you slip," he added, "you're going to end up looking like pizza."

From here, just meters from the crater, the exhaling gas sounds like the slow whoosh of a wave against a beach. It doesn't take long to set up Sims' instruments, which he powers using a motorcycle battery purchased in the town of Masaya.

Instead of describing the volcano as a fellow creature, Amy E. Nevala, the author of the article, calls the volcano "Earth's plumbing system," giving it a machine-like character, as do all the comments by those who went down into the volcano. Reinforcing a change in views of volcanoes, and inferentially of nature around us, is the photograph I took in Taiwan of an advertising shoot taking place in front of an active volcano on Datun Mountain, just outside the capital city Taipei (Fig. 6.2). The volcano is nothing more than a decorative background, not a living creature, not something fearful, just eye-catching.

In these experiences, separated by almost four centuries, we have a comparison between a pre-scientist's organic view of Earth's interior and a modernist's scientific and technology-age perception. This gives us some insight into what a

FIGURE 6.2 An active volcano used as a scenic backdrop for a fashion shoot in the twentieth century. A fashion model poses in front of Datun Mountain just outside Taipei, Taiwan. By the end of the twentieth century, a volcano was apparently viewed simply as exciting scenery, not as part of a fellow creature, as it was in prescientific times. (Photograph by the author.)

belief in an organic—living—Earth meant in prescientific times, but let's explore this idea further.

What is the essence of the organic view? An idealized (that is, perfect) organism has certain attributes. It passes through the major life stages: birth, youth, maturation, maturity, reproduction, old age, senility, and death. An organism thus has a history; what it will be tomorrow depends not only on what it is today but also on what it was yesterday. An organism proceeds through its existence in a one-way direction, passing from stage to stage, none of which can be relived by the individual. The idealized organism in this sense is like a rose, which is first a bud, then a flower that is opening, a flower in full bloom, a fading flower, a fruit, and finally a new rose plant. A rose that lived forever would no longer be organic and would no longer have the same charm and beauty.

Scientifically, such an idealized organism is subject to chance and itself a creator of chance; it is in some ways unpredictable. To someone out in the wilderness or on a country walk, such an organism can have individuality, a personality, and thus idiosyncrasies, which are at the same time charming and troublesome. Will this rose be beautiful? Will it live, or will it die prematurely? *Uncertainty is an essential organic quality*, adding to the rarity of other special individual qualities and thus to their value. Even an idealized organism lacks continuing perfection, since it eventually ages, dies, and decays. This process is natural and gives organic entities individuality, charm, and attractiveness, enhanced by a touch of sadness at the fleeting nature of organic beauty.[7]

The Idea of a Living Earth

The idea that Earth might be viewed as organic in this sense may seem quaint and strange to us in the twenty-first century, but it has been one of the predominant views throughout human history, perhaps more common than the idea of divine order, and has a much longer history than our own scientific explanation of nature. The organic view has been traced back by archaeologists and anthropologists to early cultures, pre-scientific and non-Judeo-Christian. It can be found in many contemporary non-Western "primitive" societies. In the history of the West, the organic view was important in ancient Greek and Roman thought and in Judeo-Christian thought.[8]

The organic view fit in with and helped justify the biblical stories of Earth's changes, including the loss of the Garden of Eden and the Flood. Qualities of environment that were hostile and harmful to people, both physical and biological, including natural disasters with their uncertainties, were the result of man's fall from grace. The physical hostility and chaos represented by mountains, oceans, and marshes, as well as the existence of fossils, were taken to be direct results of the Flood and of the expulsion from the Garden of Eden.

As late as the seventeenth century, theologian Thomas Burnet (ca. 1635–1715), a contemporary of the volcano-delving priest Athanasius Kircher, wrote in his influential book *Telluria Theoria Sacra* that Earth before the Flood had been "smooth, regular and uniform; without Mountains, and without a Sea." Divine order, harmony, and balance were evident everywhere. Indeed, he said, "the smoothness of the Earth made the Face of the Heavens so too; the Air was calm and serene; none of those tumultuary Motions and Conflicts of Vapours, which the Mountains and the Winds cause in ours; 'Twas suited to a golden Age, and to the first innocency of Nature." The Flood created the "ruins of a broken world." Where there had been a "wide and endless plain, smooth as a calm sea," there was after the Flood "wild, vast, and undigested heaps of stone and earth," from which developed geographic variations, oceans, mountains, and even "stagnant fens and bogs."⁹

From Burnet's point of view, the world lacked perfection in part because of human failings within the Garden of Eden; that's his answer to why a world that by definition had to be perfect was not. But he also believed in a golden age like that described by the Greek and Roman writers, and therefore in an aging, therefore organic, Earth. He argued that the discoveries of science concerning the history of the Earth could be fitted into the older views and provided further revelations of God's works. Thus from Burnet's seventeenth-century perspective, a belief in the organic Earth was consistent both with explanations offered by the new sciences and with the idea of a divine order. The wounded Earth had slowly recovered to a certain extent by Burnet's time, in part, he argued, because people had undertaken the husbandry of nature.¹⁰ In sum, we were finding our way back into our proper role within nature.

The European Renaissance forced a reexamination of ideas about nature and a renewal of the ancient controversy about the balance of nature. Many developments forced this reexamination, including discoveries of new worlds, with their wildernesses, exotic and curious creatures, and peoples living a Stone Age life; discoveries about the physical universe, making clear that humanity is not at the center of the universe and that the planets do not revolve around the sun in perfectly circular orbits; and discoveries of geology, including fossils and the composition of Earth's surface.

A controversy arose: Much of the new knowledge about nature seemed to contradict the belief in a divinely ordered universe and a balance of nature. How could these new observations be reconciled with old beliefs? One solution was to return to the organic idea of nature, which included a belief that Earth had aged and, although perhaps perfect at one time, was no longer so. Some scientific findings in the seventeenth century seemed to reinforce the organic view. For example, Danish professor Nicolaus Steno (1638–1686), considered to be one of the fathers of geology, was the first to recognize that Earth's surface is made in part of water-deposited sediments, which were laid down horizontally and later made irregular by subterranean forces. In 1671 he wrote that "all present mountains did not exist from the beginning,"¹¹ implying in its way that Earth had aged and continued to age. (One could interpret what he wrote as simply the observations of a scientist, but it is phrased in a way that is consistent with the aging of Mother Earth.)

Although Steno's reflections might not have been meant literally, sometimes the organic metaphor of Earth as a living creature was taken quite literally, as in a poem of the sixteenth century, "The Purple Island," by Phineas Fletcher (1582–1650), in which he writes, "The aged world, though now it falling shows [*sic*], and hastes to set, yet still in dying grows."

That is a description of a very organic world—living and subjected to dying. Fletcher describes his purple isle as having in its "lower region" a liver "walled in by the ribs," and "covered with one single tunicle: & that very thinne, and slight." That earth-liver "is tied to the heart by arteries, to the head by nerves, and to both by veins dispersed to both." And then: "Hence rise the two great rivers of bloud, of which all the rest are lesser streams: The first is Porta, or the gate-vein, issuing from the hollow part and is shed toward the stomack, splene, guts, and the Epiploon. The second is Cava, the hollow vein, spreading his river over all the body."[12]

We of the scientific age find Fletcher's description of Earth's analogues of bones, heart, stomach, and all of its bodily organs just plain odd, perhaps fanciful, and certainly disconnected from any way we conceive of the physical Earth or one of its islands. We have to understand that, as we saw with Father Kircher, the idea of an Earth or an island as an organic being with body parts was taken literally.

Such views can be traced back to the Greeks and Romans. A poem by Empedocles (ca. 490–430 BC) titled "Tetrasomia," or "Doctrine of the Four Elements," is considered to be the origin of the idea that the universe is made up of four elements—air, water, fire, and earth—and he is said to have believed that the sea was the sweat of the Earth. Lucretius, as mentioned in Chapter 5, was the most famous opponent of the idea of divine order and, among the classical philosophers, the most important proponent of the organic idea.

"In the beginning," Lucretius wrote, "earth covered the hills and all the plains with green, and flowering meadows shone in the rich color." Earth—that is, nature—had gone through stages in its life, just as a person does. "When the earth and air were younger, more and larger things came into being," and then "earth was indeed prolific, with fields profuse in teeming warmth and wet." Earth "deserves her name of Mother" because the rich fields were like wombs, the creatures were like embryos. There were "pores or ducts of Earth, channels from which a kind of milk-like juice would issue, as a woman's breasts are filled with the sweet milk after her child is born." But Earth had aged; "of parturition, Earth has given up like a worn-out old woman."[13] Change is recognized as inevitable, but as leading unavoidably only to decay and death.

The mortality of Earth was evident in its erosion, with the soil washed away by heavy rains, and the riverbanks "shorn, gnawed by the currents." Erosion was regarded as part of a process of losses and gains: "For every benefit, requital must be given. Earth's our mother, also our common grave. And so you see Earth is receiving loss and gain forever."[14] Clearly, the mutability and decay of Earth were recognized by the ancient philosophers. In *Mountain Gloom and Mountain Glory*, which I mentioned in Chapter 1 is a classic discussing changes in beliefs about nature through the nineteenth-century Romantic period, author Marjorie Nicolson pointed out that if

Earth was like man, and man like Earth, then one "had every reason to expect to find in the globe and in the cosmos exact analogies for the structure, functions, and processes of the human body."[15]

This idea of Earth as a living creature seems so strange to us that one wonders how it could possibly have come about. But consider this: We explain new things to other people in terms of what we and they know. The unfamiliar is described by analogy with the familiar. A classic example is William Clark's description of a badger, the Lewis and Clark expedition's first zoological specimen. The animal was unknown to Western science until one of the men on the expedition shot it on July 30, 1804. Clark described it as having a shape and size "like that of a beaver, his head, mouth &c. is like a dog with its ears cut off, his tale and hair like that of a ground hog, something longer and lighter, his internals like a hog's." He compared the hair to a bear's, with a "white streak from its nose to its shoulders."[16]

Before machines, what people knew were people—their own bodies and those of other people—and some other living things in the vicinity. Earth changed, with the daily cycle, with the seasons, from year to year. The climate changed; the seas changed. They were unlike a single rock that you could pick up and seemed always the same. A rock could not serve as an analogy to the diurnal cycle, to the seasons, to the waxing and waning of the moon. Only living things—yourself and those around you—would seem to share any of the qualities that you observed in nature. You would have no other choice but to conceive of nature, your environment, Earth, sun, moon, and the wandering planets, in terms of living things. A biological analogy, this organic imagery for all around us, would have been only natural. The organic idea was extended so that all creatures were described in terms of human beings, the creatures that we knew best. Not only was Earth a fellow creature, but each living thing shared deeply in organic qualities that we have.

For example, in *A Topographical Description of the Dominion of the United States*, published in 1776, Thomas Pownall wrote that trees in a forest "grow up, have their youth, their old age, and a period to their life, and die as we men do." A visit to a forest shows "many a sapling growing up, many an old tree tottering to its fall, and many fallen and rotting away, while they are succeeded by others of their kind, just as the race of man is."[17] In this way Earth is kept vegetated just as people continue on Earth.

New Technologies and an Aging Earth

Beginning with the Age of Exploration—approximately from the early fifteenth century to the beginning of the seventeenth century—and continuing into the eighteenth century, European explorers discovered the Americas, Australia, and parts of Africa and Asia previously unknown to Western civilization. They found strange new animals and plants and vast oceans and mountains. The invention of the microscope at the end of the sixteenth century revealed new kinds of life, while

Galileo's development of the telescope, along with major advances in the science of physics, brought new insights about the sun, the solar system, and the universe. These discoveries forced a reconsideration of the theological arguments about the existence of God. In previous philosophical writings, the perfection of Earth and the universe was taken as evidence that they were created by God—a perfect and infinite God could make only a perfect world. Now, instead, the mountains and oceans, with their fearsome irregularities, became signs of the power of God. In the eighteenth century, Joseph Addison (1672–1719), a British writer, poet, politician, and also a colleague and friend of Jonathan Swift and other writers more famous today than himself, wrote in *Pleasures of the Imagination* that he saw in ocean storms an "agreeable horror." He said that the power of the oceans "raises in my thoughts the idea of an Almighty Being and convinces me of his existence as much as a metaphysical demonstration."[18]

Proof of the existence of God had completely shifted from nature's structural symmetry to the demonstration of power in the (apparently) disorderly ocean storms. Dynamism could be godlike; a dynamic Earth, more organic in character than the earlier idea of an Earth of perfect but static symmetry, could serve as proof of His existence.

In the seventeenth century, an organic idea of beauty also underwent a resurgence. In classical Greece and Rome, order, permanence, and regularity were beautiful; this view had predominated throughout the Christian era into the Renaissance, but at the end of the seventeenth century, ideas of nature's beauty also began to change. In 1690, in *Geologia: A Discourse Concerning the Earth before the Deluge*, Erasmus Warren (1642–1718) wrote that the present world, with its irregularities, was beautiful. Nature has asymmetries and "a wild variety," he wrote. "Without its irregularities, ruggedness and inequalities, the Earth would be not only less useful and less diverse but less comely." In a book first published in 1693, *The Folly and Unreasonableness of Atheism Demonstrated from the Origin and Frame of the World*,[19] Richard Bentley (1662–1742) wrote "there is no universal reason that a figure by us called regular, which had equal sides and angles, is absolutely more beautiful than any irregular one." Sir Thomas Browne (1605–1682), an English physician and author who wrote about medicine, religion, and science (his best-known work is *Religio Medici*), justified the ugliness of the toad, bear, and elephant on the grounds of their adaptation to their needs (Fig. 6.3). In time, irregularity along with variation became beautiful.[20]

Also in the seventeenth century Andrew Marvell (1621–1678), one of the best and most famous of that century's English poets, wrote that mountains are unjust, deformed, and ill-designed excrescences.[21] By the eighteenth century, mountains were perceived quite differently. For example, John Dennis, who traveled through the Alps in 1688, wrote that the mountains were a place where one experienced a "delightful horror" and "terrible joy." By the mid-eighteenth century a descriptive poetry had emerged that glorified the wildness of nature. James Thomson (1700–1748), for example, in his poem "The Seasons," wrote that in winter he saw "Earth's universal face as one wild dazzling waste," which he viewed with "pleasing dread."

The irregular structure of the observable biological world had become beautiful.[22] The transition was not restricted to literary formalisms but included the actual appreciation, and manipulation, of nature. In the eighteenth century, Addison, for example, admired Chinese gardens more than his own English gardens because the Chinese gardens were less artificially regular and appeared closer to nature.

This change continued into the nineteenth century, when the organic perspective was celebrated in romantic poetry. The romantic perception was an important part of the major shift from classical aesthetics, which, as I have explained, were based on symmetry, balance, harmony, and structural order, to the aesthetics of a dynamic, imperfect, powerful Earth that I have just described. The aesthetics of the nineteenth century can be seen in part as a reaction to the scientific age, a rejection of the machine-dominated world and the mechanistic perceptions that accompanied the scientific advances of the eighteenth century, and in part as a product of the increased accessibility of remote regions, including rugged mountains, to more and more people, including the poets themselves. Today we have easy access to such ranges as the Canadian Rockies, where we moderns find great beauty in the rugged, glacier-carved slopes (Fig. 6.4). Many a modern hiker and mountain climber would be surprised that what today seems so obviously beautiful was two centuries

FIGURE 6.3 In ancient Greece and Rome, an elephant's asymmetry would have been seen as odd and ugly. In the nineteenth century, with a revival of the organic idea about nature and a concomitant change in aesthetics, an elephant's powerfulness gave it a kind of beauty. When I took this photograph, this old bull had a 24-hour guard in a Kenyan national park because he was believed to be the oldest elephant in the park and his tusks were the largest. Therefore he was considered a national treasure and was very likely to be killed by poachers. (Photograph by the author.)

ago considered hideous and to be shunned. In the nineteenth century, mountains became places where power dwelt in tranquillity. Shelley wrote in his 1816–1817 poem "Mont Blanc: Lines Written in the Vale of Chamouni":

> Remote, serene, and inaccessible:
> And this, the naked countenance of earth,
> On which I gaze, even these primæval mountains
> Teach the adverting mind.
> . . .
> Mont Blanc yet gleams on high:—the power is there,
> The still and solemn power of many sights,
> And many sounds, and much of life and death.
> In the calm darkness of the moonless nights,
> In the lone glare of day, the snows descend
> Upon that Mountain;
> . . .
> The secret strength of things
> Which governs thought, and to the infinite dome
> Of heaven is as a law, inhabits thee!

FIGURE 6.4 This mountain peak, in Jasper National Park, Canada, with its Athabasca Glacier, would have been considered ugly by the ancient Greeks and Romans because it is not symmetrical. But we consider this scenery beautiful and worthy of special trips to view it, as I did myself. This illustrates the change from classical requirements of symmetry to a new, organic aesthetic arising in the nineteenth century, where mountains and ocean storms became a representation of God's power, and therefore held a "terrible beauty." (Photo by the author.)

That "remote, serene, and inaccessible" mountain had a "secret strength of things which governs thought," impressing the viewer of the scene with the power and infinity of God. Change and decay, lamented by earlier poets, are an inevitable part of the "enormous performances of Nature."[23]

The nineteenth-century idea of beauty can also be regarded as an attempt to grapple with the deeper implications of the new sciences, primarily the idea that the world was held together not by structural symmetries but by universal laws that from a religious point of view could be interpreted as evidence of the wisdom and power of God.

This idea—that adaptation did not have to be symmetrical and beautiful to be the creation of God, and that God was the only way that such adaptations could come about—has become familiar in the twenty-first century with the recurrence of the theological design argument: that Earth, life, and the universe are too amazing, complex, and intricate to have come about by chance, and therefore must have had a designer. But today the design argument is taken a large step farther, using the new knowledge of DNA and many other modern biological discoveries to argue that these could not have come about by a blind, unguided process of Darwinian evolution, but must have had a maker, and that maker had to be God.

What is of interest in our current discussion is that symmetry and the kind of beauty that can arise only from symmetrical objects are no longer important parts of the design argument—at the DNA level, a balance of nature in time, space, or any other way is not required to argue against Darwinian evolution. (Oddly, this has not seemed to counter the strength of the balance-of-nature idea when it comes to discussions of environmental issues and of the relationship between human beings and nature.)

In sum, a strong theme of an organic aesthetic is clearly apparent in the Romantic poets, an aesthetic in some ways dependent on science, and in other ways dependent on rejection of the sciences. The important point here is that there was a shift from an argument that God must exist, because His world looked perfectly ordered, to the argument that God is evident in the power of natural forces—a shift from an explanation based simply on structural characteristics to an explanation based on processes and dynamic qualities. This is a key transition, which will come up again in later chapters. But looking ahead, this is a transition that the science of ecology had to make and in general did not in the twentieth century, and in many ways still has to make, from a perception of nature as a fixed structure to nature as dynamic systems, with an emphasis on *dynamic*.

Thoreau, Transcendentalists, and the Summit of Mount Katahdin

Henry David Thoreau represents an interesting transition in the nineteeth century among the perceptions of nature, since he was one of the first modern botanical scientists and modern naturalists, but was deeply intrigued by the beliefs, feelings, emotions, beauty, and the connections between people and nature.

His mentor, Ralph Waldo Emerson (1803–1882), was one of a group of intellectuals around Concord and Boston who adopted the philosophical-religious doctrine of transcendentalism. Among its precepts were the central role of biological nature in religion and human life in general, the belief that nature could be read spiritually, and that studying God in nature could substitute for reading the Bible. Another precept was that nature was benign and concerned about human beings.[24]

Thoreau, as was typical of him, listened to what his mentor and others said, mulled it over, and then went off to see for himself whether it was true. He made three trips to the Maine woods, much of which today we would call wilderness—land little if ever affected by people—and he was one of the first people to climb Mount Katahdin, 5,267 feet high, the tallest mountain in Maine and second-highest in New England. The American Indians considered Mount Katahdin sacred and did climb it, and only a few people of European descent had attempted it before Thoreau. On Katahdin he hoped to achieve a spiritual connection with benign, caring nature through direct experience with it and by reading it as one would the Bible.[25]

He began his ascent of the mountain on September 7, 1846, with a few companions, but never quite reached the summit. Instead, he reached a tableland between two peaks, later named South and Baxter, the latter the higher and the actual summit of the mountain. The tableland is much like the summit in its rawness and exposure, and there Thoreau found himself "deep within the hostile ranks of clouds."[26] The clouds blew in and out. "It was, in fact, a cloud factory," he wrote. "Occasionally, when the windy columns broke into me, I caught sight of a dark, damp crag to the right or left; the mist driving ceaselessly between it and me."[27]

Thoreau was well read in the classics, and the summit of the mountain began to seem more like an ancient, pre-Christian saga than a replacement for the Bible. He wrote melodramatically that "it reminded me of the creations of the old epic and dramatic poets, of Atlas, Vulcan, the Cyclops, and Prometheus," not of a benign, Christian god, and he felt estranged here from nature in a way he had not felt before. "It was vast, Titanic, and such as man never inhabits." Standing within this awesome setting, he, a human being, seemed less significant as nature appeared more powerful. He felt "more lone than you can imagine." In the thinner air on the mountaintop, "Vast, Titanic, inhuman nature" had him "at a disadvantage." He felt that he had lost "some of his divine faculty,"[28] one of the very things he had sought to capture and understand on the mountain.

The setting took on the aspect of a theater in which Thoreau became a player and Nature became the star. He gave Nature a speaking part:

> She seems to say sternly, Why came ye here before your time? This ground is not prepared for you. Is it not enough that I smile in the valleys? I have never made this soil for thy feet, this air for thy breathing, these rocks for thy neighbors. I cannot pity nor fondle thee here, but forever relentlessly

drive thee hence to where I am kind. Why seek me where I have not called thee, and then complain because you find me but a stepmother? Shouldst thou freeze or starve, or shudder thy life away, here is no shrine, nor altar, nor any access to my ear.[29]

On the summit of Katahdin, Thoreau discovered that wild nature was overpowering. "This was that Earth of which we have heard, made out of Chaos and Old Night. Here was no man's garden, but an unhandseled [Thoreau's word] globe. It was not lawn, nor pasture, nor mead, nor woodland, nor lea, nor arable, nor waste land."[30] Within himself, the mountain was several things simultaneously—metaphorical, almost alive, almost organic, a place not for a person to live but to leave quickly.

This was the same Thoreau who challenged Jean Louis (Rodolphe) Agassiz's belief in the spontaneous generation of plants, just sprouting from the soil. Agassiz was aware of Darwin's work on the evolution of species and did not accept that idea. He believed that species were special creations of God, and he therefore believed in the spontaneous generation of life. One piece of evidence brought up by those who believed in spontaneous generation was the way trees and other green plants appeared to spring from the ground, where there did not appear to be seeds.

Thoreau was skeptical of Agassiz's theory of spontaneous generation and, as was typical of him, did not take someone else's word but attempted to find out for himself. He made direct observations about how seeds spread by wind and animals.[31] He saw seeds carried in the fur coats of small mammals, ingested and released in the droppings of birds and mammals.[32] Instead of finding evidence to support spontaneous generation of life, he established that all trees in the forests near Walden always germinated from seeds or resprouted from existing roots, and never arose spontaneously from the soil.

Thoreau did not stop with qualitative observations. He measured distances between trees of a species. He counted and weighed seeds. He measured the distance that nesting birds and squirrels carried fruits and seeds.[33] He looked at caches of seeds in animal burrows. He watched squirrels cut twigs of pitch pine heavy with cones and move them to new locations. "I counted this year twenty under one tree, and they were to be seen in all pitch-pine woods," he wrote. He continued his observations and found that "squirrels were carrying off these pine boughs with their fruit to a more convenient place either to eat at once or store up," and that along the way some of the seeds were abandoned and sprouted, spreading the species to new locations.[34] He did experiments to determine how far a bird could move a specific seed. These studies may have made Thoreau the first American field ecologist to be influenced by Darwin's theory.[35]

Thoreau's works illustrate that a great love of nature and feeling of connection with nature can be compatible with careful, scientific research. It is a struggle that still grasps us. But most of us, unlike Thoreau, are largely unaware of its grasp, perhaps never thinking much about nature and ourselves except in an abstract sense, perhaps as a political issue far removed from us, such as the state of tropical rain forests and endangered species in some faraway place. We may think about it when we open a mailed request for contributions to an environmental organization or see a stirring nature film in the theater or on TV. Then perhaps we pause for a few moments to contemplate the fate of elephants, frogs, lions, or tropical beetles. Otherwise and other times, we may live unaware of the deep importance of the meaning of nature for us, and the chasms that exist for us if we do not find that meaning. It is a chasm perhaps like Thoreau's view from Mount Katahdin when, partway up, he noted in his journal, "I had soon cleared the trees and paused on the successive shelves to look back over the country" but the view was "almost continually draped in clouds."[36]

Most of the time, our understanding of our connection with nature is like a brief view soon draped in clouds. I am trying to help us see down through the clouds, into the chasm of emotions and feelings, to a rationalist's set of ideas that can serve for us as the meaning of *nature*, in our time, in our civilization.

Nature as Superorganisms

The organic viewpoint was not as important as the idea of divine order in the explanation of nature in the twentieth century. Closest to the organic viewpoint were the ideas of Frederic Edward Clements (1874–1945), a plant ecologist famous among his colleagues for his explanation of what he referred to as "plant associations."

Clements worked in the first decades of the twentieth century, when plant ecologists were beginning to study major patterns in time and space of the major types of vegetation, such as the oak-hickory-and-chestnut forests that extended from Connecticut into the Appalachians; the maple-beech-and-northern-hardwood forests that were found in New England; and the boreal forests, the spruce-fir-birch-and-aspen forests of the North. Clements and his colleagues believed that each of these groups occurs under a certain range of climatic conditions; given the right climate, a particular kind of vegetation would be found. Analogous kinds of vegetation could be found worldwide. In North America, for example, desert plants were mainly of the cactus and yucca families. In Africa, plants that looked like cactus were members of the Euphorbia family. In a specific kind of environment, plants evolve that are similar in form, but to Clements they were not just single species. Whole "communities" of plants seemed to him to have evolved and adapted to the needs posed by the climate.

From this perspective, Clements developed the idea of the plant formation, a collection of species occurring together, as a "superorganism," arising, growing, maturing, reproducing, and dying. Its mature phase became known as the "climax" stage, and it had parents in prior climax stages. Like other organisms, this superorganism could be an object of scientific study. Clements thought that plant formations had an evolutionary history, descending from a common ancestor. He imagined a "panclimax," a "grouping of formations on different continents whose descent from a common ancestor is indicated by possession of similar dominant growth forms."[37]

Clements's organic view of vegetation was rapidly challenged and by the 1940s appeared to have been completely dismissed in the United States, remaining only as a historical curiosity, useful in explaining to ecology students why it is an inappropriate perception. The first challenges to Clements's ideas came in the 1920s from botanists led by Leonty Ramensky (1884–1953) of Russia and Henry Gleason (1882–1975) of the United States, who made two arguments against Clements's idea of the superorganism.

First, each species responds uniquely to environmental factors, and its distribution is determined solely by its relationship with the environment, not by its interaction with or dependence on other plant species; thus the co-occurrence of two species is simply an accident of similar adaptations, not a mutual dependence.

Second, the kinds of species that dominate a landscape change continuously in space; there are no sharp boundaries between communities, and no inside and outside of a superorganism. If there are not sharp boundaries, then the superorganism cannot really exist.

To these, we need to add a third, consistent with the theme of our discussion: A superorganism is static in the relationships among its parts. By making the superorganism structurally static, Clements also connected the idea to the balance of nature and, as we will see in the next chapter, to the machine age, although he may have been unaware of these interlinkages.

Many studies in the first half of the twentieth century concerned the distribution of plants; this was a major concern of plant ecologists of the time. They sought to determine whether species are distributed in groups, as Clements's idea seemed to require, or whether each species has its own individual geographic pattern. Some of this research was carried out as late as the 1950s by Robert H. Whittaker, John T. Curtis, and Robert P. McIntosh, who showed that no two species that they studied had exactly the same distribution. Whittaker put the superorganism idea to rest with the statement that "species distributions cannot be 'independent' in the sense that they are unaffected by competition and other interrelations. These interrelations, however, do not result in the organization of species into definite groups of associates . . . species distributions are 'individualistic' in the sense that each species is distributed according to its own way of relating to the range of total environmental circumstances, including effects of and interrelations with other species."

In the second half of the twentieth century, studies of pollen deposits in lakes and forest soils reinforced the individual concept of species distributions, showing definitively that species had moved at different rates to different locations as the climate changed in the past. (This is discussed in detail in later chapters.) Thus "natural communities are not organisms."[38]

The key point here is the dismissal of Clements's ideas in the twentieth century; the organic view of nature, although one of the dominant perceptions throughout the history of Western ideas, was of minor importance and was vigorously dismissed when proposed in the scientific age.

It is also worth noting that the organic view is incompatible with the perspective of the machine age. At the end of the twentieth century, the organic view seemed to most ecologists clearly wrong and inadequate. However, the simplistic mechanistic view, which underlay the descriptions of animal populations and forest history discussed in earlier chapters, is also inadequate. This left us by 1990 without a clean, agreed-upon set of ways of thinking about how nature worked.

Oddly enough, however, the organic idea of nature made a comeback in discussions of global environmental problems and in some scientifically based writings in the twentieth century about a global perspective on life. In some of the most important and influential theoretical models used to forecast possible ecological effects of global warming, vegetation is once again being treated as collections of species that by definition move about together. Most influential in this regard was the reappearance of the organic idea in *Gaia: A New Look at Life on Earth*, by the famous British scientist and environmentalist James Lovelock.[39] Lovelock brought back the Greek word for mother Earth, *Gaia*, to represent what I and other ecologists have called the "biosphere," Earth's system that contains and supports all life.

In *Gaia: A New Look at Life on Earth*, Lovelock writes that "Gaia has vital organs at the core," and that salinity control "may be a key Gaian regulatory function," which he explains in terms of the human kidney.[40] However, his ideas are a blending of an organic view and a mechanistic view and use organic, mechanistic, and computer-based metaphors to explain the functions of the biosphere. In his book, homeostasis of the biosphere is explained in terms of the temperature regulation of an electric oven, the functioning of a cybernetic system (that is, a calculating device, a computer), and an analogy with the human body, as mentioned. Even where the organic metaphor appears strongly, as in Lovelock's writing, there is a tendency, at least in some passages, to view the biosphere only mechanistically. For example, Lovelock wrote that "the only difference between non-living and living systems is in the scale of their intricacy, a distinction which fades all the time as the complexity and capacity of automated systems continue to evolve."[41]

A problem in discussions of this kind is that it is difficult, if not impossible, to avoid presenting the ideas as though there were purposefulness on the part of life. This difficulty is a well-known flaw in scientific discussions, known as a "teleological argument"—giving a sense of intentionality to objects that cannot have consciousness, desire, and purpose. Lovelock writes that "occasionally it has been difficult, without excessive circumlocution, to avoid talking of Gaia as if she were known to be sentient. This is meant no more seriously than is the appellation 'she' when given to a ship by those who sail in her, as a recognition that even pieces of wood and metal when specifically designed and assembled may achieve a composite identity with its own characteristic signature, as distinct from being the mere sum of its parts."[42]

Lovelock's book is sometimes metaphysical, going beyond the issue of scientific observations (is the biosphere biologically regulated?) to the question of whether this regulation is purposeful, as in his discussion of the possibility that "the Earth's surface temperature is actively maintained at an optimum by and for the complex entity which is Gaia, and has been so maintained for most of her existence."[43] Whether or not Lovelock meant the organic idea of "Gaia" as only a simple and superficial analogy, it has become quite common in popular, nonscientific or meta-scientific discussions of environmental problems.

When I wrote *Discordant Harmonies*, it was my guess and hope that by about 2000 or 2010 the simple mechanistic and simple organic perspectives would seem equally silly as explanations of nature. But as I have just pointed out with James Lovelock's book—and we will find more in later chapters—in odd, ironic, and perhaps paradoxical ways, these two metaphors for nature may be making a comeback at the start of the twenty-first century. All the more important to say here that it is only our heritage of the machine age, the dominance of its metaphors, which I discuss in the next chapter, and the correspondence between the machine metaphor and the divine order, that have made the mechanistic description seem so much more plausible to us than Kircher's idea of the marriage of fire and water, or Clements's concept of a forest as a superorganism.

FIGURE 7.1 Antique logging wheels, used in the nineteenth century to haul logs out of Michigan's white pine forests, repainted and shown as if a kind of sculpture at Hartwick Pines State Forest. (Photograph by the author.)

7

In Mill Hollow
NATURE AS THE GREAT MACHINE

> Machines . . . machines . . . machines! This is the cry that, running like a leitmotif through the modern world, echoes off along the highway of the future toward a goal that cannot be other than imaginatively foreseen.
>
> —EDWARD ALDEN JEWELL (1927), NEW YORK TIMES
> *ART CRITIC AT THE TIME*[1]

Fixing Chase's Mill

In the 1980s, Chase's Mill, on the outflow of Warren's Pond in East Alstead, New Hampshire, was one of the few remaining water-powered mills in New England. Unused since the late 1980s, the old mill, with its unpainted siding of weathered vertical pine boards, stands on a site used for waterpower for two centuries. Rebuilt in 1917, the mill houses a turbine wheel, a marvel of nineteenth-century engineering. Unlike the large, picturesque, overshot or undershot wheels in romantic paintings, the turbine wheel is quite small, occupying a space less than three feet in diameter. It is mounted inside the mill rather than outside, in line with the flow of the water, just as a jet-engine turbine is mounted in the flow of air passing the airplane. Such turbine wheels are much more efficient than the older kinds and had brief popularity in the second half of the nineteenth century but were soon replaced by similar turbines connected to electric generators.

The wheel was powered by water fed to it through a long flume about 18 inches in diameter that started at a small dam beside the mill. The flume went through the mill's basement and then dropped vertically into a subbasement, where the water entered the wheel's housing. When the flume gate was opened, a rush of water started the turbine spinning, generating about 18 horsepower, which was transferred by a spinning vertical shaft to the basement. In the basement are a number of huge pulleys about three feet in diameter, mounted on long, horizontal, metal shafts and connected to one another by leather belts that transferred the power back and forth across the room, to the slap-slap-slap of the leather belts, the whir of the pulleys, the rush of the water, and the rumble of the turbine, which vibrated **139** the floor. Some belts transferred power upward once again to the main floor to run woodworking tools: a planer, a power saw, a joiner.

Heman Chase, whose family owned the mill since early in the twentieth century until his death in the 1980s, used to say that the mill basement, full of belts and pulleys, was the best place for a young person to learn about machinery and the principles of physics and mechanics. That mill and its machinery were part of his ideal of life: the nineteenth-century American village, almost completely self-sufficient, producing its own goods, generating its own power, growing its own food, and boasting a pond for recreation and waterpower. In short, a self-sustaining and repairable regenerating system, the whole community operating as smoothly as the mill.

One cold winter day, Heman and I worked to replace the flume, which after 30 years of use had rusted too badly to contain the water. Like other parts of the mill, the flume had been replaced before. On that morning, we moved in the last new galvanized pipe and sealed and taped the sections together. The mill was not only as good as new but in fact the realization of the nineteenth-century mechanical ideal: a machine that never wears out, has parts that fit together and are replaceable when they age, and are readily comprehensible in their transfer of energy from flowing water to cutting tools. On the day that I wrote this in 1987, I noticed an article in which George M. Nelson, president of the Alyeska Pipeline Service Company, was quoted as saying that in some respects the pipeline is "like an airplane, you keep rebuilding and replacing parts that need it, and the thing will go on forever."[2] The idea is well expressed by an old New Hampshire joke about two men cutting trees out in the woods. When one admires the other's axe, its owner replies, "Why, this is the best darned axe I've ever had. Had it for years and in all that time it's only needed two new handles and one new head."

In addition, an idealized machine operates according to readily understood rules and laws of nature, in ways that are readily predictable. The machine exists in an idealized universe without chance. It has no history: How it will be tomorrow can be predicted from how it is today; there is no need to know how it was yesterday, since that information is observable in its condition today.

The comforting aspects of a machine like Chase's Mill lie in its predictability, the capacity for repair, potential for indefinite continuation (as long as it is properly maintained) and timelessness. For these qualities, we give up individuality and the unpredictable charm of personality; in a sense, we also give up the notion of precious moments that characterize a living thing, moments that will never come again.

While machines cannot change themselves—they are not creative, they do not evolve—people can change them, improve them, reengineer them. This, too, is comforting in one way: We are in control of the machine and can make it do what we want. So it was with Chase's Mill. The turbine had replaced a less efficient, although more picturesque, overshot wheel, which would have been the kind of wheel at this mill when it was first built and known as "Kidder's Mill" for the owner at that time.

The turbine wheel in turn became obsolete with the development of hydro-electric turbines. If a machine is not good enough, or was good enough until a better one was invented, the old gives way to the new. At best, the old machine becomes a hobby, or a picturesque artifact to display in a museum. At worst, it is for all intents and purposes abandoned, as happened to Chase's Mill at the end of the twentieth century and the beginning of the twenty-first (Fig. 7.2). (Repairs finally got under way for the mill building, but not the mill machinery, in 2010.)

Chase's Mill may seem a quaint bit of ancient history, but in the nineteenth century it was one of the modern advances of an age that had begun two centuries earlier with Kepler and Newton, the rise of the new physics and the beginning of modern mechanical engineering. To us in the twenty-first century, the new mechanical view of the Earth seems an old and permanent one, but it is not. The mechanical Earth is an idea a few centuries old that persisted throughout the twentieth century and is still part of our way of describing our planet and its life, as pointed out in the previous chapter. The mechanical view is consistent with the idea of a divine order in most of its particulars and consequences, and thus it simultaneously reinforced, and was reinforced by, the ideal of divine order. But the mechanical view of nature as perceived in the past and still perceived by many is antithetical to the organic view of nature.[3]

The Death of the Earth

A crucial philosophical change occurred in the seventeenth and eighteenth centuries when the perspectives of the new mechanical age replaced the organic view of nature. In her remarkable and impressive book *Mountain Gloom and Mountain Glory: The Development of the Aesthetics of the Infinite*, Marjorie Hope Nicolson called this transition the "death of the Earth,"[4] meaning that the Earth was no longer regarded as an animate creature, but as a vast machine. This change in perception was primarily a result of the new physics of Galileo, Kepler, and Newton, and new, highly successful machines. Stars and planets moved like clockwork according to universal rules. Earth itself obeyed these rules. To project its position and trajectory through space, only its current state need be known. New machines made such discoveries possible: telescopes, improved clocks, machines with the same inherent qualities as the turbine wheel in Chase's Mill. The development of modern sciences, beginning with physics, led to a change in metaphors, but more profoundly to a change in explanation: from a belief in Earth as an organism created by the Great Artist to a belief in Earth as a magnificent machine invented by the Great Engineer.

Then came discoveries of the new physics that upset things. One was Kepler's discovery that the orbits of the planets are not perfect circles, and thus not ideal in terms of the classical Greek and Roman idea of beauty and perfection. Instead, they were explainable in terms of the beautiful symmetry of the laws of Newton, a

FIGURE 7.2 Chase's Mill in 2008. Perhaps symbolic of the passing of the machine age, Chase's Mill, still a pretty building where the water glistens over the spillway (A), now stands idle and unused (B). After Heman Chase died in 1988, the mill passed through a number of hands. Some tried or hoped to keep it running, but no one with the money had an interest until 2010, when some repairs to the mill's foundation reportedly started. The first mills at this dam, a gristmill and fulling mill, were built here in 1767, and several other mills were built soon after just downstream. (Photographs by the author.)

perfection in motion rather than structure. The "old truths began to lose their force, gradually becoming the language of poets rather than of scientists."[5] René Descartes (1596–1650), in the *Principles of Philosophy*, provided a major turning point, departing completely from theories of divine creation and thus a divine origin of order to develop a theory from a basis in a *mechanistic* universe. Descartes and Isaac Newton (1643–1727) built the foundation of the machine-age idea, which replaced the organic idea.

There was great enthusiasm for the machine age, as illustrated by this chapter's opening quotation and in paintings, in which a great machine is viewed as an object of beauty. It's hard to overemphasize the degree to which America and the other nations of the Western world were captivated by the machine. In a twenty-first-century biography of Samuel Clemens, Ron Powers's *Mark Twain: A Life*, Powers writes that by the late 1880s Clemens's "new god" was the machine, which "cast its politicizing spell: Clemens began to believe in the 'Machine Culture's' promise to release the energies and skills of the oppressed," and in the machine age's skilled workman as "the rightful sovereign of this world."[6] This may seem especially surprising from the creator of Tom Sawyer and Huckleberry Finn, two very human characters scrambling around in a very biological, organic world.

Indeed, the machine changed the world in many ways that we take for granted. Before the spinning wheel, a medieval invention, the traditional task of spinning fibers to make thread by hand took an estimated 50,000 hours for 100 pounds of cotton. By 1850, machines of steel had reduced that time to 135 hours.[7] Who today gives any thought at all to the effort involved now or in the past to accomplish such a task? We don't because it's so ordinary. But in the middle of the nineteenth century, these machine-age changes were happening thick and fast. If machines could be so wondrous, and if we could analyze them, as scientists and engineers were learning to do, why couldn't machines be the best—no, the only—way to explain and understand nature?

Seventeenth-Century Beginnings

The beginnings of the machine-age perspective on nature can be seen as early as the seventeenth century—just about the time William Shakespeare died—in *The Primitive Origination of Mankind*, in which Sir Matthew Hale (1609–1676), famous in his time as the Lord Chief Justice of England, wrote that the "Qualities of Natural things are so ordered, to keep always the great Wheel in circulation." That is, there is reason to believe that, like clockwork gears, nature can be kept running, and that faith in this operation of nature lies in universal laws of science. In addition, the "vicissitudes of Generation and Corruption"—that is, the processes of birth and death—are "a standing Law in Nature fixed in things." Hale recognized the prime mover of nature and life as energy, the "influxes of the Heat" from the sun as well as "the mutual and restless Agitation of those two great

Engins [*sic*] of Nature, Heat and Cold," which are the "great Instruments" maintaining the "Rotation and Circle of Generations and Corruptions, especially of Animals and Vegetables of all sorts."[8] Nature was beginning to appear as a machine driven by an engine of heat and sunlight.

The Growth of Science: Nineteenth-Century Advances

Discoveries and observations about Earth's surface and the life on that surface reinforced the mechanical interpretation. In the geologic sciences, nineteenth-century discoveries made it clear that the factors that had shaped Earth's surface could be explained from an understanding of force, energy, power, and the laws of mechanics. The intellectual power of previous centuries' physics could apparently be extended to explain how the mountains and valleys were formed, how glaciers had altered the surface of Earth in many places. Here's how that realization came about.

The climate during the past several centuries has fluctuated enough to cause glaciers in the Alps to expand and contract repeatedly to a degree great enough to be apparent to residents of the region.[9] In his wonderful book *Times of Feast, Times of Famine: A History of Climate Since the Year 1000*, Emmanuel Le Roy Ladurie uses paintings and historical records to prove that a glacier that had receded high up a mountain in the Alps in one century grew and overran towns in another, blocking a river running through the town, flooding it, and continuing to grow and knock down buildings.[10]

As I mentioned in Chapter 4, Swiss inhabitants of high Alpine valleys long ago understood not only that glaciers changed their shape and extension, but also that rocks scattered below a glacier had been shoved downslope when the glacier extended, and then deposited when the glacier melted back. They could see similar rocks at the leading edge of the glacier, attached to and moving with the ice.

Large boulders composed of one kind of mineral were found on top of bedrock of a very different kind. Granite boulders, for example, rested on limestone bedrock in the Alps, and it would be apparent to a careful observer that these boulders must have been moved there by some large, external force. Some may have believed that they were placed there by pagan gods or carried there by the waters of the biblical Flood, but others recognized a connection with the mountain glaciers—nature appeared to operate as a machine, at least in moving rocks around a countryside.

These observations by Swiss farmers, however ancient or recent they may have been, did not enter the intellectual thought of the West until the nineteenth century. It was in that century that the idea of large-scale glaciation entered scientific discussions. The idea began in 1815 when a Swiss peasant, Jean-Paul Perraudin, suggested to a Swiss civil engineer, Ignaz Venetz-Sitten, that some features of the mountain valleys, including the boulders and soil debris, were due to glaciers that in a previous time had extended down the slopes beyond their present limits. Impressed with these observations, Venetz-Sitten, speaking before a natural-history society at Lucerne in 1821, suggested that the glaciers at some previous time had extended considerably beyond their present range.

Louis Agassiz, who, as mentioned earlier, rejected Darwinian evolution, played an important role in the discovery of continental glaciation and is famous as one of the first scientific geologists. Born in Switzerland, he eventually became a professor at Harvard University, and in that way became America's first professional scientist—that is, someone who was paid for teaching science, doing scientific research, and writing about his findings. He traveled to the Alps to refute the ideas about glaciers and landscapes, but quickly became so impressed with the correspondence between the glacial debris in the mountains and the topography he had seen at lower elevations that he completely changed his mind and formulated a theory of continental glaciation. We should mention that Agassiz, arriving armed with a hypothesis and a belief—that glaciers could not have covered large areas of the land—quickly changed his mind when natural-history observations showed the contrary. In this situation, he was not one of those who stayed with assertions "in spite of appearances to the contrary."

It was soon recognized that glaciers had covered vast areas in Great Britain and North America, and the theory of continental glaciation became generally accepted, although opposition to it appeared in the scientific literature until the very end of the nineteenth century. Thus, in less than three decades, a mechanism had been suggested to explain local topographic features in the Alps and then extended as a widespread process that had occurred over large areas of the Earth. Rocky fields were no longer viewed as the aging of Mother Earth, but as the result of mechanical processes with a universal character. *Nature could be explained as a machine.*

The mechanization of life was much more difficult to imagine than that of rocks and mountains, partly because of the great diversity of life-forms. Alexander von Humboldt (1769–1859), the famous naturalist and explorer, was one of the major thinkers who in the nineteenth century attempted to confront the newly available wealth of observations of life-forms and to seek a general understanding of the order of nature. As he wrote in *The Cosmos*, his goal was "to combine all cosmical phenomena in one sole picture of nature," to find "great laws, by which individual phenomena are governed," so that he could "comprehend the plan of the universe—the order of nature," which must begin with "a generalization of particular facts, and a knowledge of the conditions under which physical changes regularly and periodically manifest themselves."[11] Humboldt was seeking to apply the scientific method to explain the diversity of biological nature.

Humboldt wrote about nature's capacity to achieve a steady state more than a decade before George Perkins Marsh did. Stating ideas similar and related to Marsh's, Humboldt wrote: "We may easily comprehend how, on a given area, the individuals of one class of plants or animals may limit each other's numbers, and how, after the long continued contests and fluctuations engendered by the requirements of nourishments . . . a condition of equilibrium may have been at length established."[12] Another proponent of the steady-state balance of nature.

Thus we arrive at the time of George Perkins Marsh, who, you will recall, believed that nature had the ability to recover from any disturbance. He wrote that nature has two key attributes. First, when "left undisturbed," nature "so fashions her territory as to give it almost unchanging permanence of form, outline, and proportion." Second, when disturbed, nature "sets herself at once to repair the superficial damage, and to restore, as nearly as practicable, the former aspect of her dominion."

The net result is nature in a steady state, at least in "countries untrodden by man," where all qualities, from "the proportions and relative positions of land and water, the atmospheric precipitation and evaporation, the thermometric mean, and the distribution of vegetable and animal life," are maintained in an almost perfect constancy, "subject to change only from geological influences so slow in their operation that the geographical conditions may be regarded as constant and immutable."[13]

This mechanization of Earth and its life-forms signaled a profound change in the perception of nature. Earth must have seemed a friendlier place to those who viewed it as a fellow creature. No wonder we value whales all out of proportion to our contact with them or their importance to our own survival, for a whale represents, in the now-inanimate vastness of the oceans, a huge and benign entity, an almost human presence, a gentle giant animating the immense and lonely oceans. If the rocks under our feet were alive and the ocean were Earth's blood, we might have less need to know the presence, beyond our horizon, of the leviathan.

A Mechanical View of Life

The rise of the mechanical worldview had several important consequences. First was recognition of the power of the new laws of physics. Second was the rise of machines: the steam engine, the steam train, the paddlewheel boat, the sewing machine, the entire Industrial Revolution. The success of machines, and their ability to transform society and improve the standard of living, reinforced the growing faith in the new sciences and the machine ideal.

A third consequence was that the mechanistic view offered a new kind of theological perspective. The wealth of new discoveries in physics and geology eventually overwhelmed arguments that there was perfect order in the observable architecture of the *physical* universe and the *physical* Earth. The awakening of interest in the natural surroundings made clear to some, particularly to students of landscape details, that a varied, asymmetrical landscape was pleasing and that the strange new creatures being discovered in the New World, Australia, and the Pacific islands were fascinating, some beautiful, some arousing strong emotions and affecting one's spirit, like the grizzly bear of North America. A new justification arose: Instead of the existence of God being proved by the perfect static structural balance and symmetry in the world, a perfectly working, idealized machine could be seen as the product of a perfect God. We find the rise of the argument, only too familiar to beginning students in philosophy, that the world is "like a watch," not only a perfect machine but also requiring a maker.

As discussed in Chapter 6, the machine age made the organic view, and poetry arising from it, seem merely quaint. Poetry moved on from Coleridge, Wordsworth, Shelley, and Keats.

The mechanistic explanation of nature developed further and continued throughout the twentieth century. For example, George Wetherill and Charles Drake described "The Earth as a Machine" in a 1980 article in *Science*. "Looked at from the purely physical point of view," they said, "the earth has often been loosely described as a 'heat engine.' It is now clear that this is not simply a metaphor but that it is literally true."[14] This change in the perception of nature was accompanied and followed by greater and more precise observations of Earth's geologic and biological qualities.

A mechanistic "nature"—except in our own age an oxymoron—would have the attributes of a well-oiled machine, including the capacity to keep operating, replaceable parts, and the ability to maintain a steady state and thus be in a balance. Births and deaths, immigration and emigration, the input of sunlight and the loss of energy as heat, the uptake of nutrients and the loss of nutrients would always happen in a way that maintained life in a constant state of abundance and activity.

A mechanistic "nature" can also be reengineered by us; we can believe it possible to tinker with nature and improve it, replace nature's equivalent of an overshot wheel with a turbine wheel. This is the other side of the coin of the mechanistic view, the side that has dominated much of our management of natural resources and the environment in the twentieth century. Not only has it been customary to use civil-engineering approaches to environmental issues, but it is consistent with the mechanistic perception of nature that, as a machine, nature is better improved by using novel engineering devices than by employing organic approaches. This point of view is consistent with much of our attitude toward the development of land and resources.

A classic example of the machine-age approach to land development is the history of flood management on the Missouri River. The Missouri is famous for its floods, and no one who has seen it, lived by it, or been on it would call it a good example of the Greek ideal of beauty in symmetry. As one writer put it in 1907, "Some people would think it was just a plain river running along in its bed at the same speed; but it ain't. The river runs crooked through the valley; and just the same way the channel runs crooked through the river. . . . The crookedness you can see ain't half the crookedness there is."[15]

The meandering Missouri, with its sandbars, drifting logs, and complex channels, was a threat to nineteenth-century steamboats, and although frequent droughts were a problem for agriculture and urban water supply, the Missouri's frequent flooding threatened emerging towns, cities, and farms. By the time *Discordant Harmonies* was published, the U.S. government had spent $25 billion to build a system of levees, walls, and other flood-control measures on the Missouri–Mississippi river system, straightening its channel with a kind of Euclidean geometric orderliness in space and smoothing out its flow so it ran like a well-oiled steady-state machine. The Army Corps of Engineers did much of the work, but state and private groups did some.

In the mid-1990s, I traveled partway down the Missouri from Omaha on an Army Corps of Engineers boat while the engineers checked the channels. One of the engineers explained their conception of what they wanted the river to be. He drew a diagram of the river with the six dams that the Corps had built. The drawing looked like a plumbing system. The six major dams were designed to hold back and control floodwaters, to release water so there would always be enough in the channels for safe navigation, and to keep enough water in the reservoirs to provide flow in years of drought.[16]

There are two kinds of dams on the Missouri: big storage dams and control dams. The storage dams are the ones farthest upstream: Fort Peck, Garrison, and Oahe. Each can store approximately 25 million acre-feet, and together they store enough Missouri River water flow to last three years even if there is no rain or snow.

"Under perfect conditions, the storage drops to fifty million acre-feet—a two-year supply—in March, just before spring runoff from the mountains," an Army Corps engineer explained. "Then we hope that the spring runoff will just be enough to fill the reservoirs back up to a three-year supply." As the upper reservoirs fill, water is released to the three lower dams, which then release water as needed so that the channel stays as close as possible to desired constant conditions. The lower dams are the valves, the upper dams the storage tanks, and the river a delivery system, much like the pumping system in a town's water supply and the system that comes into your house.

Many of the levees near cities and towns were built high along the river to prevent floodwaters from overflowing the main channel. This was the engineered river that was in place in 1993 when the Missouri refused to be tamed and flooded anyway, causing considerable damage around St. Louis and other Missouri towns. The engineered-river plan failed because of what Aldo Leopold would have called "its own too much." The flood-control levees created a paradox: The more the river is crowded in by tall levees, the faster it flows and the more erosive and dangerous it becomes. The result is a war of levees. A town or house with no levees or low levees suffers the flooding that is being avoided by the places with better protection. One way to respond to this is to do what the town of Chesterfield, a suburb of St. Louis, did: Build even higher levees, making sure they are higher than those of the neighboring towns. Of course, Chesterfield's neighbors could have retaliated, building their own levees even higher. But in the end, nobody would win. Also, ironically, the Army Corps' vision of a calm and peaceful river made so by modern engineering led to some complacency: The greater the apparent control over the Missouri, the greater the faith people had in their own effectiveness, and the less alert they were to possible dangers.

Then in 1993 the Missouri did what it had always done—flooded, big-time. The floods began in late June 1993. At St. Louis, near the meeting of the Missouri and Mississippi, the rivers rose to 49.6 feet, more than 19 feet over flood stage[16] and 6 feet higher than ever recorded. Heavy rains continued through July. At Papillion, Nebraska, a suburb of Omaha, an inch fell in 6 minutes on July 26, 1993. To the north, the Red River rose 4 feet in 6 hours. As much as 17,000 square miles were inundated.[17] Nine midwestern states suffered destruction estimated at $10 billion from the combined flooding of the Missouri and Mississippi, with hundreds of bridges washed out, more than 50,000 square miles of farmland flooded, and tens of thousands of houses estimated to have been damaged or destroyed. "This summer's war against the water has changed life along these rivers—and people will be living with its effects for months and years. In some ways, forever," journalist Richard Price wrote in *USA Today*.[18]

The alternative to a heavily mechanized river full of tall levees is to allow sections of the floodplain to flood naturally. And after 1993, some followed this idea. A 4,000-acre wetland near St. Louis, called Columbia Bottom, was converted from farmland (which had to be protected from floodwaters) into a nature preserve. The species of the natural riverside ecosystems had evolved with and adapted to change. The natural preserve would be allowed to flood, taking the dangerous floodwaters away from St. Louis. This was a more ecological and organic approach. Rather than fight the river with machines as if it were a completely controllable machine, rather than believe the Missouri should run constant and smooth in a perfect steady state, as straight as an arrow, let it be its meandering, dynamic self, changing over time, extending and contracting, metaphorically more like an organism than like a mechanical clock. Lewis Mumford (1895–1990), in several of his books, such as *The City in History* and *The Pentagon of Power*, detailed similar machine-like qualities characteristic of modern Western civilization and the ironic problems they create.[19]

Of course, the developed nations of the twentieth century were not the first civilizations to greatly alter landscapes or change rambling countryside and meandering rivers into geometrically uniform and symmetrical shapes. The ancient Romans are famous for their carefully engineered roads and their geometric cities with streets forming rectangular grids. Ancient Egypt, China, Japan, and the civilizations of Central and South America built geometric and symmetrical towns and monumental towers. But the Western nations from the nineteenth century onward were the first to develop machines on such a scale that a very large percentage of Earth's land surface could be altered, and were apparently the first cultures for which machines became a fascination, things of beauty, a kind of perfection, seen not just as a normal part of life but as an expected and necessary part of life.

Not only do we take machines for granted, we assume that all peoples at all times and places would have enjoyed and welcomed them. But in his book *The Lever of Riches*, Joel Mokyr discusses curious aspects of the ancient world. He writes that the classical world of Western civilization, especially the ancient Greeks, recognized "fully the importance of the elements of machines. . . . Yet these insights were applied mostly to war machines and clever gadgets . . . rarely put to useful purposes."

The "clever gadgets" include the well-known steam engine invented by Hero of Alexandria (AD 20–62), a sphere that spun as steam pushed out of it. Oddly, neither he nor anyone else thought of turning it into a device that could replace human and domestic animal labor, except to open temple doors. He is also said to have invented a coin-operated machine for temple holy water. In addition, Mokyr writes, Ctesibus (third century CE) "reportedly invented the hydraulic organ, metal springs, the water clock, and the force pump." Most interesting was the Antikythera mechanism found in a sunken ship near Crete that was "a geared astronomical computer machine of astonishing complexity, built in the first century BC." It seems like an analog computer in that it was apparently capable of calculating the motion of planets. But rather than extend such inventions to make life easier, they were used as playthings or had a religious role, as in predicting the motion of planets, but not a major economic one.[19]

Mokyr concludes that this was a cultural choice; basically people lacked our fascination with machines. "How could this be?" we of the modern world ask. Steven Saylor played with this problem in *Catilina's Riddle*, one of his mystery novels about ancient Rome, in which the hero inherits a farm in the countryside and builds a gristmill. He thinks his wealthy neighbors will appreciate it and want to build more like it, but they find it hideous and hateful, preferring the energy of their animate slaves. It's one of the few fictional portrayals I know of that reveals a culture that could make machines but didn't like them.[20]

Writing as I am about ways in which the machine age and its machine metaphors are inappropriate for an explanation of how ecological systems work and how we can sustain them, you might think I dislike machines completely and am suggesting we move away from them. Back to nature, I must be saying, 1960s-style. But quite the contrary.

As my work with Heman Chase in his water-powered mill should make clear, I do find classical machines things of beauty and fascination. Antoine de Saint-Exupéry (1900–1944), the famous French pioneer airplane pilot, said it right in his classic work that won the Grand Prix of the Académie Française, *Wind, Sand, and Stars*. (He may be better known today for his charming and fanciful *The Little Prince*.) Saint-Exupéry started to fly in 1926 and was one of the first to fly the mail from France to Africa. During those flights—many at night, alone with nothing but the glowing instruments and roar of the motor, the lights of the stars and occasional lights on the ground, with moonlight illuminating Earth's features—he found great beauty and, more important, the most direct contact with nature he'd ever felt. "Contrary to the vulgar illusion, it is thanks to the metal, and by virtue of it, that the pilot rediscovers nature . . . the machine does not isolate man from the great problems of nature but plunges him more deeply into them."[21]

Since the rise of modern environmentalism in the 1970s, the machine has become for many the enemy of nature and a thing that serves to separate people from nature. But Antoine de Saint-Exupéry saw it differently. "Little by little the machine will become part of humanity," he wrote. When first invented, "the locomotive was an iron monster. Time had to pass before men forgot what it was made of. Mysteriously life began to run through it, and now it is wrinkled and old."[21]

To say that a machine is not the right metaphor for nature, that machine-age mathematics is not the way to solve environmental problems, is not to reject machines and what they do for us, the way they make our lives so much easier and more comfortable than those of people who lived before the modern industrial age. Indeed, ironically, even those holding the most extreme positions in conservation and economic development share a worldview heavily influenced by the machine age. Environmental preservationists, who believe that nature should be left completely alone, argue that the machinery of nature functions perfectly without human intervention.

Once again, Antoine de Saint-Exupéry had it right: "Numerous, nevertheless, are the moralists who have attacked the machine as the source of all the ills we bear, who, creating a fictitious dichotomy, have denounced the mechanical civilization as the enemy of the spiritual civilization."[22] Some among them argue that even the scientific study of nature is an undesirable activity that is bound to affect and change nature. "Do our dreamers hold that the invention of writing, of printing, of the sailing ship, degraded the human spirit?" Saint-Exupéry asked.

Most conservationists believe in some use of natural resources by people and some intervention by people in nature, but there is a wide range of opinion about how much we should interfere with natural mechanisms. To both preservationists and engineers, nature is as malleable as heated iron. From the engineering point of view, this tractability is good and should be exploited. From the preservationist's point of view, it has allowed the abuse of nature. These two points of view seem to conflict. But looking at it from outside and being cognizant of the mechanistic paradigm, the contradiction appears more like a subtle difference in the assessment of our ability to successfully engineer that malleable and engineerable nature.

The machine-age idea of nature reinforced the main precepts of the classic balance of nature: that nature on its own maintains a steady state, that nature is a machine, independent of man, and since it runs perfectly by itself, we amateur mechanics only destroy its perfection, disrupt its function. The machine-age view of nature became so dominant that the idea of divine order declined from a "truth" to a metaphor, while the organic view, also reduced to a metaphor, was dismissed, no longer thought of as a serious notion.

Perfection Revised

The rise of the nineteenth-century scientific explanation of nature had two major effects: rejection of the idea of an animate Earth and an organic concept of nature; and refutation of the idea that the world is composed of perfect structural symmetries, physical symmetries—for example, that Earth is a perfect sphere, or that the planets move in orbits that are perfect circles. As discussed in Chapter 6, the first effect led simply to abandonment of the organic perspective. The result of the second is more subtle: Belief in aesthetically pleasing and theologically satisfying physical symmetries was replaced by belief in an aesthetically pleasing and theologically satisfying conceptual order. While the belief in gross physical attributes of symmetry, balance, and order was no longer tenable following the new observations of nature, Newton's laws created a conceptual order. Subsequently, theologians used this conceptual order to justify their belief in a perfect world where a perfect order—the laws of nature—ruled our asymmetrical and structurally imperfect world.

This point perhaps needs some explanation. An argument based on observation can make use of two kinds of evidence: structural (of form) or conceptual (of either processes or rules—that is, laws—that govern the workings of the universe). Structures, processes, and laws may each provide a kind of balance of nature. The search for this balance had begun with structure, then moved to processes and laws. This adds to the importance of Newton's *Principia*, for during a period in which an argument from structure was becoming more and more difficult to support in regard to the stars and planets, Newton offered evidence that order and proportion existed in universal principles or laws of motion.

For life, structural balance is represented by constancy in the number of organisms. For example, in the seventeenth century, Sir Matthew Hale, mentioned earlier, wrote of evidence of wisdom and purpose found in the structural balance of nature: "These Motions of Generations and Corruptions, and of the conducible thereunto, are so wisely and admirably ordered and contempered, and so continually managed and ordered by the wise Providence of the Rector of all things," that a balance is maintained, "things are kept in a certain due stay [meaning constancy] and equability." An equilibrium is evident, "a continual course, neither the excess of Generations does oppress, and over-charge the World, nor the defect thereof, or prevalence of Corruptions doth put a period to the *Species* of things."[23]

The distinction between a structural and a conceptual balance of nature is an important one, although it must be recognized that it is a distinction that we make looking backward and not an explicit one made by either the classical Greek and Roman philosophers or the Renaissance writers who borrowed so many of their ideas and arguments from the classics. Explanations of a balance of nature based on processes become more and more important as one confronts the questions that particularly disturbed the Christian thinkers of the seventeenth and eighteenth centuries: How can there exist predators, or what they called "pernicious and venomous creatures," in a perfect world made by a perfect God? The concern extended to geology and cosmology, to the form of the Earth and of the universe, which should be perfect also.

Giving up the belief in structural order, balance, and constancy in the cosmos and in geology was difficult, as illustrated by a classical question: How can a perfect world have mountains, which destroy the perfect proportions of Earth? I remind the reader that the classical answer, according to Marjorie Nicolson, was given in the seventeenth century by Godfrey Goodman, Bishop of Gloucester, who wrote that "the mightiest mountains . . . carrie some proportion to the lowest bottome at Sea . . . that God might observe some kind of proportion." This idea that the heights of the mountains were balanced by the depths of the oceans, Nicolson wrote, was "very common in the seventeenth century and continued into the eighteenth."[24]

With this chapter, we have now come full circle in our story, explaining the origin and foundations of the mechanistic view of nature, which has dominated our own time. *The modern dilemma can be understood as a conflict among three major explanations of nature: organic, divine order, and mechanical.* The organic myth, long dominant in human history, had been abandoned in recent centuries, overwhelmed and replaced by the myths of a machine-nature. By the end of the twentieth century, nature had come to resemble the devices in the old mill in East Alstead, New Hampshire. In the history of Western civilization, this perception of nature is as new as the mill's turbine wheel was in the second half of the nineteenth century. But like the industry of the machine age, which has so changed and dominated civilization, the machine idea rapidly rose to dominate our perceptions of nature.

This idea has been reinforced by the much older and more deeply ingrained myth of divine order. Both the myth of divine order and the myth of machine-nature lead to similar conclusions about the character of nature and the possible roles of human beings in nature. From the points of view of both myths, nature undisturbed can function perfectly and beautifully, and both myths allow two possible interpretations of the role of human beings: either to complete the perfection of nature or to interfere in its perfect processes. While our technology has moved us rapidly past Chase's Mill, and while sciences have also moved beyond their insights of the nineteenth century, our perceptions of nature and the science of ecology, which attempts to explain and interpret nature, have lagged behind. In some fundamental ways, ecology remains a nineteenth-century science, which continues to lead us into failures in the management of natural resources and to unsettling contradictions in our beliefs about nature and therefore about ourselves.

PART THREE

Evolving Images

To grasp the meaning of the world of today we use a language created to express the world of yesterday. The life of the past seems to us nearer our truest natures, but only for the reason that it is nearer the world of yesterday. . . . We are in truth emigrants who have not yet founded our homeland. . . . We shall have to age some before we are able to write the folksongs of a new epoch.

—ANTOINE DE SAINT-EXUPÉRY, in *WIND, SAND, AND STARS*[1]

FIGURE 8.1 One of 250 remaining whooping cranes, nesting at Aransas National Wildlife Refuge in southern Texas. This species was reduced to 14 individuals in 1938. Its numbers have varied over the years since, but the population is generally increasing. A complete census of this species has been maintained since 1938, and this made possible the first-ever mathematical calculation of the probability that the species would go extinct. The probability, discussed in this chapter, was found to be surprisingly small.

(*Source*: Jim Kruger/iStockphoto)

8

The Forest in the Computer

NEW METAPHORS FOR NATURE

Nature has . . . some sort of arithmetical-geometrical coordinate system, because nature has all kinds of models. What we experience of nature is in models, and all of nature's models are so beautiful.

—BUCKMINSTER FULLER (1966)[2]

If we were to put our ear against the wall of the Sistine Chapel choir [we would hear the music less accurately than] the entirely new performance generated by machinery from the meticulous numerical score of a digital recording.

—FREDERICK TURNER (1984)[3]

Bacteria as Computers

Bacteria, those tiny creatures that rarely enter our thoughts unless they are making us ill, continually amaze scientists with their importance to all life, with their capabilities, and with the vast differences between their lives and ours. Beginning in the 1970s, a change took place in our perception of these microscopic forms of life. This change is illustrated by two books on the shelf by my desk: an encyclopedia published in the early 1950s and a small book translated from the French and innocuously titled *A New Bacteriology*. The encyclopedia described bacteria much the same way Louis Pasteur must have viewed them a century ago, as "minute unicellular organisms, usually classified as plants. . . . [They] range from c. 1/250,000 to 1/250 in. long; reproduce chiefly by transverse fission."[4]

The old description of bacteria seems simple and straightforward: Bacteria are tiny individuals, each cell acting independently, obtaining resources, processing chemicals, giving off wastes, like tiny machines. However, Sorin Sonea and Maurice Panisset, authors of the 1983 book *A New Bacteriology*, saw them entirely differently. They wrote that bacteria "form a planetary entity of communicating and cooperating microbes, an entity that, we think, is both genetically and functionally a true superorganism." Bacterial cells are very different from the cells that compose our bodies. The bacteria, which are prokaryotes, lack a definite cell nucleus; their genetic material is scattered throughout the cell. All other living things, except viruses, have their genetic material neatly packaged in the cell nucleus, and they have other organelles, which are discrete units with specific functions. One type of bacterial cell passes genetic material to another, which, as Sonea and Panisset explain, provides a kind of communication. According to them, bacteria are not so much single-celled individuals as a vast network of cells sharing a gene pool that "may be compared to the central data bank of a large electronic communications network."[5]

An intriguing discovery in 2010 reinforces this idea about bacteria. Off the Pacific coast of South America, a huge mat of microbes was discovered that covered an area as large as Greece and seemed to be functioning together in some ways.

"Bacteria work in teams," Lynn Margulis (1938–2011), one of the twentieth century's experts on this kind of life and the evolution of early life, wrote in the preface to *A New Bacteriology*. "They contain a *constant*, stable genetic system . . . but they function in the world by acquiring and exchanging a diverse set of variable genetic systems."[6] Bacteria are regarded as resembling nothing more than memory bytes in a computer that operates at the planetary level. Here is a radically different view of life from that which could have been stated before our time, either in machine-age ecology or in the prescience of Cicero and Aristotle. It is a description that could not have been made before the invention of computers, because the metaphors on which it is based did not exist. Sonea and Panisset recognized the originality of their definition of these organisms. "Bacteria are radically different from all other living creatures," they wrote, but everywhere else bacteria are "still described and discussed in nineteenth century terms."[7]

As the contrasting descriptions of bacteria illustrate, technology is changing our perception of nature. Computers are providing new metaphors not only for bacterial life but also for our entire perception of life on Earth, from the way that we regard bacteria to the way that we view ecosystems and our entire planetary life-support system. Most important, computers have made it much easier to analyze non–steady-state systems, which were difficult to model or make forecasts about with pencil-and-paper mathematics. Another key element of the computer metaphor is simultaneity, based on the computer as a network of memory units that can function together simultaneously, so that many tasks can be carried out at the same time or at least so rapidly that they seem to be occurring at the same time. Even older computers that I was using in the 1960s and 1970s could print the results of one program while running another. Many modern computers have true "parallel processing." Do you want to calculate the growth rate of 100 bacterial populations? In a parallel-processing computer, 100 chips can do the calculations for all the populations at one time. In this and other ways, computers are revolutionizing our concept of nature, our perception of our relationship with nature, and our ideas about managing nature.

Computerized Nature

Peter Kalm, the Swedish botanist mentioned earlier who was sent to the New World by Linnaeus to collect plants, arrived in Philadelphia in 1748. One of the most articulate of the early natural-history explorers of North America, Kalm traveled through much of what is now the northeastern part of the United States and the southeastern part of Canada. In his travels, he saw the primeval forest, so much a part of our mythology about nature. He recognized that the forests were not at all uniform landscapes, as later naturalists have romanticized them. "I was told, in several parts of America," he noted in his journal, "that the storms or hurricanes sometimes pass over only a small part of the woods and tear down the trees in it." This seemed confirmed to him when he found "places in the forests where almost all the trees had crashed down and lay in one direction."

On his way to Montreal, Kalm passed through virgin forests in northern Vermont. "Almost every night," he wrote in his journal, "we heard some trees crack and fall while we lay there in the wood, though the air was so calm that not a leaf stirred." Knowing no reason for this, he suggested that the dew loosened the roots of old trees at night, or that immense flocks of passenger pigeons settled on the branches unevenly. Reading his account, one imagines him confronting the deep woods not only with curiosity but also with a sense of mystery and awe at the basic life processes of birth, growth, and death in those uncivilized vastnesses.[8]

Peter Kalm was interested in explaining the cause of what he observed, but he could only speculate, which, from our modern perspective, he did very well. Linnaeus honored him for his work by naming the mountain laurel genus *Kalmia* for him, and until recently we couldn't do much better than he did in making and using observations such as his to explain cause and effect or to predict the fate of forests in the future. Throughout most of the twentieth century, our scientific tools took us into the laboratory, where the anatomy of trees and their basic physiology became well understood. But putting the details together to create a moving-picture show of an entire forest eluded scientists for most of the century.

To understand the processes of birth, growth, death, and entire forest regeneration, one must realize that trees compete at all times for essential resources— light, water, minerals, and space. In any particular spot in a forest, the tall trees shade smaller ones, silently suppressing their growth. In this quiet competition, a tree "wins" by growing faster than its neighbors and shading them before they shade *it*. No one species wins the competition everywhere all the time. As a result of untold years of competition and adaptation, different species have evolved to take advantage of different spatial and temporal conditions, and each species is best adapted to a specific range of environmental conditions. Some are adapted to the conditions immediately following a catastrophic clearing, like those Peter Kalm observed. In northern New England where Kalm walked, pin cherry is such a species. Its seeds are spread widely by birds and mammals that eat its fruit. The cherry seeds don't sprout in the deep forest shade, but wait until large trees fall.

Then, on some signal from the environment, all the cherry seeds sprout within a year after the clearing. If soil conditions are right, the seedlings grow into a dense stand of cherry trees that are close in age. The young trees grow rapidly, but live a short time. Such rapid growth and short life are typical of trees adapted to the early stages of a forest's development.

Since the seedlings of pin cherry cannot survive in the shade of their parent trees, a stand of cherries is temporary in any location, slowly giving way to trees whose seedlings can persist in deep shade. In New England, sugar maple, beech, and red spruce are typical shade-tolerant trees. In contrast to the pioneer types, they tend to be slow-growing and long-lived. They are the dominant trees of old (long-ago disturbed) forests, where Kalm heard the old trees fall in the night.

This process of change in the species of trees that dominate the landscape over time as a forest develops is an example of ecological succession, nature's melody of the forest played against the changing chords of storms and fires and short-term climate changes, all of which are heard against the grander, ponderous themes of glaciation and soil changes. The process of forest succession is real, a pattern repeated over time. Once known, it can be remembered like a melody and replayed in our minds as long as the instruments, the species, remain the same. The idea of succession is important to our understanding of nature and our management of natural resources; a problem in our management of forests has arisen because we have incorrectly projected the hypothetical end point of succession, believing that nature's melody leads to one final chord that sounds forever.

Not only do the growth, reproduction, and death of trees respond to light intensity, moisture, temperature, and nutrients, but trees, in turn, strongly affect these environmental factors in a locality, and the survival of individual trees depends on their interactions with neighbors and on their species, size, age, and vigor. The interactions are complex, and it is difficult to predict their consequences among even a handful of species. In the past, an understanding of such consequences, when it existed at all, lay only in the minds of a few experienced naturalists.

One such naturalist was Murray Buell, the well-known Rutgers professor I mentioned earlier who was my mentor in ecology and died in the 1970s. Every Wednesday morning Murray, his wife, Helen, and a botanist friend, John Small, went for a walk in Hutcheson Memorial Forest. In the late 1960s, when I was the caretaker of the forest, I sometimes went along too, and on many other occasions Murray and I would walk through these and other forests trying to understand the relationships among living things, and between living things and their environment in the woods.

Murray never lost his youthful enthusiasm for and curiosity about nature. One morning, we came upon a walnut tree in Hutcheson Forest, an unusual tree in those woods. "Now, Dan," he asked, as he had on many similar occasions, "why do you think that tree is growing here?" We talked for a while, deducing what we could—from the soil, the surrounding species of trees, and the shape and form of the individual stems—about the history of the woods: whether it had been cut or plowed, when it had last been cleared, what the requirements for walnut trees might be. Big limbs low to the ground told us that some of the trees had once grown in an open area. A rough soil surface told us that the soil had never been plowed.

By the time we left, I believed we had the answers. Walnut trees are uncommon in the area because New Jersey is too cold for them to grow well, and they can survive only under the best of conditions. They are only moderately tolerant of shade, so the old walnut before us must have sprouted when there was a small clearing in the woods, possibly created by the death and fall of a large old tree. We talked about the kind of soil that best suits walnuts, and about other factors that might have limited or promoted this species. But in the end, there was no way we could prove to ourselves that we were right. We couldn't plant another walnut and wait 50 or 100 years to see if it survived and grew, what pests attacked it, and what other trees competed with it. We couldn't build mechanical trees that would mimic the natural forest and give us an answer about the validity of our insights into natural history.

At the same time, I was working at Brookhaven National Laboratory in an experimental forest that was being subjected to radiation as part of a study of the possible environmental effects of nuclear war. Brookhaven is a high-tech place, and its forest was scrutinized with the most advanced devices available, including the first digital recording system ever used to collect information about a forest. This system recorded observations about the forest in a format that could be read directly by a computer and thus be processed rapidly and efficiently. It was itself a precursor to a computer, but was also an early transitional device and in many ways familiarly mechanical. It had been jury-rigged at the laboratory from big surplus telephone-system relays and stepping switches that made loud clicking noises. Standing in front of these devices, it was easy to see the parts move and to understand how the mechanisms operated. The data were recorded as a series of holes punched on paper tape; hence the trailer in the woods where the whole machine stood was filled with the rat-a-tat-tat of the mechanical paper punch, and the floor was covered with little paper disks from punched holes. At the end of a day, the paper tape was taken to a computer, which attempted, sometimes successfully, to read the data.

Computer programming was quite new to biologists, and not many ecologists had tried it. But it occurred to me that Murray and I might have found a way to test what we believed by writing down what we knew about forests in computer code and seeing whether our explanations could produce an imaginary "forest" that would grow and change over time, just like a real one.

A few years later, in 1970, I began to work with Jim Wallis, a hydrologist, and Jim Janak, a theoretical physicist, both at the IBM Thomas J. Watson Research Center in Yorktown Heights, New York, to develop a computer program that would mimic forest growth.[9] In computer code, we wrote down what we believed happens between the environment and trees in a forest: how the growth of a tree changes with the amount of light it receives; how this amount of light changes with the size and number of competing trees nearby; how the growth also changes with the amount of water in the soil; and so forth. Each idea could be written down mathematically because the long history of laboratory study of trees provided mountains of quantitative information about tree growth and reproduction.

Development of the computer model of a forest went through several stages. We talked for days and then set down our ideas about forest growth based on our assumptions and knowledge. This produced a "conceptual" model—a model of ideas, not of quantities—that was translated into a set of mathematical equations, which in turn were translated into a computer code. We created the computer model step by step. First we wrote a program that grew a single tree, and made sure that this worked properly. Next, we elaborated the program to grow a group of trees competing with one another under a fixed environment, and made sure that worked. And finally, we added more complexity so that the program grew a forest under changing environments.

During the first step, writing a computer program that could grow a tree, we confronted a problem that has continued throughout my career: lack of basic data. The computer program gave us many characteristics of its hypothetical tree, including the number of leaves on it. Jim Janak and Jim Wallis asked me, as the ecologist, to find out whether the estimated number of leaves was accurate. I spent two weeks in the Yale School of Forestry's special library and found only two references where somebody had actually counted the leaves on a tree: a small fruit tree in a greenhouse at Cornell University that had about 10,000 leaves, and a large forest-grown tree that had about 100,000 leaves. Our computer trees had numbers of leaves in the right order of magnitude. By the end of 1970, we had created a computer program that grew trees of many species under a changing environment.

Up to that time, it had been a truism in ecology that this science was mainly empirical and therefore data-rich, but had little theory. More often than not, however, the opposite has turned out to be the case. Zoologist Kenneth Watt published a paper in 1962 that made the point that ecology suffered from too little data and too much (wrong) theory, but not many ecologists believed it.[10] In every research project I have done, and every ecological environmental problem I have tried to help solve, basic and essential data have always been lacking.

This may seem like an aside, but it's not. If there are rarely sufficient data, and sufficient interest in those data, to make a truly scientific test of the theory, it's no wonder that the scientific study of nature has been captured by mythologies. Thoreau was frustrated by the same problem, as I discuss in my 2001 book *No Man's Garden: Thoreau and a New Vision for Civilization and Nature*.

Since 1970, our computer program has been shown to mimic forests quite well. (For details about it, see my 1993 book *Forest Dynamics: An Ecological Model*.[11]) The forest model simulates ecological succession accurately and realistically; its projections about the characteristics of a very old, undisturbed forest match early surveyors' descriptions of forests in New England, including some observations that were not widely known or widely accepted among modern ecologists. For example, it predicted that a very old forest in New Hampshire would have many more spruce trees at low elevations than are found today in the White Mountains National Forest, and more than foresters and ecologists believe would grow. However, all of that forest, except for a few very small areas, had been clear-cut at some time since European settlement. It was commonly believed that the woodlands that had regrown closely resembled the original forests, but I found reports of measurements made in the uncut forests at the turn of the century showing that spruce had been more common than it is in the present-day younger forests, just as the model predicted.

The model even reproduced forest qualities that I didn't know. David Smith of the Yale Forestry School, the author of a standard book on silviculture, agreed to test the model by asking me to make it grow any kind of forest stand he wished. He suggested that we grow a model forest for 50 years at an elevation of 2,500 feet in New Hampshire and then cut all the trees that had trunks larger than five inches in diameter. The computer did this and projected that the result would not be a catastrophic clearing. Forest growth continued, but only of those species found in mature stands; the imaginary forest was not opened up enough to allow the pioneer trees to grow. Exactly right, Smith said. It was new to me, however.

In the more than four decades since its inception, this forest model has been tested widely around the world, and it is now used by many scientists, from Sweden to New Zealand and from tropical forests to the most northerly forests of Canada. We continue to work with it and improve it, and it continues to relentlessly confront us with the consequences of our assumptions. Over the years, it and we have become more accurate in our projections.[12] (As I write this in 2012, Australian scientist Michael Ngugi of Australia's Queensland Herbarium and I showed that a version of this computer program successfully forecasts changes in a forest Down Under.[13])

Computer simulation was getting started in many fields in the 1960s, but, unlike our forest model, many of the approaches were borrowed from engineering, especially engineering methods called "systems analysis." Those early computer models were based on the assumption that any system under study has a steady-state condition and tends to seek that condition—one of constancy and stability, like that of a well-made mechanical device. In addition, the systems-analysis models assumed that the processes and the steady state could be described in simple mechanical terms. Thus, they were simply reassertions of the older mechanistic view and didn't mimic nature well.

Nevertheless, steady-state computer models have had a major influence since the 1970s. Jay W. Forrester, a computer engineer who in the 1950s invented random access memory, a major advance in computer hardware, was one of the most influential pioneers in this kind of steady-state modeling—inventing and developing what he called "system dynamics."[14] This approach was used as the basis for the Club of Rome's 1972 book *Limits to Growth*, which was the first major attempt to use computer simulation to forecast the effects of people on nature and was widely influential in the late twentieth century. Although the authors saw the book as forecasting a Malthusian future—exponential growth of populations, especially the human population, and linear growth of resources—they made fundamental steady-state assumptions. In the preface, they wrote:

> For example, a population growing in a limited environment can approach the ultimate carrying capacity of that environment in several possible ways. It can adjust smoothly to an equilibrium below the environmental limit by means of a gradual decrease in growth rate, as shown below. It can overshoot the limit and then die back again in either a smooth or an oscillatory way, also as shown below. Or it can overshoot the limit and in the process decrease the ultimate carrying capacity by consuming some necessary nonrenewable resource, as diagramed below. This behavior has been noted in many natural systems. For instance, deer or goats, when natural enemies are absent, often overgraze their range and cause erosion or destruction of the vegetation.[15]

At the level of metaphor, *Limits to Growth* repeats and reinforces the ancient balance-of-nature idea with its first interpretation of the role of man in nature: Nature achieves a perfect balance until people get into the act and, because of exponential growth, destroy that balance. Whether or not it was the intention of the authors, whether or not it was an idea so deeply embedded in their way of thinking that they were unable to formulate the dynamics of the biosphere with people in it in any other way, the result—the "output," to put it in their terms—was to reinforce the ancient belief that had underlain the West's approach to nature.

People generally have a peculiar perception of computers and computer models, believing in the power of computers either too much or too little. In the late twentieth century, when computer simulations of aspects of the environment were first developed, there was great skepticism about their reliability. Oddly, during the first decade of the twenty-first century, computer models of a wide range of environmental characteristics, from climate to whales, are accepted as the truth at face value, often believed more by the public and by scientists than empirical evidence. We have to accept these models for what they really are. They are, in one sense, an exact and unyielding demonstration of the implications of one's assumptions about how nature works—in my case, the demonstration of the assumptions and perceptions that Murray Buell and I shared when we stood before the old walnut tree in Hutcheson Forest. Assumptions have a tendency to slowly bury themselves in our unconscious, becoming in effect myths.

The computer relentlessly confronts us with these assumptions and their implications. When we make nature act like a machine in a computer program, it does so exactly, and the results are quite unnatural. When we follow the knowledge and long experience of an ecologist like Murray Buell, add to it information gained by many scientists in laboratory studies of photosynthesis and the growth of trees, and then translate these data carefully into computer code, the model works. As George E. Yule observed 90 years ago, "If you get on the wrong track with the Mathematics for your guide, the only result is that you do not realize where you are and it may be hard to unbeguile you." You have to be on the right track, Yule said, for logic and mathematics to be useful. "To find the right track you must exercise faculties quite other than the logical—Observation, and Fancy, and Imagination: accurate observation, riotous fancy, and precise imagination."[16]

Such models can help us avoid two traps of the past that I have emphasized repeatedly in this book: continuing to believe in myths about nature even when they are clearly contradicted by facts, and believing two mutually contradictory ideas about nature at the same time (for instance, that nature is constant and nature is not constant). At the practical level, these models can help us synthesize what is too complex for our minds to combine working alone. But the computer cannot help us out of these traps if we begin our construction of a computer program with those traps assumed and built in as necessary to the internal workings of the model's system and a necessary output. As I explained earlier, this is my primary concern about the climate models, the general circulation models. When computer models are based on assumptions that contradict nature, they become a kind of pseudoscientific forecast, more like methods used by astrologers. This failing happens not only with ecological models but also with stock market forecasting methods, as was demonstrated in recent years. This is the great irony of computer models.

The computer model of forests illustrates the second way that computers are changing our perception of nature. By allowing us to mimic nature realistically, computers let us return to the intricacies of natural history in our scientific explanations, in our theories, in our tools for making projections, and thus in our management of forests and other resources. On the downside, computer models that use scientifically wrong assumptions and are data-weak can hide their failings among the details of lengthy computer language codes. Thus computer modeling is two-faced, and can serve us either well or ill.

In the 1970s, a revolution began to take place in computer simulation, which has now become so commonplace that we are not often aware that what we are being told as news by the media is the outcome, or relies on the outcome, of computer simulations. So it is with weather forecasts, in part empirical, using satellite remote sensing (also made possible by modern computers and computer simulation programs). Corporations that offer personal financial planning make use of computer models, often of a very simple kind. One of these companies offered to provide me with forecasts of how my investments would do, taking into account and averaging the results of more than 400 different forecasts. This was using a fairly simple computer model.

Farmers can use a computer simulation of crop growth that takes into account current weather patterns and the farmers' past activities and suggests when they should plant, irrigate, and fertilize. Knowing that bacteria act like data banks and that we can mimic a forest in a computer, we no longer need deny the complexities of nature in our theories and ideas.

Prediction and the Balance of Nature

At dusk one summer in the 1970s, far away from the hum of computers, a pack of wolves began howling near our camp at Washington Harbor on Isle Royale. There is something eerie and fascinating in the calls of the wolves, which begins with the howling of a single adult, who is soon joined by its packmates. The sounds build to a crescendo, and at times the wolves call in minor thirds. The result is almost musical, and the calls are always evocative, threatening yet fascinating, primeval, and wild. With the day's fieldwork done, we decided that evening to take advantage of the chance of the moment and follow the calls to try to get a view of the wolves.

As darkness fell, we set off along one of the trails that led back into the forest. We took flashlights along, but used them only occasionally; under these conditions their brightly focused beams only blinded us, and we made our way much more easily by the dim light of the moon. We hiked a mile or so, up and down, past dense stands of trees familiar to us during the day, across a stream, and finally up a small ridge on which we stood and waited and listened. We knew that we could not predict exactly where the wolves would go or determine very precisely from their calls just where they were. Chance and uncertainty were inherent in our choices and in our understanding of the behavior of wolves in the wilderness. But that unpredictability added interest to the night. We did not find the wolves, but standing on the ridge in the darkness and listening to the fading sound of their calls, we enjoyed a deep sense of the wild that rewarded us as much as a view of the wolves might have.

Our failure to find the wolves was the result of two kinds of limitations on our ability to predict where the wolves would go next: our lack of knowledge about the wolves, and just plain luck. Additional knowledge might have helped us in many ways. The calls of the wolves penetrated the forest but didn't give us a very good idea of the direction they were coming from. A scientific instrument that could determine the direction from which a sound came could have helped us estimate the location of the sound and would have increased the likelihood of predicting the wolves' next steps and then finding them. A better understanding of wolf behavior, such as how wolves move through the forests, also would have helped. Our ability to reason about wolf behavior and to understand causes and effects was limited. More studies of wolves might have led to a computer model of their behavior, from which we might have more accurately projected their location. But no such models existed—and if they had, back then we would have been able to operate them only in a windowless, air-conditioned room, of little help to us as we walked in the darkness.

With today's iPhones that have GPS capabilities, we might have accessed an app that would have used a computer simulation of wolf-sound travel and wolf behavior, taking into account the surrounding terrain, to forecast in real time where we should go next for the statistically best chance of encountering the wolves.

But no matter how good our forecasting methods were, there would always be some ambiguity about where the wolves were and where they would be next, because their behavior, as well as ours, always involves some kind and degree of chance and uncertainty. In short, there was also luck, even the chance that we and the wolves might cross paths without planning on our part. My colleague Peter Jordan, who had worked on the island for years, had that kind of luck just once, when he met a wolf crossing a trail in the middle of the day. Such luck was not with us that night.

It is an old philosophical question whether luck, as just described, really exists, whether chance or randomness is inherent in nature or whether there is merely an appearance of chance that results from our incomplete knowledge of causes. Here the key issue is prediction, which is at the heart of science and of the way we can deal wisely with our surroundings. The idea of the predictability of nature will remind some readers of the famous doctrine of the eighteenth-century astronomer and mathematician Pierre-Simon Laplace (1749–1827): If a mind were as large as God's and supplied with exact information about the position and velocity of every particle in the universe, then all future states of the universe could be predicted exactly. From Laplace's perspective, the human sense that things happen at random is simply a consequence of our ignorance of the state of the universe and the limitations of our reasoning capacities—the result of uncertainty. Laplace believed that the universe was deterministic, that there was no chance, only our ignorance.

Laplace and the wolves confront us with two fundamentally different concepts of reality: that chance arises simply from ignorance, and that chance arises from truly inherent randomness in nature. To Laplace, our failure to find the wolves that night on Isle Royale stemmed simply from our lack of knowledge about them and the rest of the wilderness. Similarly, our inability to accurately predict the life and death of trees in a forest would be solely the result of our imperfect knowledge and limits on our ability to use it. Computer models, by enabling us to make better use of our knowledge, help reduce Laplace's kind of uncertainty. In a nature with inherent randomness, which is referred to as a nature with "risk," the probability of an event may have specific causes and be well determined. When this is so, whether and when something will happen can be calculated rather accurately and computers can help us with the calculations, but only as to the probability of its happening

Most who have thought and written about nature throughout the history of the West have accepted implicitly or explicitly Laplace's point of view, rejecting the possibility of inherent randomness in nature. The classical nature of divine order and the nature of the machine age were not at all like seeking to find a pack of wolves at night on Isle Royale. Classical and machine-age nature was reliable, predictable, and comfortable, although not exciting; everything was in its place, and every future event exactly calculable.

For some thinkers, there were only two choices: either this completely knowable world with everything completely determined, or chaos, a world of complete randomness without any order, without any cause and effect, the sound and fury signifying nothing. This is the world of A. J. Nicolson, quoted at the beginning of Chapter 2: "Animal populations must exist in a state of balance for they are otherwise inexplicable."

This classical myth of determinism created a world of brightness and seeming clarity, which was in fact a faint shadow that had little correspondence to real nature. Beyond, in the wilderness, where the bramble and not the rose grew, where the wolves called in the night, there was the reality of another nature—of movement, sounds, and chance, but a nature that is predictable in its own way, which is intrinsically different from what we used to think and requires a change in our fundamental ideas about nature.

It is worth restating that a world that is like a watch that not only operates smoothly and is stable in the classical sense, but also is absolutely and completely predictable. When we watch the gears moving inside an old-fashioned grandfather clock, we can see which gear teeth will mesh next, and we can immediately view the realization of this small prediction. The gears are moving according to inexorable laws of physics. This is called a deterministic process; the status of the clock at any time in the future can be predicted exactly from present conditions, as long as the clock keeps operating.

Certainty and predictability are complex ideas in themselves. In the old adage "Nothing is certain but death and taxes," the certainty of each is quite different. Taxes are a deterministic certainty; both the event and its time of occurrence are certain—every April 15, U.S. federal income taxes are due. Death, on the contrary, is a stochastic certainty: The event is certain, but the time of its occurrence is not. Death in this sense involves risk; taxes do not. Philosophers may argue whether each death is ultimately the result of myriad deterministic causes at an ultimate level. But to each of us, at the level that we observe ourselves and our surroundings, and to each of our fellow creatures, at the level that they sense and respond to their surroundings, there is no way to distinguish the timing of that final event from the result of a series of chances.

So it was with our attempt to find the wolves at Isle Royale, and their effort to avoid us. The rustling of the wind in a branch might have disturbed either the wolves or us and changed our speed or direction. For the trees whose growth I had sought to mimic with a computer model, that same wind might have affected the fertilization of a tree's flower or the direction in which a seed fell to the ground.

The argument between determinism and chance, which needs to be delved deeper among ecologists and all those who hope to do well for nature, is familiar to physicists. Whether chance is an essential quality of the universe was the focus of intense controversies in twentieth-century science, especially among the physicists of the 1920s. Albert Einstein could not accept the idea that the universe functions fundamentally from a set of probabilities. The controversy led him to make one of his famous statements: "I shall never believe that God plays dice with the world."[17]

Other physicists had found that events involving fundamental particles can be explained by assuming that there are chance events—events that occur with a certain probability. In 1927, quantum physicist Werner Heisenberg (1901–1976) published what has become known as the uncertainty principle: that the position and velocity of a particle cannot be simultaneously determined with complete accuracy. Although either can be measured to any degree of precision, the more accurately one is measured, the less accurately the other will be known. Although where a particle would be and how fast it would be going could not be predicted with exact certainty, a probability of the occurrence of an event could be written down.

The profound philosophical arguments that arose from the development of quantum theory in the 1920s opened up the possibility of a very different perception of the physical universe: the universe as fundamentally stochastic to some degree. With this as a background, it may be easier for us to accept the idea of chance in our perception of biological nature; at least it should be easier, but it hasn't been to a large degree. Because computer programs can so easily be made to mimic chance events, computers reinforce this metaphor and offer in themselves a new basis for the metaphor of nature as a set of probabilities.[18]

The philosophical issues are more difficult for the physicists than for the ecologists. In the forests of Isle Royale, infrequent severe storms are an important cause of the death of trees. From a tree's point of view, if one can use that expression, the timing of the occurrence of such a storm is unpredictable. The effect of the storm on the tree's survival and on the evolution and adaptation of trees in a forest is a result of events that cannot be distinguished, at the level of perception and response open to trees and other living things, from a truly probabilistic event. One can argue whether the processes that lead to the development of a major storm are at an ultimate level deterministic or stochastic—whether they are like a clock or like dice. But these events occur at a rate and timing that, given the capacities of wolves and trees to detect their surroundings, are indistinguishable from probabilistic events. At the level that organisms respond to and affect their environment, the world is one of risk, predictable only in terms of probabilities. Nature as perceived by living things is a nature of chance.[19]

Although scientists generally rejected the idea that nature at the level perceived by trees and wolves might play dice, rapid advances in our understanding in recent decades hinted that randomness might be inherent in nature. For example, in the second half of the twentieth century new techniques allowed the reconstruction of Earth's climate for incredibly long periods, as I discussed earlier. It became possible to estimate the total amount of ice that had existed in glaciers, and ice volume is an index of how cold or warm the climate is. These histories of ice volume were subjected to statistical analyses. About one quarter of the variation in climate for the past 700,000 years could be explained by changes in the direction of Earth's tilt toward the sun, but the rest could be accounted for only by random variations, referred to as "red noise." Nature seemed to be throwing climatic dice.[20]

What I have described here is a kind of uncertainty principle for ecology: Since life and ecological systems that sustain it inherently involve chance, there is always going to be a limit on the accuracy with which we can predict anything about nature. This may sound rather obvious, especially to anybody who has gambled, or has done extensive hunting or wilderness hiking. But many who are attempting today to make forecasts about our environmental future will quite likely be distressed and try to dispute this uncertainty of nature. In the years since the publication of *Discordant Harmonies*, this is one of the things that has disappointed me most—the failure of those doing the forecasting to understand that we live in a world of chance and that chance must therefore be a characteristic of environmental forecasting methods.

The Erratic History of the Whooping Cranes

The problems posed by the threatened extinction of endangered species require a consideration of risks and probabilities. The whooping crane, the tallest bird in North America, once lived throughout most of the continent. Hunted, subjected to disruption of its habitat by land development, and extremely shy of people, the crane retreated with European colonization of North America. By the mid-nineteenth century, there were perhaps somewhat more than 1,000 remaining, but the population declined rapidly, and by the turn of the twentieth century no more than 100 cranes were left. Although the whooping crane, along with other birds, came under the protection of the 1916 Migratory Bird Treaty between the United States and Canada, the crane population continued to decline and seemed to be vanishing. Where the remaining cranes nested in the summer or fed in the winter was a mystery.[21]

Things began to look up for the crane in the late 1930s. The last wintering grounds of the bird were discovered in the Blackjack Peninsula of Texas. In 1937, the Aransas National Wildlife Refuge was established there, and an annual count of the entire population, which distinguished the number of adults and newborn, was begun. This practice has continued ever since, providing a unique population history. The breeding ground remained a mystery for about 20 more years. In 1955 the last nesting area for this bird was found in Wood Buffalo Park, Northwest Territories, Canada, an area east of the Canadian Rockies in remote wetlands.

The fate of the whooping crane seemed to hang in the balance for many years. The population declined to ten adults and four young in 1938. But then a slow recovery began that has continued since. An interesting aspect of this slow recovery was its erratic pattern; the population rose to 26 individuals in 1940, fell back to 15 in 1941, rose to 34 in 1949, fell back to 21 in 1952, and so on to the then most recent counts (Fig. 8.2).

With these rises and falls of the whooping crane population over the years, the question naturally arose about the birds' chances of extinction. In the early 1970s, Roy Mendelssohn, Richard S. Miller, and I analyzed the history of the whooping crane population and estimated this risk. The existence of a complete census of adults and young compiled over more than 30 years made this analysis possible. Instead of viewing the population pattern as a classic deterministic growth curve— the exponential or logistic—we viewed it as exhibiting true randomness, known formally as a "linear stochastic birth-death process." With this approach, we could calculate the chance that this erratically upward-trending curve might wander downward to zero—one way to calculate a probability of extinction. Our calculation showed that the chance of extinction was amazingly small. Given the population of 51 birds in 1972, the calculated chance that the curve would drift downward to zero by 1992 was only 5 in 1 billion. To my knowledge, this was the first such calculation.

FIGURE 8.2 Whooping crane population change appears to illustrate some randomness in nature. This shows the history of the whooping crane population from its low of 14 individuals in 1938 to more than 50 by the 1970s, when we made our calculation of the probability of extinction of this species.

(*Source:* R. S. Miller, D. B. Botkin, and R. Mendelssohn (1974). The Whooping Crane [*Grus americana*] Population of North America. *Biological Conservation, 6,* 106–111.)

This calculation was made with very specific and demanding assumptions, however. The critical assumption was that causes of variation in the previous 30 years would continue into the future and be the only causes of variation. In reality, one good sharpshooter could have eliminated the entire species, as could the introduction of a new disease or another new kind of catastrophe in the summering or wintering grounds. However, contrary to the assumption of the fragility of such small populations, the whooping crane seemed to have almost no chance of extinction unless faced with a new catastrophe. In fact, the species has continued to increase, reaching 247 individuals at Aransas National Wildlife Refuge in October 2009 (Fig. 8.3) and a total of 352 in the wild. (Since the census began, the cranes have been reestablished on flyways other than the Aransas to Wood Buffalo National Park.)[22]

The kind of analysis done for the crane is different from those done earlier in the century and illustrates a part of the change from a perception that populations follow exact, mechanistic patterns to acceptance of the essentially stochastic quality of population change over time. This idea allowed the calculation of the

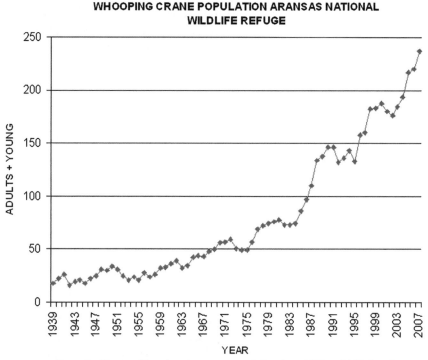

WHOOPING CRANE POPULATION ARANSAS NATIONAL WILDLIFE REFUGE

FIGURE 8.3 The entire history of the whooping crane population, from 1938 to 2009. Consistent with our 1970s forecast, the whooping crane population migrating between Aransas National Wildlife Refuge in the United States and Wood Buffalo National Park in Canada has not gone extinct but rather has continued to increase substantially, reaching 247 adults in 2009. (Data courtesy of Tom Stehn, Whooping Crane Coordinator, U.S. Fish and Wildlife Service, Aransas National Wildlife Refuge.)

chance of extinction. Acceptance of such erratic, random qualities appeared to become more common in the study of populations toward the end of the twentieth century and continues into the twenty-first century, illustrating one of the transitions at the heart of this book, a transition that can lead us to acknowledge the discordant harmony of nature. I'll return later to whether this transition has occurred in any large way.

This kind of calculation of the probability of extinction of an endangered species, based on observations of the population's change over time, would seem an obvious thing to do. In the mid-1990s, I was teaching a course about ecological forecasting at George Mason University, and asked the students to do a worldwide search, including use of the Internet and direct contacts with researchers, to locate other calculations of the same kind. I was surprised that they could find only a few, less than 15 other cases, and that the U.S. Fish and Wildlife Service had data sufficient to do such a calculation for only one species, the Puerto Rican parrot, not for any of the major "headliner" species like the spotted owl or salmon. Today, in the second decade of the twenty-first century, the number of species analyzed in this way remains very small.

Chaos and Well-Behaved Chance

> But fortune, which has great power in all matters and most of all in war, causes great shifts in human affairs with just a little disturbance.
>
> —Julius Caesar[23]

We have a tendency these days to believe all or nothing. Perhaps it is because, drowning in information and saturated with visual and audio media, we seek slogans and sound bites as if they were adequate substitutes for knowledge, understanding, scientific conclusions, and wisdom. The all-or-nothing approach to the concept of chance in nature has given rise to a popular and in a sense fashionable belief that there is a kind of chance in nature that leads only and invariably to complete chaos in very much the way that ancient Greek and Roman philosophers meant it—without form, structure, cause and effect.

But we know from our own lives that there is a wide variety of chance and risk. Take a game of dice. While each throw has an unknown outcome, overall the game has a very regular, reliable, predictable behavior: Over many throws, the number seven will be the most common result. If you want to make a long-term bet and are conservative in your approach to risk, you would always pick number seven. The game is well behaved because the probabilities are fixed over time and independent of anything in the environment except, of course, cheaters with loaded dice. This is the most predictable kind of chance: a process that has some inherent randomness, but whose probabilities—chances of what will happen—are fixed forever and independent of everything else.

Next in "well-behaved" randomness are processes where the probabilities change with specific external (in the case of our interests, environmental) factors. The chance that a pack of wolves will travel at all depends in part on the weather. For these kinds of phenomena, the more we understand the connection between that thing of interest to us (like whether wolves will run toward us) and the specific influencing factors, the more precise our predictions can be. But once again, we cannot know perfectly, exactly, the future of any specific series of events. We could say that on average, given a certain set of weather conditions, a pack of wolves will do such and such, and give an estimate of all the other things a pack could do under those circumstances. But we would still be running in the forest in the dark to a certain extent, hoping that the pack we were trying to meet up with was doing exactly what that average pack would do. As with a game of dice, the best you can say is "Good luck." But to make the point more precisely, *luck* and *good luck* can be precisely defined, just as the number seven can in a game of dice.

From here we can move downward to less and less well-behaved probabilities. The next kind would be a process where the external probabilities that affect the chances of what we are interested in change over time themselves in ways that are subject to chance. And so on downward until one reaches what I will call complete chaos, an imaginary world where there is no cause and effect and that world is therefore without form, structure, and anything that we would call "understandable" in ordinary terms—like if it's cloudy it's more likely to rain than if it's sunny.

When it comes to discussions about chance in nature, current popularity goes to "chaos theory." Two mathematical discoveries in the second half of the twentieth century led to the development of chaos theory. The earliest was a very curious finding by Edward Lorenz (1917–2008). Lorenz, who began his career as a weather forecaster and later joined the faculty at MIT, tried to use a computer program to predict the weather and was surprised to find that very small differences in the initial quantitative conditions produced very different results. This came about originally because the relatively crude computers of that time only tracked numbers that had a small number of decimal places. But from the results of these initial computer models, he concluded that very small differences in initial conditions could have huge effects on the outcome.

Lorenz's most famous description of this is known as the "butterfly effect." The idea is that the tiny flapping of a butterfly's wings before a storm developed could affect the intensity of the storm or even whether it amounted to a storm. This idea has been extended in science fiction to the possibility that this small flapping of a butterfly's wings a long time ago could change history. This was striking and in some ways an amazing conclusion that has become influential. The important question for us is whether nature operates in this way. Chaos theory has been applied in medicine for such problems as eye-tracking disorders among schizophrenics, and certain heart conditions. Geologists have applied it to consider the predictability of earthquakes.

My own work with ecological computer simulations gives a different story about the possible importance of Lorenz's chaos theory. The simplest way I can put it comes from my flying experience. When I was a pilot in training, I made many, many practice landings, each of which created a different amount of turbulence—the equivalent of a butterfly flapping its wings. When I was doing very well, the plane would land very smoothly, generating minimal turbulence as the wheels settled to the ground. But if I was having a bad day, I might bring the plane down not very smoothly, with even a little bounce on touchdown. I'm sure that if the butterfly effect is real, I personally have had a huge effect on the weather in Cape Cod and Santa Barbara, where I did most of my flying. But as I'm sure you'll agree, that doesn't seem likely at all.

Nor was the forest model we developed that chaotic. Instead, the response of the (computer's) forest to any initial conditions was to be influenced by them, but in a comparatively well-behaved way. So, whether there is going to be a butterfly effect depends on the entire set of cause and effects that are inherent in the system.

The second mathematical discovery, which led to a different meaning but was also called chaos theory, occurred in the second half of the twentieth century. This began with a discussion of a peculiar deterministic equation—that is, an equation that did not have chance associated with it. Mathematicians S. A. Woodin and James Yorke showed that there is an equation expressing changes in populations that involves no chance events at all but one with which you cannot calculate the size of the populations for all times in the future simply from current conditions. Instead, the future size must be calculated step by step from one time period to the next; beyond that, there is an appearance of unpredictability. This equation became one of the primary methods to calculate "random" numbers for computer programs, including the kind of randomness associated with many video games. This is because, for all intents and purposes, it is very difficult to tell the output from this equation from a truly randomly picked number. This is unlike a clockwork mechanism whose position at any time in the future can be predicted exactly from its present position. Robert May, a mathematical ecologist, used this equation as the basis of a discussion of predictability and chaos in populations.[24]

As I have explained, for nature involving risk, there are three possibilities in regard to prediction. Two of them lead to a well-defined kind of predictability: probabilities that are constant over time, such as those determining how dice will fall; and probabilities that change over time in response to environmental conditions and are therefore predictable within some range. The third is probabilities that change in a random fashion, uncorrelated with anything, and would lead to a chaotic nature from the point of view of living things. When the chances of birth and death remain constant over time and these are the only factors that matter—that is, the only factors that are allowed to vary in the computations—a population of trees, for example, will eventually achieve a constancy of numbers, although the fate of any individual remains subject to chance. After a very long time in such a forest, the relative importance of different species of trees would become fixed even though the location of individuals of any one species could not be predicted exactly. When the probabilities of birth and death change over time, the outcome is more complex. If the changes are direct responses to environmental and biological conditions, then it is possible to project what will happen, but the projections are generally too difficult to make with pencil-and-paper mathematics.

For example, the chances that a seed will germinate may increase as rainfall increases and as sunlight increases and as the abundance of animals that spread the seeds increases. But each of these can vary independent of one another, and one may increase while the others decrease. Except in very special cases, the only way to make forecasts about the future condition of such a forest is to make many, many calculations one by one. If you were king, you would give a large number of people pads and pencils and assign each of them, in groups, to make certain of these calculations. You, meanwhile, would have to organize the whole thing and make sure the calculations were done in the correct order. In effect, you and those working with you would be functioning as a biological computer. Today, we do this with modern computers and refer to it as "computer simulation."

New Ways to Observe Nature

I said earlier that the chances of finding wolves in the night could have been improved by three things: first, better observations of the present (the location of the wolves when they first called); second, better understanding of cause and effect and better ability to use that understanding to make predictions (the factors that led the wolves to call and then move through the forest in a particular direction); and third, the ability to make predictions that involve chance. Computers help us with all three. Perhaps seeking to find wolves on Isle Royale at night will not seem such a dark search if we know that these devices can be of help to us.

As mentioned, our search for the wolves might have had greater success if we had had better means of direct observation. Computers are also playing a direct role in changing our ability to observe nature, which in turn influences our perception of nature. Computers are at the heart of many of the devices used today for measurement and observation. Even by the late 1980s this was beginning to happen. High in the remote and mist-draped mountains of Taiwan that extend well above 10,000 feet in Lala Shan Nature Preserve, rainfall has been measured since 1987 by computer-based methods, using chips invented in the United States and produced in Japan. At many other remote locations, data on environmental factors such as rainfall and wind speed and direction were collected and stored on computer chips or in a computer format on tape cassettes. The locations were then visited periodically to collect the information.

As early as the 1980s, the potential for producing electricity from wind energy was being assessed in California using computer-based monitoring.[25] Yet another example, the chemical characteristics of the environment can be analyzed by methods that require computers, or components derived from the computer industry. The level of pollutants can now be measured in place in the lagoon of Venice by a probe that relies on such computer-based devices. Remote sensing, which we discuss in later chapters, is possible only with the help of computers and computer-derived devices.

Today, of course, remote measurement has become much more sophisticated. A scientist can go into the field—into a remote forest or a ship at sea—and with a handheld device make measurements that are automatically radioed to a computer in an air-conditioned and centrally heated room. The location and time of the observation are automatically included. All made possible by computers.

These are just a few examples of the major changes in observational abilities that are rapidly entering science and management.

New Mythology

Making environmental predictions that involve chance requires not only new techniques but also a change in our myths about nature. This is yet another way that computers are changing our perception of nature. Until the advent of modern computers, it was not possible to make extensive predictions that involve stochastic processes. Acceptance of the idea that we might benefit by viewing nature as characterized by chance and randomness is a deep and unsettling change. On the surface, at the level of visual perception of the environment, the computer model of forest growth is simply an expression of generalizations based on careful observations of nature. But at a deeper level, that program, operating through the rapid processing of numbers within the silence and darkness of microchips, expresses a new perspective on nature, rapidly moving shadows of the ponderously slow reality outside in the sunlit forest.

A key aspect of this deeper perspective is the metaphor of many events considered simultaneously in a connected network, which we met before as the definition of bacteria. It allows us, for example, to perceive the forest as made up of many individual trees growing, taking up air, water, and nitrogen, producing seeds, and dying—all these processes occurring simultaneously, interconnected yet independent. Another key aspect is the inclusion of chance as a fundamental aspect of life and death. In the computer model of forest growth, whether a tree dies and whether seeds sprout are determined by the computer's equivalent of the roll of dice. The chance of survival depends on two factors. One is inherited: a genetic potential expressed in terms of the maximum longevity of the species. The other is environmental: how well the tree grew during the past year. If a tree's growth falls below some crucial minimum in a year, the tree has a much larger chance of dying when the dice are rolled. Reproduction of trees also involves an element of chance: The number of saplings of a species that are added in a year in the forest model is determined by the computer version of throwing dice.

The computer's ability to handle chance occurrences is one of the reasons that a computer program has been so helpful in considering forest growth. Old-fashioned pencil-and-paper mathematics cannot take us very far in projecting birth and death when these events involve chance occurrences that depend on many factors.[26]

Nature that is inherently risky may seem less appealing than nature that is completely deterministic. Superficially, nature characterized by probabilities seems less likely to be in balance; random events might throw the otherwise balanced system out of whack for no "purpose" or "reason" except the very probabilities that govern the events. In contrast, nature without chance seems of necessity to be readily predictable.

Computer Chaos Becomes Our Reality

Let us step back a bit to the discussion of our beliefs about nature, and the role of images, metaphors, symbols, and mythology in influencing those beliefs. One of the oddest results of Lorenz's chaos theory, to me at least, is that it has become our reality. The output of a (primitive) computer program posed an idea that has taken hold widely, entering mainstream beliefs and ingrained with the media. For example, in the film *Jurassic Park*, one of the characters, presented as having a mathematics background, keeps reminding the others that nature is chaotic, and the collapse of the park and the attack on the characters by the dinosaurs is thereby attributed to chaos theory. In this case, how did we explain the world around us? Not from what we knew about our own bodies, as was the case before modern machines and science. Not from what we knew about machines, the dominant nineteenth- and twentieth-century metaphors, but from the output of a computer program.

As I have suggested in this chapter, the computer has made possible quite a large set of new images, symbols, and metaphors that are quickly becoming the source, or at least one of the sources, of our explanation of how nature works. In the case of the whooping crane, the understanding and application of a very well-behaved kind of randomness seems helpful to us and to our conservation of the cranes.

Although this nature of chance may seem less comforting than a predictable clockwork world, it is the way we find nature with our modern means of observation, and therefore it is the way we must accept nature and approach the management of resources. Managing from the comfort of a believed-to-be deterministic world when one lives in a world of chance is like following the beam of a flashlight at night on Isle Royale: What appears in the beam is very clear, but one is likely to stumble and fall over the roots and rocks that lie just outside of it. Instead, once we accept the idea that we can deal with these complexities of nature, we begin to discover that the world of chance is not so bad, that it is interesting and even intriguing now that we understand that chance in our world is not complete chaos, that seeking the wolves at night involved some probability, predictable itself, of our finding them. Thus we must accept nature for what we are able to observe it to be, not for what we might wish it to be. Accepting this perception of nature, we discover that we have the tools to deal with it. And once we realize we have the tools, this new idea of nature takes on its own appeal, just as the game we played seeking the wolves at night became a pleasure, filling us with a sense of the wild.

The flapping of butterfly wings seems to lead us in the wrong direction—wrong in that it is not a good description of how biological nature behaves. Whether these new computer-born images, symbols, and metaphors are to our and nature's benefit or disbenefit is not clear and, like all else in a world of chance, it depends. . . .

Living and Nonliving

The distinction between life and nonlife used to seem simple and clear, but it is becoming obscured by modern technology. This is another way that computers are changing our perception of nature. While computers have made machines seem lifelike, other fields that rely on modern technology are blending the mechanistic and the organic from the opposite direction. Molecular biology and genetic engineering are opening up organic entities to engineering and making life seem machinelike. In this way, genetic engineering also obscures the distinction between life and nonlife because scientists can manipulate life in ways that, even a decade ago, people believed could be done only with nonliving things. We could design a new automobile or transfer an engine from one car to another, but we didn't know that we could transfer genes from one life-form to another. Today, genetic engineering has become so common and easy that there are hobbyists creating new varieties within species and perhaps even new species, and apparently doing it without very much understanding of how ecological systems work or of the history of well-meaning intentional introductions of exotic species. As I will discuss later, this is probably the single major untalked-about threat to nature.

That it is now possible to do genetic engineering affects us at the level of image, symbol, metaphor, and myth. Does that make life more or less machinelike in our deeper perceptions of nature? Once again our technology in the second decade of the twenty-first century is blurring the distinction between living and nonliving.

In Western civilization, as late as the eighteenth century, the distinction between life and nonlife was not made in the way we understand it. The strong distinction is a result of the machine age, the recent industrial and mechanical world. Primitive peoples who believe in animism would not understand it. Objects that we, as inheritors of the ideas of the nineteenth-century Industrial Revolution, conceive of as nonliving—rocks, mountains— earlier peoples believed were animated. Thus the blending of the ideas of life and of nonlife brings us back to a perspective more in harmony with beliefs held throughout most of human history. For some, this will seem another strange irony of modern times.

Nature as wilderness, the out-there that has played such an important role in Western ideas throughout the centuries, seems now fundamentally different from what it seemed before. Wilderness is a nature of chance and complexities that we need no longer fear as unknowable or completely unpredictable. The most novel of our tools, the computer, is helping us grasp what we have feared to seek.

FIGURE 9.1 Earth from the Moon, one of the iconic images of the late twentieth century, changing forever how we think of life and its environment. (Photograph, of course, by Apollo astronauts, courtesy of the U.S. National Aeronautics and Space Administration.)

9

Within the Moose's Stomach
NATURE AS THE BIOSPHERE

It is by no means for nothing that [the uninhabited parts of the Earth] come to be. . . . The sea gives off gentle exhalations, and the most pleasant winds when summer is at its height are released and dispersed from the uninhabited and frozen region by the snows that are gradually melting there.

—PLUTARCH (CIRCA AD 75)[1]

The relationship to the living matter of the earth and to its decomposition products must form the central theme . . . because this relationship is responsible for the most remarkable feature of the atmosphere in contact with the liquid and solid materials at the earth's crust, namely, the fact that the atmospheric gases in contact with water, do not represent a mixture in thermodynamic equilibrium.

—G. E. HUTCHINSON (1954)[2]

An Afternoon on Isle Royale

One summer afternoon on Isle Royale, I came across a moose in a shallow pond that lay back from the shores, sheltered from Lake Superior's chilly winds. I stopped to rest my feet and view the scene, which was peaceful, calm, and pleasing. Feeding placidly, the moose seemed alone in the wilderness, breathing oxygen from inert, lifeless but life-giving air and digging through dank, lifeless mud for a succulent morsel of green water plants. Yet looks are deceptive, and here, after our long journey through time and over Earth's surface and among its many creatures, it is appropriate to take a lesson from Aristotle and view the "mean and despicable" creatures, for with this inquiry we can, at long last, begin to reinterpret nature for our time.

The moose—with its sagging belly and long, spindly legs that looked too thin for its stocky body—seemed an ungainly creature in the wilderness. Who would design such a creature? It has none of the attributes of classical beauty; instead, a face without merit; a drooping hang-jaw and protruding lower lips; a hard, gummy palate instead of upper front teeth. The moose has inspired none of the rhapsodies

183

to the grandeur or sublimity of nature, so important to Wordsworth. Only in the autumn does a bull moose, with its large antlers, present any semblance of nature's magnificence—viewed, that is, from a distance; close up, the comical reappears.

If I had been hiking with another twentieth-century biologist versed in the Darwinian theory of evolution, he could have explained to me how well adapted a moose is to its environment, despite the superficial ugliness of the adaptations. Its lips, for example, are wonderfully adapted for pulling up underwater plants and tearing leaves from twigs: The lower front teeth push against the upper gums so that only the soft, edible leaves are removed, leaving the twigs on the plants. I knew from past experience that when disturbed the moose's long, spindly legs can propel it over the waters at surprising speed, enabling it to flee or chase an interloper like me with great agility through the thickets and over dead logs, much faster than I could run.

Watching the peaceful lakeside scene under the bright blue sky against tea-colored waters sheltered in the forest, I was reminded of those ancient questions that we have been considering in this book. All seemed constant beneath the sky, unchanging and permanent, without the slightest suggestion of the inevitable death of the moose or of the population explosions and crashes of its ancestors, without a sign of trampled vegetation or of water lilies and yew forced to the brink of extinction on the island by this ungainly herbivore. Although it seemed dependent on the wilderness—the air, water, soils, and plants—the moose appeared to have little influence on its immediate environment or on the rest of nature beyond the island.

In this still life, the grazing moose seemed to be alone, but in fact he was not. As an ungulate, a true ruminant with a four-chambered stomach, the moose carries within its intestines an intricate array of symbiotic microbes, as do its many ruminant relatives, which include the rest of the cervids (the family of northern deer), domestic cattle, and many of the big-game animals of the African plains and savannas. Its stomach teems with microbes, 1 billion per cubic centimeter, performing tasks that the moose cannot do for itself. Who would have designed a moose this way, of necessity harboring in its gut vast numbers of creatures on which its life depends, unable to make the enzymes to digest the vegetation that, in the northern wilderness, is its only food?

An ungulate's rumen is so complex that a textbook in animal physiology called it "an ecological system in dynamic equilibrium" with inputs (of food and saliva) and outputs (of those materials that the moose can digest).[3] Of the many species of microbes in the rumen, the most important to the moose are anaerobes, bacteria that can live only in an oxygenless atmosphere. There are other species of bacteria and other unicellular organisms that feed on the bacteria—predators and prey—all growing, reproducing, feeding, and dying within the gut of this large mammal.

The anaerobic bacteria are truly symbiotic with the moose, enabling the moose to survive and in turn depending on the moose for their own survival. A large fraction of the vegetation swallowed by the moose consists of cellulose and complex carbohydrates, which it cannot digest. The bacteria release enzymes that digest cellulose and other carbohydrates, breaking down these complex compounds into simple sugars that might make a good energy source for the moose, but it doesn't get a chance to use them. The bacteria take up the sugars and use them as their energy supply before the lining of the moose's intestines can absorb these compounds. In the oxygenless atmosphere of the rumen, ordinary respiration, which requires oxygen, cannot take place. Instead, the bacteria ferment the sugars and give off fatty acids and other acids as waste products of their metabolism. Of these, acetic acid, the acid in vinegar, is the most important. These acids become the moose's food; they are absorbed by the walls of the rumen.

The moose, like all mammals, produces urea in the process of digesting its food. Urea is 47% nitrogen, an element essential for all living things and sometimes difficult for the moose to obtain in sufficient quantity, especially in winter. The excretion of urea is a loss of this valuable resource. But instead of eliminating all the urea, as we do, the moose is able to transfer part of it through its blood to its saliva; when the saliva is swallowed, the urea returns to the rumen. There, the bacteria are able to convert the nitrogen-rich urea, as well as ammonia (produced in other reactions), to proteins. The nitrogen in the urea is important to the bacteria, since it fertilizes their growth. The bacteria also synthesize vitamins necessary to the moose. Thus, the moose can live on what would otherwise be a very poor diet, low in protein and vitamins, too low for its survival if not for the bacteria in its gut. No scurvy for the moose, because it hosts all these bacteria.

The bacteria digest about 50% of the plant material taken in by the moose and convert about 10% of the rest to methane, which is belched and released to the air unused by the moose. The remaining vegetation passes down the digestive system, along with a large mass of the bacteria (about 15% by weight of the original mass of vegetation taken in by the moose). The moose digests these bacteria as high-quality food and is also able to digest about half of the remaining vegetation by its own digestive processes.

That moose and all ruminants, including domestic cattle, release large quantities of methane was known to farmers but little known to urban dwellers in the twentieth century. G. Evelyn Hutchinson (1903–1991), whom I believed was the greatest ecologist of the twentieth century, wrote about this in a scientific publication that has one of the most amusing titles in science ever: *The Biogeochemistry of Vertebrate Excretion.*[4] With the rising concern about possible global warming, beginning in the late 1980s, many people became aware of this release by ruminants because methane is one of the greenhouse gases.

The moose contributes many things to its symbiotic relationship with bacteria. It provides the food for the bacteria and maintains an environment with a constant temperature of approximately 102°F. Along with vegetation, it swallows saliva, which is slightly basic (slightly alkaline) and neutralizes acids produced by the bacteria, helping to maintain an acidity in the rumen within a range acceptable, perhaps even optimal, for the bacteria. The bacteria live, grow, and multiply in a protected environment, with a special atmosphere and a steady input of food. The moose removes fatty acids that are wastes of the bacteria and would otherwise poison the microbes and threaten their survival in the rumen. The rumen, in sum, is an ecological system overwhelmingly biologically produced and controlled.

In some ways, the moose's rumen is a miniature model of the biosphere, our planet's entire life-support system. Like the biosphere, the rumen includes and sustains life. But the rumen is also a system under biological control and in a steady state within a certain range of conditions: as long as the moose is alive and healthy. In these qualities the stomach of a moose brings us back to the ancient questions about the characteristics of nature without human influence, and the question of whether nature is in a balance. The microbes in the moose's gut live in an environment that seems bizarre to us and certainly unsuitable for human beings. There is little oxygen and in some parts of the intestine no oxygen at all. Noxious gases, such as methane and ammonia, abound, produced by the microbes and the moose's own digestive processes. The fluids are acidic, and the microbes withdraw nitrogen from the air (nitrogen is the most abundant molecule in Earth's atmosphere) and convert it to nitrate, nitric acid, ammonia, and other small and unpleasant chemicals.

The microbes that live in the rumen would die in the air outside, the air that I found so clear and pleasant on that sunlit afternoon by the pond. Oxygen would kill them. But in the pond-bottom that the moose kicked at and waded through as it fed, there were similar bacteria that also can survive only where there is no oxygen; they live beneath the still waters of the pond, where poisonous oxygen cannot penetrate. Nitrogen-fixing bacteria that live in the root nodules of leguminous plants, such as alders on the pond's edge, also cannot survive in an atmosphere rich in oxygen. For them, the nodules provide a home, an *ecos*, protecting them from the oxygen that is so deadly to them.

Imagine that you are one of those bacteria, except that you and your colleagues are intelligent, thoughtful, and conversational. Your universe would be the horizon that you could see within the moose's stomach. A Henderson among you (remember his book *The Fitness of the Environment*, discussed earlier)[5] might speculate about how and why the environment was so well suited to the life that all of you bacteria enjoyed, and how its remarkable steady state came about. You might develop a theory about the constancy of nature, a balance of nature. That would unfortunately be interrupted by the death of the moose, but before that catastrophic ending, your group might have developed a deterministic differential equation model of your biosphere, and since it assumed and demonstrated a steady state, and since you were only as intelligent as a bacterium, you might have believed that it proved you were living in a constant nature that would persist forever.

The Debris in the Bay of Bengal

Earth is a peculiar planet when compared with its nearest neighbors, Venus and Mars, even though the three planets are similar in several ways. They are within a factor of 2 in their distances from the sun and within a factor of 2 in their diameters.

Since they are similar in distance from the sun and in size, one would expect them to be similar in many other ways, including the composition of their atmospheres. However, scientific observations in the latter part of the twentieth century, both from space probes sent to Mars and Venus and from Earth-based telescopes, revealed that their atmospheres are very different from Earth's.

The atmospheres of Mars and Venus are more like each other's, composed mainly of carbon dioxide; nitrogen is the second most important constituent, and oxygen is a rare and minor component. In contrast, Earth's atmosphere is 79% nitrogen and 21% oxygen, while carbon dioxide makes up only 0.03%. The atmosphere of Venus is very dense, while that of Mars is very thin.[6] How could this be? It's a question that has been long debated. It turns out that the relative abundance of the primary constituents of Earth's atmosphere differs from those of both Venus and Mars because of the effects of life on Earth's surface.

Whether or not life affects its environment globally to any significant degree is an argument whose resolution is important to the questions that we have been pursuing throughout this book: What is the character of nature undisturbed by human beings? What is the effect of nature on people? What is the effect of people on nature?

Whether living things have important effects on Earth's geologic characteristics has been argued in what we would call scientific terms since the seventeenth century. Today it may seem obvious that life changes its environment globally, a simple truism, because we are bombarded with information that makes this case for many human activities that seem to be affecting the environment globally, especially with respect to global warming. But until the end of the twentieth century, the notion that life had global environmental effects was generally considered ludicrous by leading scientists.

The change is in part a result of late-twentieth-century observational technology, but it is also—and we should not ignore this—in part a result of a change in myths, metaphors, and perceptions, which in turn are influenced by technology, bringing us full circle back to the photograph of Earth from the moon, which appears at the beginning of this chapter. That photograph could not have been possible without modern science and technology, hence without modern rationalism. But it probably also would never have come about if there were no people like Robert Goddard (1882–1945), the inventor of the modern rocket, and Edwin Eugene ("Buzz") Aldrin (1930–), one of the first two men to set foot on the moon, people with great imagination, literally "flights of fancy," a sense of adventure and beauty, of thrill and of life. The rational fed the nonrational, which fed the rational, which in turn gave us the photograph of Earth from the moon that has forever changed our feelings and beliefs, our idea of what is beautiful, important, meaningful, pleasant, amusing.

To appreciate this change in understanding and belief, it is helpful to consider some of the classic scientific arguments about the influence of life on the global environment. The possible influence of life on air, water, soils, and rocks, on the shape and persistence of Earth's landforms, on rates of erosion, and on the climate was actively discussed in the eighteenth and nineteenth centuries, when scientific study of these phenomena was beginning. In the early nineteenth century, when geologists were starting to formulate the first scientific understanding of the processes of erosion, they began to ask what forces opposed the slow degradation of the land and what created new landforms.

Adam Sedgwick (1785–1873), one of the founders of modern geology (and also one of Charles Darwin's professors and friends), argued that vegetation is the primary force opposing erosion: "By the processes of vegetable life an incalculable mass of solid matter is absorbed, year after year, from the elastic and non-elastic fluids circulating round the earth." This material is "thrown down upon" Earth's surface so that "in this *single* operation there is a *vast counterpoise to all* the agents of destruction [Sedgwick's italics].[7]

Sir Charles Lyell (1797–1875), another of the fathers of modern geology, dismissed Sedgwick's argument as "splendid eloquence." Lyell's ideas are important to us because of the important role he played in establishing geology as a science. He wrote the first major book in English on geology, *Principles of Geology*, first published in 1830. The book's subtitle was *An Attempt to Explain the Former Changes of the Earth's Surface by Reference to Causes now in Operation*. That's something we continue to try to do.

Lyell's ideas were based on the evidence available to him at the time, as well as the worldviews of the time, the metaphors and beliefs about nature that were acceptable and seemed viable. He thought the idea that life might play an important role in large-scale Earth processes was not worthy of serious consideration, but required attention nevertheless because "such an opinion has been recently advanced by an eminent geologist" (meaning Sedgwick). Otherwise, Lyell wrote, "we should have deemed it unnecessary to dwell on propositions which appear to us so clear and obvious."[8] It is worth emphasizing that what was "clear and obvious" to Lyell is that life could not have any important effect on the global environment now and never did. It's important to emphasize because this will now seem so strange to twenty-first-century popular culture, where rock-and-roll musicians, asked what they could do to help the world, suggest a contribution to oppose global warming.

As a counter-example, which he hoped would put the issue in the grave forever, Lyell considered the materials washed down annually into the Bay of Bengal, where, as it appeared to him in 1831, "what remains, whether organic or inorganic, will be the measure of the degradation which thousands of torrents in the Himalaya mountains, and many rivers of other parts of India, bring down in a single year." This was an extreme case, with an abundance of rainfall and vegetation and a high rate of erosion. If you couldn't make the case for the geologic importance of life here, you couldn't make it anywhere, Lyell argued. The forces of physical erosion were so strong in this region, he believed, that the role of vegetation "can merely be considered as having been in a slight degree *conservative*," merely retarding erosion, but not acting as a constructive or an "antagonist power" against the forces of physical erosion. Because it represented an extreme, Lyell generalized from this case that vegetation can never play an important role in countering erosion.[9] *Strike number one about life affecting something large-scale about the environment.*

Lyell also rejected the possibility that organic matter represents a large and increasing fraction of the material on Earth's surface. Organic soil, which he called "vegetable mould," was "seldom more than a few feet in thickness" and "often did not exceed a few inches," and as such represented a small portion of Earth's material. Furthermore, from the knowledge he had available in the eighteenth century, organic deposits did not appear to have increased in past geologic periods.

Lyell and his contemporaries had come to recognize that mountains are formed by uplifting generated by forces deep inside the Earth. Mountains and other areas that had been subjected to such uplifting were observed to be generally denuded of soils, implying that they did not retain soils as they rose. Lyell concluded that soils are not conserved during these processes or over long time periods. Moreover, the growth rate of vegetation is not correlated with the abundance of organic debris; peat, for example, is common in the Far North, where vegetation grows slowly, and uncommon in the tropics, where vegetation growth was believed to be greatest. From these lines of evidence, Lyell concluded that vegetation-derived organic matter in the soil could not be important to the creation of landforms or the prevention of erosion. *Strike two against life as a global environmental force.*

Life seemed to Lyell to have played a minor role, if any, in the cycling of chemical elements necessary for life. He knew that animals and plants withdraw elements from the air, land, and waters, but he argued that their death results in only a small and very slow return of these chemicals to the environment. "If the operation of animal and vegetable life could restore to the general surface of the continents a portion of the elements of those disintegrated rocks, of which such enormous masses are swept down annually into the sea, [then] the effects would have become ere now most striking; and would have constituted one of the most leading features in the structure and composition of our continents."[10]

The effects were not visible or striking in Lyell's time with the scientific methods of observation available. *Strike three against life's global environmental role.* Life seemed incapable of countering the major changes in the structure of Earth's surface, and the role of life in affecting the nonliving environment appeared minor. Instead, Lyell concluded that "igneous causes" must have provided the "real antagonist power" to "counterbalance the leveling action of running water." That is, processes deep in the Earth must offset erosion. Of the materials required for life, "fresh supplies are derived by the atmosphere, and by running water," he wrote, "from the disintegration of rocks and their organic contents, and from the interior of the Earth, from whence all the elements before mentioned, which enter principally into the composition of animals and vegetables, are continually evolved."[11] Life seemed dependent on the environment, like the moose alone in the wilderness, but appeared to have little effect on that environment, unlike the community of bacteria teeming symbiotically within the moose's gut.

Lessons from a Dead Mouse in a Jar

In the history of modern science, both before and since Lyell, it has been common to dismiss the role of living things in determining many of Earth's features, including its atmosphere, because, just as Lyell had argued, the total mass of living things is only a tiny fraction of Earth's mass. Using modern estimates, the mass of all living things on Earth is a few hundred billion metric tons, perhaps as much as 1,000 billion metric tons.[12] The total mass of the Earth is about 6 trillion billion metric tons. (For those of you familiar with the American or British ton, 2,000 pounds, don't worry; by coincidence, a metric ton and an American-English ton are almost the same size.) We living things add up to less than one-billionth of the mass of Earth. If all of life were evenly mixed with the mass of Earth in a giant blender, the concentration of living things would be less than 1 part in 1 billion—at the border of detection by the most sophisticated twentieth-century analytic techniques. The total biomass is tiny even in comparison with the mass of the atmosphere, which is estimated to be 5 million billion metric tons, so that if the biota and the atmosphere were evenly mixed, life would add up to a few millionths of a percent of the mixture. Talk about being made to feel small and unimportant!

But it is a mistake to attribute importance simply to the amount of material.[13] And this is the mistake that Lyell made. He failed to recognize that a small amount of life could have chemical reactions at a rate high enough to affect the biosphere. His mistake is analogous to saying that the weight of the gasoline in your car's gas tank—about 6 pounds per gallon, so about 120 pounds in a 20-gallon tank—is a small fraction of the total weight of the car, which could weigh 4,000 pounds, and therefore could not play any meaningful role in propelling the car.

That animals and plants have had some impact on the makeup of the atmosphere, despite representing a small fraction of Earth's materials, was recognized first in the eighteenth century by the English scientist Joseph Priestly (1733–1804). Famous as one of the first modern scientific chemists, Priestly was also a Unitarian minister and a friend of Benjamin Franklin. One of his contributions was the book *The History and Present State of Electricity, with Original Experiments*, probably stimulated by his friendship with Franklin.

Most important for us, in 1772 Priestly presented a paper to the British Royal Society titled "On Different Kinds of Air." He discovered that green plants grow better in a bell jar in which a mouse has died than in one with ordinary air, while mice, in turn, receive something of value to them from plants, transferred through the atmosphere.[14] He had found that during photosynthesis plants give off what we now call oxygen, which is required for animal respiration (and plant respiration as well), and that mice give off what we now call carbon dioxide, released by respiration and required for photosynthesis. Priestly's experiments provided an early clue that *although life represents a small mass compared with Earth, life can have an important effect on processes that occur in the biosphere.* But the time when the importance of these biological processes for the biosphere would be recognized lay in the future.

Ancient Nature

Bacteria are an ancient form of life, the oldest form of life known to us that still exists on Earth. Just how ancient became known in the 1950s. In 1954, paleobotanist and Harvard professor Elso Barghoorn (1915–1984) and his geologist colleague, University of Wisconsin professor Stanley Tyler (1906–1963), made a microscopic examination of rock called the Gunflint Chert taken from the Canadian Shield several hundred miles north of Isle Royale.[15] The Gunflint Chert was dated using radioisotopic techniques quite new at the time. These measurements indicated that the rock was 2 billion years old. Using tools that were a twentieth-century invention, Tyler and Barghoorn cut extremely thin sections of the chert, thin enough to be transparent to light and to be viewed under a microscope. These thin sections revealed fossils of prokaryotic organisms—bacteria or their ancient relatives—that appeared to be much the same as modern forms, perhaps exactly the same species as occur today.

Fossil records discovered since 1954 suggest that 2 billion years ago Earth was "teeming with prokaryotic microbial life."[16] Indeed, they seem to have been much more abundant and diverse then than they are today. The greatest diversity of prokaryotes occurred about 900 million years ago. As mentioned in Chapter 8, bacteria differ in fundamental ways from all other life-forms. They lack a cell nucleus, which is found in all forms that evolved later (except viruses). Although the prokaryotes have certain primitive features, they are chemically sophisticated and can carry out chemical reactions that none of the structurally more sophisticated life-forms, called eukaryotes, can.

Curiously, the 2-billion-year-old microbes found in the Gunflint Chert appear to have lived in an atmosphere more like that in the muds beneath the still waters of a pond at Isle Royale, or in the stomach of a moose, than like that of present-day Earth. Several kinds of evidence suggest this is true. On the ancient Earth, certain species that are ancestors of modern bacteria were the only photosynthetic organisms. Like all other photosynthesizers, they took carbon dioxide out of the air and gave off oxygen.

The Gunflint microbes, in the 1950s the most ancient life discovered, turned out to be comparatively recent arrivals. The most ancient fossils known today, and therefore the earliest known forms of life on Earth, were discovered in the 1980s by Professor Stanley Awramik of the University of California, Santa Barbara.[17] Dated as having formed about 3.5 billion years ago, they appear to be made up of single-celled organisms that grew joined together in long filaments and formed mats. They look very similar to modern stromatolites, mats of blue-green photosynthetic bacteria that can be found today in quiet marine bays, such as Shark Bay, Australia (Fig. 9.2).[18] The fossil mats have been distorted by time's long travail, which subjected them to repeated heating and pressure, bending and twisting them to create odd shapes that sparked the curiosity of paleontologists, who described them as

A

B

FIGURE 9.2 (A) Ancient and (B) modern stromatolites. (Courtesy of Stanley M. Awramik.)

looking like fossil cauliflowers. Within the matrix of the fossilized filaments of the bacteria are pebbles with a peculiar feature: Their surfaces are made of chemically reduced minerals—minerals that are not oxidized and could not have developed in an atmosphere with oxygen.

A pebble is a product of erosion, a bit of rock broken off from a larger surface, perhaps by the action of freezing and thawing water, and then smoothed by wind or water, exposed on all sides to the atmosphere. The outer surfaces of the ancient pebbles within the stromatolites would have been oxidized if the air in which they formed had been like the present-day atmosphere. The pebbles provide evidence that Earth's atmosphere 2 billion years ago and earlier lacked oxygen and, at least in this important characteristic, was very different from the modern atmosphere.

Global Nature

How could it be that nature, long believed to be perfect and unchanging, once had an atmosphere so "unnatural" to us? When some of the photosynthetic bacteria were buried but did not completely decompose, a slight imbalance resulted in the uptake and release of oxygen, which slowly built up in the atmosphere. In this way, early life, in the form of these lowly, seemingly insignificant mats, began to change the entire planet. Free oxygen in Earth's atmosphere is the result of more than 3 billion years of photosynthesis and is therefore a product of life. Like the atmosphere in the rumen of a modern moose's stomach, Earth's atmosphere has been fundamentally altered by life.

By the end of the twentieth century, it had become clear that life had altered not only the air we breathe but also the rocks we stand on. Perhaps nothing seems less lifelike than a piece of steel or iron, a steel girder or a chunk of iron ore, but the origin of iron ores is intimately connected to life and its history on Earth. The rocks that form the major economically important deposits of iron were laid down between 2.2 and 1.8 billion years ago, deposited because photosynthetic bacteria released oxygen into the oceans.[19] A vast amount of unoxidized iron was dissolved in those ancient oceans.

Unoxidized iron is much more soluble in water than oxidized iron. Oxygen released into the ocean as a waste product by photosynthetic bacteria combined with the dissolved iron, changing it from a more soluble to a less soluble form. No longer dissolved in the water, the iron settled to the bottom of the oceans and became part of deposits that were slowly turned into rock. Over millions of years, these deposits formed thick bands of iron ore that are mined today from Minnesota to Australia. After most of the dissolved iron had been removed from the oceans, the oxygen produced by photosynthesis began to enter the atmosphere, leading to another major change in the biosphere, converting it from a reducing to an oxidizing environment.

From the chemistry of rocks such as stromatolites and iron-ore beds, geologists have been able to reconstruct the history of Earth's atmosphere. The early atmosphere, before the emergence of life, was composed primarily of hydrogen, methane, and ammonia. Approximately 3.5 billion years ago, about the age of the earliest fossils, there was a shift to an atmosphere with free nitrogen and carbon dioxide, which was followed by an increase in the concentration of oxygen and a decrease in that of carbon dioxide. If Earth had remained lifeless, the concentration of carbon dioxide and nitrogen in the atmosphere would have remained high, as would the concentrations of hydrogen, methane, and ammonia. Although oxygen would have increased as a result of the activity of sunlight, it would not have increased to the concentration found today.[20] Carbon dioxide would have been ten times more abundant; free nitrogen, ten times less.[21]

Early life on Earth altered not only the chemistry of Earth's atmosphere but also Earth's heat budget and surface temperature. The latter is a result of energy exchange and physical characteristics of the surface. Energy is received from the sun, and a very small amount is generated from Earth's core (by radioactive processes that heat the center of the Earth). Energy is lost to space by radiation from Earth's surface. The hotter an object, the more rapidly it radiates energy and the shorter the wavelength of the predominant radiation. A blue flame is hot; a red flame is cooler. Our sun, with a surface temperature close to 10,000°F (5,505°C), emits most of its energy as visible and near-infrared light ("near" meaning the shorter infrared wavelengths, nearer in length to those of visible light). Earth's surface and the surfaces of plants and animals, including ourselves, are so cool that energy is radiated predominantly in the mid- to far infrared, which is invisible to us.[22]

If an object is cold and gives off less heat than it receives, it will warm up. But as it warms, it also radiates heat more rapidly. As a result, for any constant input of energy, a physical object will eventually reach a temperature that will allow it to radiate energy at the same rate that it receives energy.

The rate at which Earth's surface radiates heat depends on its average "color" and temperature. A perfect emitter of heat is called an "ideal black body," and black surfaces radiate heat much more readily than white surfaces. A white planet would therefore radiate heat very differently than would a black planet. Changes in the amount of ice, the amount of cloud cover, and the distribution of algae and vegetation over the Earth alter the reflecting and emitting characteristics of the planet's surface. Their absorption of light changes with the seasons. Marine algal mats can change from light and highly reflective in one season to almost black and highly emitting in another. Some algae produce sediments such as calcium carbonate, which, when pure, is chalky white and has a different reflective characteristic than the sediments produced from a lifeless surface. Even a single-cell layer of algae spread over a large area of water could greatly alter the rate of energy exchange and therefore the temperature of Earth's surface. The same is true of vegetation on the land. Ice reflects 80% to 95% of light; a dry grassland 30% to 40%; a coniferous forest 10% to 15%.[23]

In the last decades of the twentieth century, climatologists and planetary scientists estimated that a 1% change in the amount of sunlight reflected by Earth would cause approximately a 3°F change in the average temperature of Earth's surface.

All of this suggests that life has greatly altered its environment at a global level over the history of Earth, and this explains in part the "fitness of the environment" that Lawrence Henderson observed in the early years of the twentieth century. The environment appears "fit" for life because life has evolved to take advantage of the environment and, conversely, has greatly changed the environment at a global level. Living things clearly have a much more important effect on our planet than was believed before the last few decades of the twentieth century.

In sum, evidence of life's great influence on the Earth can be found not only in the fossil record but also in comparing Earth with other planets in the solar system.

The Planetary Theater and the Biospheric Play

By the late twentieth century, some geologists, along with some ecologists, were beginning to view life as an integral part of very long-term global geologic processes.[24] Peter Westbroek (1937–), professor of geology at the University of Leiden, in the Netherlands, summarized this new perspective in his landmark book *Life as a Geological Force: Dynamics of the Earth*, published in 1991, just after *Discordant Harmonies*.[25]

Westbroek begins by pointing out that the theory of plate tectonics, which became accepted by geologists in the 1960s, offered a new image of Earth's bedrock as a group of huge plates in very slow but constant motion and with very slow currents within the materials that make up the rocks. The idea that the continents must have moved was suggested in the sixteenth century by Abraham Ortelius (1527–1598), a Flemish cartographer who made the first modern atlas, *Theatrum Orbis Terrarum* (*Theater of the World*). He realized that the west coast of Africa and the east coast of South America looked like they fit together, and theorized that they had once been joined. Benjamin Franklin was another who believed in this possibility.

The first modern scientific proponents of moving continents were the American geologist Francis B. Taylor (1860–1938),[26] who proposed it in 1910, and the German geologist Alfred Wegener (1880–1930), whose 1915 book *The Origin of Continents and Oceans (Die Entstehung der Kontinente und Ozeane)* was the first influential discussion of the idea.[27] But this theory was vehemently opposed by most scientists, who said there was no mechanism that could move the massive landforms, and it was pretty much put to rest in 1943 by an article by the famous American zoologist and paleontologist George Gaylord Simpson (1902–1984).[28] So it was no small matter when geologic knowledge advanced to the point where the movement of continents could be explained. Note that the theory of moving geologic plates was dismissed as impossible when first proposed but is now widely accepted.

As a result of plate tectonics during Earth's history, the location and orientation of the continents have changed, resulting in major alterations in the distribution of life, providing opportunities for the evolution of life-forms and leading to the extinction of others. Where plates collide, one plunges under the other, bringing its material to Earth's deeper regions, where it is heated and subjected to intense pressures and becomes molten. Where plates separate beneath ocean waters, fresh molten material is released from the deeper Earth. Over extremely long periods, material is recycled, carried down from the biosphere by the continental plates and then returned to the biosphere at the sites of seafloor spreading. This new understanding is built not only on the discovery and acceptance of the idea of plate tectonics but also on modern chemical methods that have allowed chemists to better understand the origin and causes of the composition of Earth's rocks.

Not until the 1970s did geologists begin to focus on the global cycling of chemical elements, and it soon became accepted that, as Westbroek wrote, "each of the chemical elements is channeled along a characteristic maze of cyclic routes through this global system. . . . The geochemical cycles of different elements are intertwined to form a network of interconnected routes." It was a perception of a dynamic Earth, deep and superficial at the same time, a geologic dance.

Following from these developments, the stage was set for a reconsideration of life's role in this immense play, and by the end of the 1980s Westbroek was able to conclude that "life is a geochemical process, a self-perpetuating organization of geochemical cycles that emerged from the abiotic fluxes in the early Earth. . . . In geochemical terms, life is a localized pattern of circuitous detours and transformations, maintained in this circulatory system. In the biosphere the fluxes of materials and energy are vastly accelerated; they are knitted together into transient webs of extreme complexity."[29]

In ways such as these, Westbroek concluded, life and the rock cycle were intimately connected to each other. The rock cycle has been greatly changed by life, as the origin of iron ore, discussed earlier, illustrates. In addition, Westbroek writes, biological evolution has "favored the emergence of powerful biological 'mining' mechanisms" by which living systems obtain the chemicals they need from rocks, water, and air. For example, organic acids produced in forest soils leach chemical elements from rocks. The process of biological evolution has also led to biochemical mechanisms that remove toxins from the biosphere. Mercury, which is toxic to all living things, occurs in some soils. Bacteria take up mercury from the soil and produce methyl mercury, a gas that is readily lost from the soils. As a result of these processes, "supplies of raw materials," made available by geologic processes, are "efficiently exploited" by life, while "toxic substances are removed."[30]

The geologic evidence as described by Westbroek suggests that life was indeed a global force, an actor in the biospheric play.

The image that emerges is of the biosphere made up of four dynamic parts—rocks, oceans, air, and life—each with its own characteristic ranges of movement and rates of change: rocks changing in composition most slowly, the oceans much more rapidly, and the atmosphere more rapidly still. Each part affects the others, and the inherent differences in their movement and rates of change can create complex patterns in time and space. Modulating all these changes is life. Many chemical reactions happen more quickly when living organisms are involved, and new pathways are established for the transfer of chemical compounds from one part of the biosphere to another.

Acts and Scenes in the Biosphere's Biography

Since the origin of life, a number of major events in biological evolution, in addition to the "invention" of photosynthesis and the resulting release of oxygen and removal of carbon dioxide, have changed the rest of the biosphere—the atmosphere, oceans, soils, and rocks. These are one-way events: Once they have taken place, the biosphere cannot move backward from them to previous conditions as long as life is present. In this way, the biosphere has a history, a one-way directional change over time. Among these one-way events were the evolution of oxygen-breathing organisms; of cells with true nuclei (eukaryotic cells, the cells of animals, fungi, plants, and some other life-forms); of calcium-containing skeletons—us vertebrates—and shells; colonization of the land by plants and then animals; and the evolution of flowering plants and of *Homo sapiens*.

Here's one such "invention" that I have found especially intriguing. Several hundred million years ago there was a lot of silicon floating around in the oceans and on sandy shores (most sand is silicon dioxide), but living things were not making use of it. Since silicon can be the basis for many kinds of useful structures, as we have learned in modern times, we can think of this silicon as a resource without a user. Then tiny single-cell diatoms evolved, a kind of brown algae that made a hard shell out of silicon, exploiting a new resource. In geologic time scales they quickly became successful and many species evolved. Today they are among the most important and abundant algae.[31]

Each of the major biological inventions has three stages. First, there is a kind of biological breakthrough in evolution; new opportunities are opened up by the evolution of a group of species, which evolve rapidly, "taking advantage," so to speak, of the new opportunities. Next, the new forms of life cause some kind of change in the biosphere. Finally, other life-forms evolve to "take advantage" of the altered environmental conditions; evolution takes place within the new environmental conditions of the biosphere, which provide a new set of problems and opportunities.[32]

The evolution of photosynthetic organisms is an important example of this process. Organisms that could carry out photosynthesis were not the first to evolve on Earth. They were preceded by forms that could obtain energy from compounds in the primitive oxygenless atmosphere and waters, using organic compounds that could have built up by nonbiological processes. (Research at the end of the twentieth century and continuing today shows that some organic compounds are surprisingly widespread throughout the universe, produced outside of and without living organisms.[33]) Free oxygen in the cells of photosynthetic organisms, and in the atmosphere outside, was highly toxic to them. They simply dumped this toxic waste outside their cells to get rid of it. Over billions of years, the early photosynthetic organisms poisoned their world for themselves, forcing their modern descendants to live in obscure corners of the modern Earth, in the stomachs of ruminants, the muds beneath still pond waters, the root nodules of alders and other legumes, the guts of termites.

However, oxygen is a high-energy element; when it combines with unoxidized compounds, a lot of stored energy is released. Organisms that could use oxygen to "burn" their biological fuel could use energy rapidly. This ability has had many advantages. An oxygen-using individual can move faster for a longer time, and its internal structure and biochemistry can be more complex. The kinds of life most familiar to us, including animals and flowering plants, require a high rate of energy use and could not have evolved in an oxygenless world. But free oxygen produced by life presented both a problem and an opportunity: It was lethal to many existing forms of life, but led to the evolution of species that could make use of this energy-releasing quality. Given the presence of oxygen-using organisms, a whole new evolutionary "game" was possible.

From an ecologist's point of view, looking at Earth as a large-scale habitat, living things developed something new, which created a new toxic waste that they simply dumped into the environment, which polluted the world but created new opportunities, which the next set of living things evolved in the presence of and made good use of. In this way, there was no simple going back, running the machinery of global life in reverse, like turning the hands of an old-fashioned mechanical clock backward.

To summarize where we have come so far: Biological evolution has led to global changes in the environment, which in turn have led to new opportunities for biological evolution. In this way, a long-term process of change has occurred throughout the history of life on Earth, which is an unfolding, one-way story. A machine is not a good metaphor for this system. You can stop a steam engine and start it again later. You can move the wheels and levers and gears backward to some point and then restart the engine. But you cannot turn the biosphere back from one of its major evolutionary steps to a previous one. Instead, the new emerging history of the biosphere is reminiscent, metaphorically, of the organic idea of Earth described in Chapter 6. Like an idealized organism, the biosphere has had an origin and has passed through major stages. Like an idealized organism, the biosphere has had a history, and what it will be tomorrow depends not only on what it is today but also on what it was yesterday. Like an organism, the biosphere proceeds through its existence in a one-way direction, passing from stage to stage, none of which can be revisited.

In the nineteenth century, Lyell considered the possibility of a global balance of nature in terms of uptake and loss in the cycling of chemical elements through living things. He thought a balance was possible in theory and could in theory be affected by life, so that the supply of necessary "hydrogen, carbon, oxygen, azote [nitrogen] and other elements" might be obtained from the "putrescence of organic substances" and the release of the elements to the atmosphere. This would imply that "vegetable mould would, after a series of years, neither gain nor lose a single particle by the action of organic beings." Although Lyell concluded that this was "not far from the truth," he believed that most of the organic matter that washed down from the land to the sea became "embedded in subaqueous deposits" and would "remain throughout whole geological epochs before they again become subservient to the purpose of life."[34] The persistence of life seemed to Lyell to be made possible only by a continual release of chemicals essential to life from Earth's interior, not by the recycling of the elements by the biota. He believed that life depends on Earth, that life itself cannot create a balance or constancy in the supply of the elements on which its survival depends.

Three Schools of Thought About a Balance of Nature

Broad acceptance of the possibility that life, and especially our species, affects the global environment has spawned a return to the belief that nature could be in balance. Today, there are three schools of thought about a balance of nature at the global level. One is that the biosphere is in a steady state, exactly as was assumed to be true throughout most of the twentieth century for nature at a local level. The second is that life acts as Earth's thermostat, requiring and creating constant conditions, like a moose's physiological feedback mechanisms that maintain its body temperature near 102°F, even if not achieved precisely. The third school of thought is that the biosphere is always changing, and life is changing with it, beyond the ability of life to act as Earth's thermostat, and it is this very quality of dealing with, adapting to, change at the planetary level that has allowed life to persist.

Let's examine each of these views more closely.

FIRST SCHOOL OF THOUGHT: OUR WOBBLY EARTH IN A STEADY STATE

I have written extensively so far that Earth's climate is always changing, but I haven't focused directly on what causes these changes. Although the climate is always changing, it isn't changing entirely at random. For starters, there are three long-term patterns of climate change: cycles of 20,000 years, 40,000 years, and 100,000 years.

Our spinning Earth is like a wobbling top following an elliptical orbit around the sun. As it does so, three kinds of changes occur. This spinning top makes a complete turn every day, of course, but because it wobbles, it is unable to keep its poles at a constant angle in relation to the sun. In our time, the North Pole points to Polaris, the North Star, but this changes as the planet wobbles. The wobble (think of where the North Pole points) makes a complete cycle in 26,000 years. As a result, 13,000 years ago—before the first human civilizations, but when people were wanderers who likely could have navigated in some sense by the stars—Earth's North Pole pointed as far away from Polaris as it can go, 47°F. Then the star Vega, in what we call the constellation Lyra, was Earth's north star. And Vega will again be the north star by 14,000 CE.[35] Not only does Earth wobble on its spinning axis, but the tilt of wobble also varies over a period of 41,000 years, like a top on its way to eventually falling over.

In addition, our planet circles the sun in an elliptical orbit, not a perfect circle but a stretched one, and the shape of this ellipse varies. Sometimes it is a more extreme ellipse, with Earth's farthest distance from the sun much greater than its shortest distance from the sun. At other times it is closer to a circle, so its distance from the sun doesn't change very much over a year. This change in the shape of the orbit occurs over 100,000 years.

In the 1920s, Milutin Milankovitch (1879–1958), a Serbian astronomer, realized that climate change might have to do with the way Earth revolved on its axis and rotated around the sun. Milankovitch knew about the three climate cycles and the variations in Earth's spinning, its tilt, the tilt's wobble, and the variation in the shape of its path around the sun, and he made calculations showing that these correlated well with what was known at that time about the climate cycles. If this is true—and it is—then there is no way that Earth's climate could be constant, and therefore no way that there could be a climatic steady state.

The big surprise to me since 1990 is the reemergence of scientific research and publications and public and political policies that assume and require that the biosphere is in balance, that a global balance of nature does and must exist, including changes in climate. People have seen the picture but have not heard the music, nor recognized the dance.

What I wrote in the preface to *Discordant Harmonies* is worth repeating here: Peeling back this surface layer of our widely held ideas and broadly accepted observations reveals another, perhaps closer to the level at which our minds function. It is a level of two fundamental worldviews that some classical philosophers have characterized metaphorically as that of Apollo, god of light and knowledge, and that of Dionysus, god of music and dance, the classical opposition of the analytic and the artistic—in modern parlance, left brain and right brain.

The Apollonian view, which includes the rational, characterizes modern science as it is supposed to be and is at its best. Metaphorically, the Apollonian perception dominates science and technology. However, it does force us to focus on solvable problems, which have often been viewed as requiring us to freeze nature conceptually so it can be analyzed. At the same time, our thoughts are mixed with the Dionysian, as I have tried to show, in the way that we have thought about, made laws and policies about, and attempted to manage, conserve, and preserve animals and plants. We approach these partially rationally but also we are still involved with the intuitive and emotive, the Dionysian, which deals with nature through motion, sound, and action.

The major scientific theories that focus on global warming, and on the fundamental dynamics of Earth's great life-supporting and life-containing system, are predicated on a balance of nature. The huge computer models of climate, known as general circulation models, are steady-state models. They use deterministic mathematical equations to calculate the transfer of energy and matter in the atmosphere; and given a fixed input series (like the standard non-greenhouse-gas scenario used for comparison, which uses the weather from 1960 to 1980 or from 1960 to 1990 as a repeated and therefore constant time series), they give a fixed output. For any change in input, they give a new steady-state forecast. It always leads to a balance of nature.

More important is the conclusion that not only is Earth's life-support system out of balance but also that this is bad. In 2008, space scientist James Hansen wrote in a scientific paper with several colleagues that "realization that today's climate is far out of equilibrium with current climate forcings [energy imbalances, to put the term simply] raises the specter of 'tipping points,' the concept that climate can reach a point where, without additional forcing, rapid changes proceed practically out of our control."[36] I will return to this in Chapter 12.

Most of the methods used to forecast possible biological effects of global warming add another layer of steady-state, balance-of-nature theory. For example, computer models that forecast what will happen to individual species and groups of species—ecological communities and ecosystems—start with known temperature conditions where a species or group of species is known to exist today. They then take the results from a climate model that forecasts where those temperatures will be in the future, and the species or group of species is mapped onto those new locations. The underlying assumption is that the present relationship between temperature and the presence of a species is a steady-state relationship—the true, perfect, and necessary one—and therefore it is the one formulated. That a species may happen to be passing through a nonoptimal or unusable range or location at the time of observation is not considered. Yes, this relationship is assumed to be fixed forever—contrary to what is well known from Darwinian evolution and ecology, the species has no ability to adapt. Everything is in a fixed relationship to everything else.

An assertion in 2007 by the Intergovernmental Panel on Climate Change that 20% to 30% of species may be threatened with extinction because of global warming—and just a few-degree change in average atmospheric temperature at that—arises from these layers on layers of balance-of-nature assumptions.[37]

We need to mull over this question: If the balance of nature never existed and does not exist, then how can theories based on it provide a realistic forecast and a secure foundation for policy?

SECOND SCHOOL OF THOUGHT: LIFE AS NATURE'S THERMOSTAT

The second school of thought is that life acts as Earth's thermostat, requiring and creating constant conditions, like physiological feedback mechanisms in a moose that maintain its body temperature near 102°F. This is consistent with a machine metaphor, but can be extended to have aspects that are organic. The belief in nature's thermostat is powerful and influential today.

The first section of this book provided evidence that constancy and a steady state do not exist for nature at all levels smaller than the biosphere, and that we have to abandon the machine metaphor, abandon the idea that the entire range of ecological phenomena, from individual populations to entire forests, are like watches, car engines, or steam engines.

With what can we replace that metaphor? For a start, we can listen to Lewis Thomas (1913–1993), a famous physician and immunologist, called by his colleagues "the father of modern immunology and experimental pathology" at a symposium held in his honor in 1982.[38] In his book *The Lives of a Cell: Notes of a Biology Watcher*, which was considered a classic of nature writing, Thomas asked, "What is [the Earth] *most* like?" and answered, "It is *most* like a single cell."[39] In the book's final chapter, "The World's Biggest Membrane," he responded to the then-new photographs of Earth from space, those great iconic symbols of the space age, by writing that our blue planet is "the only exuberant thing in this part of the cosmos." He compared the atmosphere to the membrane of cell. For a cell to stay alive, it has to "be able to hold out against equilibrium, maintain imbalance, bank against entropy." The linkage to the planet comes when he writes, "and you can only transact this business with membranes in our kind of world." He brought back, probably without knowing its history, the ancient organic metaphor for nature.

How far can we push a new organic metaphor as Thomas suggested? How much is the biosphere like a single living cell, or, for that matter, like the internal organs of a large mammal? James Lovelock is a modern proponent of the second school of thought. You will recall that Lovelock proposed the "Gaia hypothesis," using the term as shorthand for a hypothesis that "the biosphere is a self-regulating entity with the capacity to keep our planet healthy by controlling the chemical and physical environment." More specifically, he lists several of "Gaia's principal characteristics," which include "vital organs at the core" and a "tendency to keep constant conditions for all terrestrial life."[40]

Lovelock's "Daisyworld" computer model has become an archetypal explanation of Earth's thermostat. Lovelock and Andrew Watson wrote a small computer program that had just two varieties of plants on Earth, white daisies and black daisies. They were identical except for their color and their absorption, radiation, and reflection of light. As a physical object warms, it radiates more light waves. Physicists will tell you that a perfectly black body absorbs 100% of the electromagnetic waves (which include light and heat waves) it receives, and is also the perfect radiator of these waves. A perfectly white body that reflects 100% of the electromagnetic waves that fall on it is also the worst radiator of energy. When it was cold, the black daisies did better because they absorbed more sunlight and kept warmer than the white daisies, reproduced abundantly, and took over most of the surface. But when the computer model warmed the Earth past a hypothetical ideal "normal," the white daisies, reflecting most of the sunlight and therefore better avoiding overheating, did better. As the white daisies took over, the planet's surface cooled. The result was a highly stable system that was both a classic balance of nature, directly analogous to the logistic growth equation applied to populations since the nineteenth century, and an organic regulator, thus a throwback to both the machine and the organic nature metaphor.[41] This model is useful as a way of illustrating how a classically stable system functions, but it is not helpful if we think of it as an explanation of the stability of Earth's nature, because there is so much more going on simultaneously, so many kinds of time-lags and other features that complicate how nature responds. (I will return to this idea in Chapter 11.)

Thomas's and Lovelock's ideas are examples of one of the oddities, ironies, and contradictions of modern environmental science and its applications. As it was in the twentieth century, so it is today: We record the message, but we ignore the meaning. The glass has hardened for us. We still see nature as the ancients believed it to be. But unlike those in the seventeenth century, we don't think of Earth as having a liver, heart, arteries, and veins. Ironically, herein lies the greatest danger: that we retreat to the completely unscientific and use what is called scientific knowledge to defend that retreat, covering our scientific tracks, so to speak, as we hurry backward while turned forward.

THIRD SCHOOL OF THOUGHT: THE GREAT DANCE OF THE GLOBAL SYSTEM

> When something is suggested, or some evidence is produced, the first response [of scientific colleagues] is "It can't possibly be true." And then, after a bit, the next response is "Well, if it's true, it's not very important." And then the third response is "Well, we've known it all along."
>
> —Dr. Jonas Salk[42]

Followers of the third school of thought argue that the biosphere is always changing and that it is this very quality at the planetary level that has allowed life to persist. This school would take Charles Elton's classic 1930 comment about wildlife, "The balance of nature does not exist, and perhaps has never existed," and extend it to the biosphere. It would take Elton's explanation—"The numbers of wild animals are constantly varying to a greater or lesser extent, and the variations are usually irregular in period and always irregular in amplitude. Each variation in the numbers of one species causes direct and indirect repercussions on the numbers of the others, and since many of the latter are themselves independently varying in numbers, "the resultant confusion is remarkable."—and extend it to Westbroek's conclusions about chemical cycles, and say that these too vary, that one affects the other, perhaps even extending to Elton's final observation that "the resultant confusion is remarkable."[43]

This is the idea that, I argue, we are forced to accept if we look at the biosphere completely rationally, and it should have been our guide, our set of rules, from 1990 onward, but it has not. We will look at how this can come to the fore in the final section of this book.

One reason that this third school of thought has been largely avoided is that it seems to bring up the fear of complete chaos, which I discussed in Chapter 8—the disturbing possibility that life and its environment are in total chaos as defined by the ancient Greek philosophers: without cause and effect, without rules or structure. As I discussed in Chapter 8, some, in fact, believe that any system can be only one way or the other, completely stable or completely chaotic—a point we will return to shortly.

What Does Time Matter to a Moose? Some Thoughts About Nature's Hierarchies in Space and Time

Let's return to our moose quietly grazing in a shallow harbor at Isle Royale, and let us consider some deeper philosophical questions. That complex discordant harmony, the ways that life has affected the environment, and been affected by that environment, over several billion years, but with changes happening at many different time intervals, raises a curious question, perhaps most easily stated as "What does time matter to a moose?" It's like the old New Hampshire joke about what the university agricultural extension agent, full of new scientific knowledge, said to the chicken farmer.

"You know, if you kept your chickens penned up, rather than let them run around all this acreage, it would take a lot less time for them to get to marketable size."

"Oh, hell," replied the farmer. "What's time worth to a chicken?"

We can rephrase this conversation in terms of the moose at Isle Royale and what environmental changes affect it. We can ask the same question about the entire moose population on the island. Thinking from a moose's point of view, an animal that lives about 18 years, what matters directly is what is happening in its environment during that here and now, when it is living. What created that environment, an integration of many kinds of events, each with its characteristic rate of change in time and over space, would not be in the moose's senses or consciousness, only the resulting environment. This leads to two ideas. One is a hierarchy of interacting causes that I just referred to as a complex discordant harmony. Scientists have noted that in some cases—some argue in most cases—the longer the time interval of change, the larger the area involved. The second idea is that if you stand back far enough, everything becomes an unchanging point.

Consider Isle Royale's geological history, which goes back about 1.1 billion years ago when alternating layers of sedimentary rocks, mainly conglomerates and sandstones, and volcanic lava flows were laid down. The sedimentary rocks are of a kind that usually form along a sea or ocean coastline. These layers were eroded and the process was repeated a number of times. These were laid down horizontally, of course, but with plate tectonic movements over the ages the layers were broken (geologically speaking, they were faulted) and tilted in a southeast-to-northwest line. Volcanic lava, the hardest rock, most resistant to erosion, formed the ridges, the island's hilltops, while sandstones, the most easily eroded, formed valleys, and the conglomerates, with a hardness in between that of lava and sandstone, formed hillsides or higher valleys. The formation of the sediments, their conversion into rocks, their later erosion, faulting, and tilting took place over very long time span, many millions of years.

A skeptic about the imbalance of nature could point out that if you viewed Earth from the moon or Mars or some other body far off in the solar system, Isle Royale would appear as a tiny spot on a small planet, and the same could be said about the island's changes over time—if you take the perspective of a billion years, what happened over a century or a thousand years or ten thousand years would disappear, and one could say that, from long time-and-spatially distant view, the island was a constant thing.

This could be extended as well to climate. The climate in the second quarter of the twentieth century (1931–1956), as recorded at Grand Marais, Minnesota, the weather station nearest to the island, reached the July highest of 94°F and the January highest of 48°F. The lowest July temperature was 28°F; the lowest January temperature was minus 34°F. But 10,000 years ago and more, the island was much colder, covered repeatedly by continental glaciers, which both added their own kind of erosion and left behind deposits of soils and rocks that formed lakes and hills. However, viewing the climate from far enough away and over a long enough time, we could calculate an average value for the entire existence of the island, and claim that this was therefore the "normal" and that the island therefore would seem to have had a constant climate.

These observations led in the last quarter of the twentieth century to a series of inquiries about a hierarchy of space and time and how such a hierarchy could affect nature. Some general rules were formulated, which became the foundation for "hierarchical theory," and became known as hierarchical structure: Events at very small spatial scales tended to occur rapidly; events at very large spatial scales tended to occur very slowly.

WHAT IS THIS DISCORDANT HARMONY OF NATURE?
MORE LIKE BACH OR CHARLES IVES?

Taking hierarchical theory into account, one path to answering the question "what is the character of nature?" is to understand what is discordant about its discordant harmonies. The paths we have taken so far show that life and its environment have changed each other over many scales of time and space, as short as the few seconds a moose takes to crop the top off a mountain ash sapling so that it cannot grow straight, and as long as the many millions of years that photosynthetic bacteria took to oxygenate Earth's atmosphere. If these variations over many scales of time could be converted to sound vibrations, a complex harmony and disharmony would resound, perhaps more like a composition by Charles Ives or improvised modern jazz by Dave Brubeck (1920–), or the current jazz style of Wynton Marsalis (1961–), who mixes jazz with the blues and some qualities of classic music, or the intricate rhythms and harmonies of some West African folk songs, like one of my favorites, "Sambo Caesar."[44] So one answer to the question "What is the character of nature?" is that nature like a discordant kind of music—modern jazz, blues, folk music and classical, all mixed together. Perhaps we can rephrase our question about our role in nature as "Can we hear, and more than hear, can we enjoy, that music of our planetary sphere?"

To Summarize . . .

Life and the environment affect each other at a global level. Together they form a planetary-scale system, the biosphere, that sustains life. The idea of the biosphere, and the growth in understanding it, can be traced back to several books published in the first few decades of the twentieth century, including Henderson's *Fitness of the Environment*, and books published in the 1940s and 1950s by a very small number of scientists, including G. E. Hutchinson in the United States. However, it wasn't until the 1970s and well into the 1980s that this idea gained momentum, helped by the popularization of photographs of Earth taken from space by the Apollo astronauts.[45]

Just as the images of nearby planets, appearing as objects something like Earth, with surface features and moons, affected ideas about the universe in the eighteenth and nineteenth centuries, so in the late 1960s photographs from space showing Earth as a cloudy blue marble floating in inky blackness became profound images for our time. The pictures of our living planet floating alone, a unique cosmic island, evoked in the public consciousness strong feelings about the fragility of all life on our planet, in a way that could not have been possible in times past—not to Cicero, with his Earth-centered, divinely ordered cosmos; not to Galileo and Newton, with their exact calculations of the continual sweep of planets through the solar system; not to Wordsworth, with his challenges to a powerful nature.

Perhaps more than any other single image or any single event, these photographs of Earth have changed our consciousness about the character of life, the factors that sustain it, and our role in the biosphere and our power over life. Those images from space have radically altered our myths about nature. The power of this image is demonstrated by its repeated use in recent years in many contexts, to the point of being almost a cliché—if it weren't such a powerful picture.

But behind this image, what are the characteristics of the biosphere? It is something like the stomach of a moose.

The idea that emerged late in the twentieth century was that life affected the environment globally and had done so for several billion years. Life changed the chemistry of the atmosphere, rocks, and soils, and changed the rate at which chemical elements were exchanged from one part of the biosphere to another. This was a revolutionary idea, whose implications are only beginning to be understood, held back in large part among scientists by the continued assertion that the biosphere, in the short run, is and must be in a steady state, in balance, unless affected by human action.

The blending of new metaphors, the extent to which the biosphere is like a computer or like a rose, and the influx of some of the older machine-age ideas into these newer ones represent a task for scientists to debate during the next decades, whether they recognize the underlying metaphors or not. That the issues have come to this stage is in itself a fundamental change in the perception of nature. The biosphere is very different from a machine. Earth is not alive, but the biosphere is a life-supporting and life-containing system with organic qualities, more like a moose than like a water-powered mill. The biosphere is not a mystical organismic entity contraposed to rationality, but a system open to scientific analysis and a new kind of understanding because of new knowledge and new metaphors.

In its dynamic qualities, its one-way history, and its complexity, a new Earth is revealed. Like a pond on a quiet afternoon at Isle Royale, the reality of nature is revealed not by what is seen in the stillness but by what is perceived within.

PART FOUR

Resolutions for Our Time

FIGURE 10.1 Morph the Moose replaces Smokey Bear as the symbol of the new learning and knowledge way about nature. (Artwork by Jeff Thomas. Trademark and copyright 2002 Daniel B.Botkin)

10

Fire in the Forest
MANAGING LIVING RESOURCES

> People are here (and in many other places) in regard to wood, bent only
> upon their own present advantage, utterly regardless of posterity . . . [they
> take] little account of Natural History, that science being here (as in other
> parts of the world) looked upon as a mere trifle, and the pastime of fools.
>
> —PETER KALM (1750)[1]

By now it should be clear that the old ideas of steady-state populations, ecosystems, and the biosphere are not correct for nature. You may wonder whether that
leaves us with any concepts to replace the outdated ideas of carrying capacity,
maximum and optimum sustainable yields, and optimum sustainable populations. The good news is that there are new ways to think about nature, and the replacements can be not only just as formal mathematically but also even more
useful in establishing policies and regulations and in conserving and managing
nature and our natural resources. (The more formal aspects of this are discussed in
the postscript at the end of the book.) More important to the nonscientist, these
new ways of thinking about nature can help us understand our relationship with
nature, where we fit into it, what actions to take, no longer on the outside but as
part of that always-changing environment. By adopting the new view of nature
that I am proposing, we can come to terms with nature and our relationships to it
in a non-hysterical way, without the extreme polarization that characterizes the
viewpoints on so many environmental issues.

As I wrote in the first chapter, we interact with nature in two ways: emotionally and rationally. I mean "emotional" in the largest sense of all that is outside of
rationality—our folkways, our myths, our spiritual feelings that arise from deep
within us, our religious sensitivities. Both the rational and the emotional are
important. We get ourselves into trouble when we confuse the two, letting emotions determine what we tell ourselves are rational decisions and actions, and believing that rationality can substitute for the emotional. In this and the next
chapter, we focus on getting our uses of the rational and emotional correct, or at
least better than we have in the past. Therein lies the path to successful action.

Monuments of Living Antiquity

Soon after they were discovered in the Sierra Nevada of California near Yosemite Valley in 1852, giant sequoia trees came to be regarded as natural monuments, curiosities of nature to be dismantled and then reassembled and displayed in museums as a kind of natural sculpture.[2] As historian Al Runte has written in his landmark book, *National Parks: The American Experience*, the giant trees, along with Yosemite Valley itself and the geysers of Yellowstone, were viewed in the mid-nineteenth century as America's answer to the paintings, sculpture, and other trappings of European culture, an American contribution to the world, and as symbols of antiquity connecting the past and the present.

No "fragment of human work, broken pillar or sand-worn image half lifted over pathetic desert—none of these link the past and to-day with anything like the power of these monuments of living antiquity," wrote the American explorer and surveyor Clarence King after viewing these trees in 1864. When the famous journalist Horace Greeley saw sequoias in 1859, he wrote that they "were of very substantial size when David danced before the ark, when Solomon laid the foundation of the Temple, when Theseus ruled in Athens, when Aeneas fled from the burning wreck of vanquished Troy."[3] Greeley, impressed, linked the sequoia trees not to historical facts but to Greek mythology and Judeo-Christian Bible stories, suggesting the kind of juxtaposition of rational and nonrational that would later get us into trouble when we tried to conserve and manage such wondrous natural resources.

With these perceptions, it's no wonder that when Yosemite, with its famous Mariposa Grove of sequoias, became a national park in 1890, the sequoia "monuments of living antiquity" were managed as though they, like sculpture, would persist indefinitely as long as they and their environment were undisturbed.[4] Until the second half of the twentieth century, Mariposa Grove was therefore managed to protect the forests from all disturbances.

It may have seemed reasonable at the end of the nineteenth century to assume that the survival of trees several thousand years old required undisturbed environments. But by the 1960s, after decades of protection, the sequoias were not regenerating in protected, undisturbed stands. The trees produced seeds, but the seeds didn't sprout. Naturalists noticed, however, that the seedlings did sprout where dirt roads had been cleared through the forest, allowing sunlight to reach the soil surface and scraping the soil clear of litter. A comparison of photographs taken between 1859 and 1932 showed that the undisturbed forests were no longer as open as they had been, but were becoming crowded with white fir, a species whose seeds can germinate, and whose seedlings survive, in the dense shade of the sequoia groves. In 1964 Richard Hartesveldt, a scientist studying the impact of tourists on the sequoia groves, realized that giant sequoias might rely on fires to regenerate.[5] Although fire had been believed to be important for some kinds of comparatively short-lived vegetation, Hartesveldt's was an extraordinary discovery: that even the largest and one of the longest-lived of all organisms requires disturbance to persist.

In 1968 the National Park Service established a new policy that allowed controlled burns in national parks; and in 1970 the first controlled fire was lit in sequoia groves.[6] The fires had to be managed very carefully because a huge amount of fuel in the form of dead branches, twigs, bark, leaf litter, and so forth had built up within the forests during the decades of fire suppression. Managed improperly, or started by lightning or by accident, a fire might therefore become unnaturally intense and either spread too far too quickly or have undesirable effects within the forests. Recognition of the necessary role of fires in the sequoia forests continues. The Sequoia & Kings Canyon National Park Fire and Fuels Management Plan: 2011 Annual Update states that a "natural-like fire regime will play an integral role in preserving park landscapes."[7]

The century from the 1860s to the 1960s, during which sequoia reproduction declined in natural stands because of fire suppression, is a mere flick of a page on the calendar in the life of these trees, so artificial management of the sequoias for this length of time hardly makes a difference in the long run. Likewise, short- and medium-term climate change of the kind that has drawn so much attention in recent discussions of global warming, like the Medieval Warm Period and the Little Ice Age, have had little if any effect on these huge, long-lived trees.[8] This is not the way we have been threatening this species—we have threatened it by clear-cutting large stands and destroying its habitat, leaving the sequoia, which has highly specific habitat requirements, with only a small remnant of its former dominion. Hence the acceptance of controlled burning and of the necessity for disturbance even in sequoia forests is an example of a transformation in attitude that must occur so that we can conserve, save from extinction, and make the best use of natural resources.[9]

To Burn or Not to Burn: Managing for Change

Management of the sequoia is reminiscent of the problem of Kirtland's warbler discussed in Chapter 4. The change in management policy to allow some fires to burn or to set controlled fires to protect the warbler's habitat illustrated the beginning of a change in underlying assumptions about nature. The old ideas were simply not working, and controlled burning was one of the first examples of movement in a new direction. The move away from the Bambi and Smokey the Bear images of forest fires to a new view of fires as a natural and desirable part of the patterns in at least some kinds of forests, shrublands, and grasslands began, to the best of my knowledge, in earnest in the United States in the early 1940s. The change has come slowly, beginning with the realization that at least some forests, such as those in dry climates where fire was common as in the American Southwest. However, episodes of large fires in famous parks repeatedly set the process back, but controlled burning is becoming a widely accepted part of managing forests and shrublands. Still, managing forests that need to burn—managing with fire—is among the hardest things to do for our living resources, and success has been mixed.

THE FAMOUS YELLOWSTONE FIRE

The largest fire ever recorded in Yellowstone National Park since its designation as a national park in 1872 occurred in 1988, two years before the publication of *Discordant Harmonies*. It burned more than one-third of the park, a total of 1,240 square miles (3,213 km²), an area equal to 80% of Rhode Island and almost 90% of Long Island, NewYork. Even though by 1988 many ecologists had accepted the important role of fire in forests, the long suppression of fire in the park, part of the Smokey the Bear philosophy, had created a superabundance of fuel that needed only a very dry summer for a large fire to be likely. And that's what happened.

Looking backward, the National Park Service issued the following explanation in 2008:

> In the 1950s and 1960s, national parks and forests began to experiment with controlled burns, and by the 1970s Yellowstone and other parks had instituted a natural fire management plan to allow the process of lightning-caused fire to continue influencing wildland succession. . . . In the first sixteen years of Yellowstone's natural fire policy (1972–1987), 235 fires were allowed to burn 33,759 acres. Only 15 of those fires were larger than 100 acres, and all of the fires were extinguished naturally. Public response to the fires was good, and the program was considered a success. The summers of 1982–1987 were wetter than average, which may have contributed to the relatively low fire activity in those years.[10]

The 1988 fire was not only the largest but also was believed at the time to be much more damaging than an occasional fire that swept through forests frequently cleared of fuel by past fires. The fire began in late July. On August 20, the worst day, 120,000 acres burned.[11] As usual, politics entered into the discussion as another example of yesterday's answers to tomorrow's problems. On September 13, President Ronald Reagan said he had questions about the National Park Service's policy of allowing certain forest fires to burn unattended, and said his administration was going to reassess the policy.[12] He compared the destruction to the effects of the 1980 Mount St. Helens eruption. The Department of Defense sent more than 2,300 soldiers, 22 military aircraft, and 105 pieces of other mechanical equipment to Yellowstone.

The next day, September 14, then interior secretary Donald P. Hodel said that the National Park Service was going to reverse its policy of letting nature take its course—which had meant doing nothing before, during, or after such a fire. The changes would include some reforestation, such as along stream banks and roads, and feeding some elk and bison so they would remain in the park and not wander outside, where they might get into trouble. Hodel apparently viewed the choices as all-or-nothing: complete interference or complete do-nothing. This was consistent with the faulty idea that people either had total control and dominance over nature or should stay completely out of nature's way, the same dichotomy that plagued most attempts to conserve, manage, and harvest natural resources in the twentieth century.

According to the *New York Times* the next day, September 15, "George T. Frampton Jr., president of the Wilderness Society, a conservation group concerned with public lands, said that before making any decisions on what actions to take after the fires, a thorough study should be conducted by a National Academy of Sciences panel. Only then should decisions be made on fire-management policy and remedial action like revegetation."[13] In other words, put off any action by appointing a committee to discuss what might be done at some undefined time in the future. Meanwhile, let the fire burn.

A few days later, on September 18, nature writer Alston Chase managed to get some useful thinking into the *New York Times* in an op-ed piece titled "A Voice from Yellowstone; 'Neither Fire Suppression nor Natural Burn Is a Sound Scientific Option.'"[14] Chase, a careful and thoughtful thinker about people and nature, was the author of an excellent book, *Playing God in Yellowstone*, and had been president of the Yellowstone Association, a nonprofit organization that listed itself as promoting research and education in the park.

The key issue, Chase wrote, was the 1972 National Park Service policy requiring that fires posing no "threat to life, structures or outside resources . . . naturally occurring (i.e., lightning-caused fire) . . . be allowed to proceed and run their course." Reminiscent of the story of Hutcheson Memorial Forest that we visited in Chapter 4, Chase observed that "before Yellowstone was established in 1872, its northern range, according to studies, was swept by fire every 20 to 25 years. Some of these fires, scientists believe, were set by Indians—suggesting that a truly 'natural' regime would include fires set by humans."

But just as fire suppression was practiced in New Jersey after the Mettler family bought the land that became Hutcheson Memorial Forest, so too, "beginning in 1886"—once the park was established—"park authorities followed a regime of total fire suppression." As a result (as noted a few pages ago), a lot of dead wood accumulated in the forest, providing such a huge amount of fuel that when a fire finally got going, it might be much worse than the more frequent fires that occurred before the park was established. Instead, Chase argued, the park service should establish "aggressive programs of prescribed burning [that is, planned by people and carried out by people according to plan] before more land is engulfed by fire."

Despite initial fears, Yellowstone's 1988 fire didn't result in terrible, irreversible damage, and the effect on wildlife turned out to be relatively mild. Elk have been the most abundant large mammal in the park. According to the National Park Service, of an estimated elk population estimated at 12,000 to 18,000, 345 elk died. After the fire, the elk population remained within that same general range. It declined only after wolves were reintroduced into the park, after a long absence, in 1996.[15] Among the other wildlife, 36 deer, 12 moose, 6 black bears, and 9 bison died in the fire.[16] Today, the National Park Service estimates that approximately 15,000 to 22,000 elk winter in the park, consistent with populations before the 1988 fire.

The Park Service also estimates that 3,000 bison currently live in the park, and that at the end of 2011 there were an estimated 98 wolves.[17]

The Park Service reported that "in the several years following 1988, ample precipitation combined with the short-term effects of ash and nutrient influx to make for spectacular displays of wildflowers in burned areas." Fire-dependent lodgepole pines—one of the pine species whose cones open and release seeds only during a fire—burned and released huge quantities of seed, from "50,000 to 1 million per acre," and the restoration of the forests began quickly.

It Isn't Easy Being Green and Burning Forests Carefully

Ecologists had been calling for this kind of planned-burning policy since the 1940s. My major professor, Murray Buell, told a story about a plant ecologist who began arguing for controlled, prescribed fires in wildlands in the 1940s. Finally he was successful and got the chance to light a fire in a preserve in Arizona. But as luck would have it, that fire got away, and the plant ecologist disappeared, never to be heard from again—at least not by Murray.

Losing control of a "controlled" fire has always plagued controlled burning, either as a possibility, frightening neighboring landowners, or as a reality. In the 1980s, the then head of the California Department of Forestry told me why his department would no longer have anything to do with controlled burns. They had worked out an elaborate program with the U.S. Forest Service to start "controlled" fires that could spread both to state-owned forest and to national forests in Southern California (there are four large U.S. national forests between Santa Barbara and Long Beach). It wasn't easy getting the plans approved—private landowners, not surprisingly, didn't like the idea of a fire that might get away and burn down their homes. But finally a plan was approved and a fire was lit.

Once again, as luck would have it, the fire got away, and it did burn some private land. This led to court battles in which the courts decided that nobody could sue the federal government. Their reasoning was that the federal government is us, and we can't sue ourselves. However, for reasons unclear to us, the courts also decided that citizens *could* sue the state, even though it, too, was supposed to be a government for the people and by the people. So the state of California ended up paying all the damages, and vowed never to get into that situation again.

Careful Wally Covington

Wally Covington, professor of ecology at Arizona State University, Flagstaff, has taken a much more careful, detailed, and thoroughly thought-out approach to managing forests that used to burn often but had been prevented from burning since European settlement. Wally focused on ponderosa pine forests near Flagstaff, one of the kinds of forests that evolved with and are adapted to fire, and whose major plants require it. Given that fuel had built up in the ground for decades, setting fires and hoping for the best was out of the question. Instead, he undertook careful studies of what the presettlement forests were like: the number of large trees per acre, the openness of the ground, and the amount of dead organic material on the ground that could fuel a fire. Then he hired field crews to go into these woods and cut out many small trees whose abundance was not characteristic of the presettlement forest, and also to remove, by hand, dead leaves and organic soils that had built up since the last fire and were also not characteristic of ponderosa woodlands familiar to the Indians and nineteenth-century cowboys.

Only after a forest area was reset to its smaller fuel load were fires lit, and they burned the way fires burned before European settlement—swiftly, lightly, keeping the ponderosa woodlands open, beautiful, and, with sufficiently frequent fires, capable of looking that way in the future. Note that this was not planning for a forest that was always the same, not for a perfect balance of nature, but for forests that changed within a range of conditions, that people liked and considered natural.[18]

We can take away several things from stories of the Yellowstone fire of 1988, the California controlled burns, and the careful resetting of ponderosa woodlands in Arizona. First, nature as we know it and like it requires change. Second, if you try to suppress change, you come out much the worse. Third, if you introduce change after people have worked so long to stop it that the landscape has been altered, plan it carefully, don't act blindly. Fourth, correcting past mistakes isn't easy or cheap. It seemed easy in the twentieth century to say "Just leave nature alone," meaning don't do a thing. But that approach created situations that were far from easy to correct. In a huge park like Yellowstone, even a fire that covers a very large area might not be very destructive. But in smaller areas with complex land ownerships crowded together, watch out. Our relationship with nature, although part of our existence, our heritage, our culture, our past, and our present, isn't something that can be continued, nor cared for, thoughtlessly. We tried to conserve nature the easy way in the past century, and it just didn't work and won't work. We can't just ignore our connections with nature because we think we are separated from it by our cities and suburbs, our cars and roads, our computers and cell phones. Nature is still right there, and it's going to burn a hole in your shoe if you don't take care.

Old-growth, Clear-cutting, and the Politicization of Forests

By the 1990s, the bright lights of the forest fires demonstrated, for once and for all, that forests evolved with, were adapted to, and required change of specific kinds; that there were some kinds of changes and degrees of change that were necessary and there were extreme actions that damaged forests. It was clear by that decade that too harsh harvesting on the one hand and Smokey Bear on the other pushed forests to an unsustainable and undesirable condition; that people had for thousands of years affected forests and that there were beneficial roles we could play. It would seem that the time had long passed when the simplistic belief in the necessary constancy of nature would have faded, and that the idea that people had always damaged nature and should only stay out of it would have passed away as well. But instead, with the increasing popularity and publicity of environmental issues, three strong opinions arose, dividing those fascinated by—some in love with—forests. One group argued strongly for the preservation of old-growth, the popularized, politically correct term at the time for forests completely undisturbed—that is, the forest that had been called the forest primeval, the virgin forest, and climax forest. One experience that stuck in my mind was a discussion with a staffer at one of the major philanthropic foundations that had previously supported my forestry research. I called to ask if the foundation would like to consider funding another project, but the staffer said no, that the foundation's new goal for the national forests was to "lock them up," meaning no logging of any kind on those federal lands. I could never verify that this was a stated goal of that foundation, but it typified the times.

Another group, in part in strong reaction to the first, argued that professional foresters knew best how and when to log, and it was only the ignorant who believed otherwise. These were what the "old-growth" group called the old-time foresters.

And then there were those of us who have gone into the study of forests as a scientific pursuit, with the realization that measuring, monitoring, and methods of quantitative forecasting are necessary for people to find their proper role in forestry, and that forests could be both used and conserved. Ironically, while forest research was undergoing great advances, public debates provided forces in the opposite, ancient direction.

Among the scientific advances were experimental clear-cuts at several national forests. Most famous are H. J. Andrews in the great Douglas fir lands of western Oregon, led for many years by forest ecologist Jerry Franklin, and Hubbard Brook in the White Mountains, New Hampshire, an experiment conceived and directed by limnologist Gene Likens, plant ecologist Herbert Borman, and U.S. Forest Service scientist Robert Pierce (Fig. 10.2 and Fig. 10.3). These scientists sought to find out the effects of various kinds of clear-cutting in real time, following standard scientific methods. Both projects have been additionally important because of the kinds of forests involved—the Douglas fir dominated forests of the Pacific Northwest and eastern deciduous forests and northern boreal forests—are among the major forests of North America in area and in recreational, economic, and conservation terms. Douglas fir has been one of the major commercial species and the focus of debates over habitats of salmon, spotted owls, and marbled murrelets. The kinds of forests found at Hubbard Brook form the largest contiguous forestland in the eastern United States, covering an area almost twice the size of Maine. These are important for water

FIGURE 10.2 An experimental logging of Douglas fir on the Olympic Peninsula, Washington. (Photograph by the author.)

supply to many cities and towns, habitats for many popular wildlife from moose to bald eagles, and have been the focus of conflicts over timber harvest, land development for vacation homes, and recreation.[19]

In a classic experiment at the Hubbard Brook Experimental Forest in New Hampshire, an entire forested watershed of approximately 50 acres was clear-cut in 1965 and an herbicide was applied to the ground to prevent regeneration. Then in 1984–1985 a second watershed was clear-cut, but without herbicide treatment. This was one of a very few experiments with a whole watershed, and therefore of great importance in enlarging the scale of ecological research and putting that research on a firm empirical foundation, following the traditions of the scientific method. Immediately after the experimental clear-cutting at Hubbard Brook, nitrate in the streams leaving the watershed increased considerably. At first, the nitrate reached concentrations exceeding public-health drinking-water standards, but that lasted only a few years; afterward, nitrate was readily retained by regrowing vegetation. Since then—45 years for the first logged watershed, an unusually long time for the careful study of a single watershed in a forested area—the regrowth of vegetation, the flow of water from this watershed, and the eroded materials and dissolved chemical elements carried by the water have been monitored. Both clear-cut watersheds have regrown considerably, typical of natural regeneration of forests in this region of New England and generally for forests everywhere.

FIGURE 10.3 A view of northern forests of the U.S. Northeast, the kind studied in the Hubbard Brook Experimental Forest, famous for their beautiful fall colors. (Photograph by the author.)

Experiments also consider alternatives to clear-cutting, including strip-cutting, in which narrow strips are cut horizontally on a hillside. For example, at the Hubbard Brook Experimental Forest, a third watershed was strip-cut so that every third strip was cut in one year and the remaining strips left for later clear-cutting. This practice had many benefits: Erosion was retarded because the intact forests acted as buffers against it; mature seed-bearing trees were near to each clear-cut because each cut was of moderate size.

Other projects advanced ecological science, such as the Long-term Ecological Research Project, funded by the National Science Foundation to provide long-term measurements over time at representative ecosystems. This was something that had long been needed, and that Henry David Thoreau would have appreciated.

Being one of the scientists in the middle, I had friends and colleagues on both the oldgrowth-only and commercial logging sides, and was saddened to hear the condemnation of each by the others. In the 1990s, I was invited by Nils Christoffersen to visit the Wallowa Resources and see their work in the Blue Mountains of eastern Oregon. One day we went out with some of the old-time foresters to view their attempts to improve logging methods. Removing logs from forests with heavy mechanical equipment like bulldozers damaged soils, sometimes creating major problems for tree regeneration. To avoid this, the loggers were using horses and light, large-wheeled log carriers, and were doing the best that they knew how to harvest timber sustainably. After visiting several of their experimental sites, we sat in the eastern Oregon sun for a picnic lunch, and the old-timers opened up. One of them, a man well into his fifties or sixties, asked why so many people opposed them when they were trying so hard to do a good job, and burst into tears.

Experiments like those at Hubbard Brook show that clear-cutting can cause immediate but comparatively short-term degradation of soils on steep slopes in areas of moderate to heavy rainfall. In contrast, where the land is relatively flat and rainfall light, and where the desirable species require open areas for growth, clear-cutting may be the preferred harvesting strategy. In the forests of eastern North America, a number of species that require open areas to regenerate, such as white and yellow birch, are useful commercial timber trees, a woodland filled with birches is a pleasant and desired part of the New England landscape. Also in relatively flat land with a well-drained soil that is reasonably fertile and not too shallow, the early-successional species regenerate naturally, and clear-cutting stands of certain sizes may lead to a landscape that is both economically productive and aesthetically pleasing. If the clear-cut area is too small, not enough light reaches the ground to stimulate sprouting of the birches. If the area is too large, it is opened up to more intensive erosion, and the amount of reseeding may be low if not enough seed trees are left nearby.

In sum, whether clear-cutting should be used depends on many factors. There is no one size that fits all. Landscape painters, like those of the famous Hudson River School of painting, often used variety and complexity as an important quality of beauty. Our new perspective takes a lesson from them, and we can appreciate diversity in the landscape, seeking to maintain several kinds of forest conditions simultaneously. We can have some nature preserves in which there is no direct active interference by people; other nature preserves whose goal is to save a particular species, such as Kirtland's warbler, and within which many direct actions are taken; and land that is subject to a variety of other uses, including commercial harvests.

This is not to say, naively, that there are no threats to resources such as forests. Of course there are. The primary threat is widespread cutting without regard for regeneration. Large areas that were once white pine stands in Michigan have never recovered from their first logging, which began in the late 1840s and continued until about 1920. It is estimated that 19 million acres of white pine were cut. Only a few remnants of the original forests remain; one of them is at Hartwick Pines State Park in central Michigan, where one small stand that originally occupied approximately 90 acres was abandoned uncut in 1873 at the time of an economic crash. It was uneconomical to log that small stand after the economy recovered, so the area remained intact and was set aside as a park. In November 1940 about half of the original stand was blown down in a storm, and now only about 50 acres remain as a virgin white pine forest. Fortunately, as of this writing in 2012, those acres remain as old growth. Some of the white pine trees in the park reach a height of 150 feet.

White pine is an early-successional species—it comes into large openings where there is abundant sunlight. But like its large, distant relatives on the West Coast, the redwood and sequoia, it lives so much longer than we do that the forests it dominates seem primeval to a casual visitor. If you walk through the old pines at Hartwick Pines State Park and look carefully, you will find hemlocks and trees of other species characteristic of older stages in forest succession making their slow appearance below the towering pines. While some Michigan white pine cutover areas have regrown to forests, large tracts in Michigan called "stump barrens" have never regenerated. They remain as open fields where reindeer moss and other lichens, bracken fern, grasses, and shrubs grow low to the ground. In the mid-1970s, I visited many of these areas and tried to determine why forests had not regenerated. An old red oak in one of the stump barrens near the town of Pellston, Michigan, provided a small clue. Gray old stumps and partially decayed gray logs lay on the ground among the grasses and lichens, but under the shade of that old oak were a half-dozen saplings of white pine, distinctive because they were the only saplings in the field. Perhaps the oak protected the saplings, but from what?

The field lay at the base of a hill created thousands of years ago as the glaciers retreated, depositing sands along an ancient lake shore. The base of that hill was subjected to downhill drainage of cold air, in the autumn, winter, and spring, when cold air, being denser than warmer air, moved down the slope, making the base of the hill colder at night than the top. Near the border of a tree species' range, early spring frosts or late fall frosts can kill young trees. Frosts are more likely near the ground because the surface loses heat by direct radiation. It is warmer at night under the branches of a large tree, just as it is cooler during the day in the shade of a tree. These factors would have been accentuated at the base of the hill.

It appears that white pine reached its northern limit during a very warm period about 4,000 years ago, the warmest period since the end of the last ice age until the present. The more recent climate of the late nineteenth and early twentieth centuries was cooler than that earlier climate, but forests have considerable inertia. Mature trees may have shielded white pine seedlings and saplings from frost. Once the mature trees were removed, such protection would have been lost. This is not to say that the forests had always been white pine stands for the past 4,000 years. As the climate warmed and cooled, the northern limit of the white pine would have responded with a change in distribution. But this would have occurred very gradually in comparison with the logging of the immense area, which took place in less than 100 years. Rapid logging of a vast area gave the white pine and other tree species little time to respond to the changes.

Failure of the forests to regenerate may be also a result of the unnaturally intense fires that followed the logging. The loggers took only the main trunks of the trees and left the rest—limbs, twigs, leaves, stumps. There was little concern about fire, and when fires accidentally started, the unnaturally large amount of wood left on the ground by the loggers produced fires intense enough to destroy the organic matter in the soil, and left too few seed-bearing trees. These explanations are based on informal observations, not rigorous scientific experiment. But whatever the exact explanation, which is not yet known, the fact is that much of this region has failed to recover. Rapid clear-cutting of such a large area in such a short time, with little care taken for treatment of the soil and the intensity of fires (not simply the existence of fires), led to undesirable results. In this situation, clear-cutting should not have been carried out as it was. But modified clear-cutting—smaller cuts scattered among intact stands, with care for the soil and avoidance of erosion—could have been part of an ecologically sensitive approach.[20]

Forest Plantations Can Help Conserve Nature

Once we get away from the idea that only untouched nature is good, many possibilities open up. One of the most striking for me was the conclusion that I and environmental economist Roger Sedjo, of Resources for the Future, reached in 1997: that widespread use of forest plantations could help in the worldwide conservation of nature, including the conservation of a lot of biological diversity and of the many roles of forests in the biosphere.[21]

We started with the observation that, on average, planted forests produce ten times more harvestable timber than natural forests do. Then we looked into the world use of forest products, in construction and to make paper and other fiber products. There is a lot of wood used—1.5 billion cubic meters annually, enough to cover the entire state of Georgia with plywood 3/8 inch thick (a typical width used in construction)—and this consumption had been fairly constant over a number of years, as had the production required to meet these needs.

We found that this much timber could be produced from the annual growth of 20% to 40% of natural (unmanaged) forests, but only 2% to 4% of Earth's forested area would be needed to produce this from plantations. Suppose our estimates were wrong by a factor of 2, a very liberal estimate of our error. Then at most 8% of the world's forested area would be needed. This would leave more than 90% of all the forested land for other uses—conservation of endangered species, protection of watersheds, recreation, wilderness, whatever.

When we first proposed this, however, it met with considerable opposition from some of the major U.S. environmental organizations, because plantations were not "natural." Since 1997, most of the nongovernment environmental organizations have come around to understanding that plantations can play a useful role. This is part of the change in ideas about nature, and our relationship with nature, that has to take place.

In Summary

Following from our new ideas about nature, our role in conservation is active and responsive to nature's needs as well as our desires. Done properly, harvesting can serve conservation as well as utilization, and the goals of utilization and conservation can be part of one integrated approach. Under the old ideas, management for conservation and management for utilization (such as harvesting fish, discussed in the next chapter, and cutting forests for timber, discussed here) appeared to be different and generally incompatible goals. From an old preservationist perspective, nature undisturbed achieved a constancy that was desirable and was disrupted in an undesirable way only by human actions. From an old utilization perspective, the forest was there to cut, take apart, replace, and put back together as one chose. If nature was like a watch, then one had to choose between the stereotyped preservationist's approach and the stereotyped engineer's approach: Appreciate the beauty of the watch and use it to tell time; or take it apart and try to improve it or else use the parts for something else.

When Julius Caesar was having difficulty defeating some of the southern Gauls in what is now near the Dordogne Valley in France, he took the time to find and cut down the Gauls' sacred forest grove, and, according at least to legend if not detailed history, he was then able to defeat them. He used what we would call psychological warfare, destroying the Gauls' spirit and their will to win by destroying something natural and essential to them. In the Babylonian myth, Gilgamesh becomes a hero by going into a forest, dark and frightening, and cutting some of it down, admitting light and making it safe.

In modern times, most of us don't consider forests sacred in these ways or if we do, it comes across as being opposed to any modification of nature by us. Sacredness and the beauty of nature became a line in the sand, crossed in favor of it by those who saw themselves as saviors of nature, attacked as unreasonable, antirational, and as the source of an uneconomic and therefore unviable approach to nature. But we are still part of nature, and its importance is deep within us.

The failure here is seeing the nonrational only as opposed to the rational and the two as incompatible. But a theme of this book is that both are important to people, deeply so. Accepting the dynamics of nature, which then places us within nature and as a part of nature, allows these two qualities to exist within our minds and spirits simultaneously.

FIGURE 11.1 Early-twentieth-century fishing for salmon on the Columbia River. American Indians had caught salmon for centuries. People of European descent began intense fishing there right after the Civil War, and fishing intensity increased through the first half of the twentieth century, as this early photograph illustrates.

(*Source*: Russell Lee, Library of Congress, Prints & Photographs Division, FSA/OWI Collection, [LC-USF33-013146-M2].)

11

Salmon in Wild Rivers and Grizzlies in Yellowstone
MANAGING WILDLIFE AND CONSERVING ENDANGERED SPECIES

The primary objective of [marine mammal] management should be to maintain the health and stability of the marine ecosystem. Whenever consistent with this primary objective, it should be the goal to obtain an optimum sustainable population keeping in mind the carrying capacity of the habitat.

—U.S. MARINE MAMMAL PROTECTION ACT OF 1972

The Secretary [of Interior] is not required to make, or require any State to make, estimates of population size in making such determinations or giving such advice.

—THE U.S. ENDANGERED SPECIES ACT OF 1973[1]

Gone Fishing

In 1995, two friends from Portland, Oregon, invited me to join them over the July Fourth holiday on a trip to their fishing cabin along the Deschutes River, one of the larger tributaries of the Columbia River, flowing into the Columbia about 100 miles upstream from Portland. I had just lost my wife to cancer, and the two men thought this might be of help to me. I took them up on the invitation immediately, even though I had done little fly-fishing.

Their cabin was an abandoned old railroad building, probably manned originally by a switchman years ago, not far upstream on the east shore of the Deschutes, and not far from an abandoned railway line. To get there, we had to drive over a bridge to the west shore of the river and carry our gear down to a rowboat tied to a small floating dock. The Deschutes is a very fast river, probably flowing about six or more miles an hour when we arrived, as it drains off Bachelor Mountain, a high, isolated peak of central Oregon that in winter is a major ski center. To make it across to the dock opposite, my friends rowed as hard as they could upstream, fighting the current, until they edged the rowboat to the center of the flow, and then let it coast down to the dock. It was a maneuver that took patience, skill, and knowledge about the river, the boat, and rowing, within and part of nature; within and part of boat technology.

We spent three days at the cabin, a rough sort of place with plain boards across the tops of low cabinets as beds and a collection of sleeping bags as blankets. Both friends were highly skilled fly fishermen, and they spent a lot of the three days trying to introduce me to it. They explained that a salmon could see both within and above the water, and was believed to be able to see a person and his shadow many feet away. You had to sneak up on the fish from downstream, using dense vegetation as camouflage. They further explained that, like all streams used by salmon, the Deschutes had a complex flow made up of ripples (shallow sections of fast-moving water) and pools (areas of deeper and slower-flowing water). A salmon liked to wait at the upstream end of a pool, where it took little energy to stay in place, and let the rapid flow in the ripple bring food down to it. Also, salmon were very smart about what was food and what wasn't. An artificial hand-tied fly not only had to look like the real thing, it had to behave like it in the water, and it had to land on the water exactly like a real fly would. This meant that the fisherman's cast had to be perfect—a graceful arch up that led the line past all the surrounding and overhanging vegetation that could (and often would) snag it, and a graceful, lifelike drop just upstream from the salmon. Then it had to float downstream exactly lifelike.

Clearly, fly fishermen gathered a lot of exacting knowledge over many decades, perhaps centuries if you count fishing in Great Britain and Europe, and the fishing itself was a great skill.

Although they never actually said so, my friends were intent on my catching a fish. And on the last day, late in the afternoon, I actually did—whereupon one of my mentors hurried to the other and shouted, "Dan didn't get skunked!" By the end of the trip, I considered myself extremely fortunate to have such friends, let me tell you.

Some years later, I was fortunate to go fly-fishing again, this time on the Anchor River in Alaska with another friend, who was a professional hunting and fishing guide. This marked the high point of my fly-fishing career, so far at least, because I caught a 20-pound coho with my fly-fishing gear. But I had a lot of help. My friend took me to the mouth of the Anchor River just as the tide was about to come in, explaining to me that this was when the salmon liked to start upstream, for the obvious reason that the flow was with them. He handed me a rod and reel and tied a fly he had made himself and chosen for exactly this moment and location. Then he instructed me to stand on a shoal—a slightly submerged sandbar—and told me the exact moment to cast. And catch a salmon I did, little credit but much fun and deep feelings for me.

There was no doubt in my mind that these three friends, the two on the Deschutes and the third on the Anchor River, understood a great deal about salmon—their biology, their feeding, their habitats, their ecosystem, all the kinds of natural-history knowledge that one would need to conserve and manage salmon and any other kind of wildlife.

What puzzled me at that time, having only recently written and published *Discordant Harmonies*, is the failure of this kind of direct experience and observational knowledge to get deeply immersed in the management, formal conservation, and oversight of fisheries, wildlife, and endangered species. This still puzzles me today, except when I put it into the context of our society's folkways, as described by David Fischer in his landmark book *Albion's Seed: Four British Folkways in America*. As I explained in the first chapter, Fischer writes that societies develop from their folkways, those "normative structure of values, customs, and meanings that exist in any culture."[2]

Folkways are persistent and dominate thinking and action today, just as they have over many generations. As best as I can understand our decisions about environmental management, what we are following are sports and learning folkways. Once again, what we have assumed to be purely scientific derives substantially from a nonrational aspect of human existence. We think and act as people with in a society, with our social history and habits.

To Count or Not to Count?

A species is threatened with extinction when its numbers drop so low that some single environmental change could kill the entire population, or when, even if it is abundant at the moment, large-scale changes taking place in its habitat and ecosystem are likely to force its numbers to zero. It's hard to imagine a determination that a species is endangered without an estimate of its abundance. Thus a major challenge in conserving and managing wildlife is determining how many individuals there are and what the population size has been in past. These are the primary pieces of information, but surprisingly they are rarely obtained. It has puzzled me all these years why such basic information is so rarely available.

One reason may be that the U.S. Endangered Species Act hedges on this point. It states explicitly that "the Secretary [of the Department of Interior] shall base the determinations and advice given by him . . . with respect to wildlife upon the best available biological information derived from professionally accepted wildlife management practices." And elsewhere the Act states that "the terms 'conserve,' 'conserving,' and 'conservation' mean to use and the use of all methods and procedures which are necessary to bring any endangered species or threatened species to the point at which the measures provided pursuant to this Act are no longer necessary. Such methods and procedures include, but are not limited to, all activities associated with scientific resources management such as research, census. . . ." But the Act also states specifically that the Secretary "is not required to make, or require any State to make, estimates of population size in making such determinations or giving such advice."

It isn't easy to count wildlife, especially in the wild. Species have evolved to be camouflaged for obvious reasons, and therefore most wild animals are not very visible. Many also have large habitats in difficult terrain. Even so, this is not the only difficult observational problem that science has had to solve since Galileo, so you would think there would be solutions—or at least more solutions than we currently have at hand.

It's hard to avoid the conclusion that knowing the present and past abundance of wildlife hasn't seemed all that important. Perhaps this is because if you believe in a balance of nature, then you believe that nature will take care of itself, as David Sheldrick believed about Tsavo (Chapter 2), and trying to count populations becomes just another of our bungling interferences. I've run into this argument directly over the years. Early in my career, in the 1970s, a conservation organization asked me to help settle a controversy between scientists who were studying whales and some conservation groups who opposed those studies. At one meeting, a representative of the New England chapter of the Sierra Club was especially outspoken against the research, which involved putting tiny radio transmitters on the tails of some of the great whales. He said it would harm the whales in some way or another and interfere with their lives, and shouldn't be done. I asked, as mildly as I could, whether we might not gain some useful information about whales that would help us better understand them and their populations, and wouldn't it be better ten years in the future, say in the late 1980s, for us to know more about whales than we knew today? "No," was the reply, "that's a problem for the people alive ten years from now to deal with."

This extreme antirationalist assertion was not general Sierra Club policy, just the views of one of its members at that time, and I've encountered it in milder forms in other contexts and with other people, including scientists.

In the mid-1990s, John Calhoun, director of the Olympic Natural Resource Center (ONRC) on the Olympic Peninsula, Washington State, asked me to help with issues there about salmon and forests. The Clinton administration had established new rules for the conservation of salmon, based on the idea that the condition of forests along the streams where salmon spawned was important to the successful reproduction and survival of young salmon. Specifically, that mature trees were necessary; their roots and logs created the best shape of the streams for salmon reproduction and growth of the young; the vegetation along the stream provided habitat for some of the species that were food of the young salmon; and leaves that fell into the streams provided food for other species that were, in turn, food for young salmon.[3]

The new policies created protected zones along the streams: No logging was allowed within a distance equal to twice the height of trees characteristic of the locale. Since there are many streams in that cool and rainy peninsula, by the time you had walked two tree-lengths away from one stream, you were likely to be within two tree-lengths of the neighboring stream. In other words, the policy basically stopped logging throughout much of the Olympic Peninsula. Some of my colleagues whose scientific work I admired and still admire were happy with this result. It was as if they still believed in the old learning and knowledge way, an absolutist nature devoid of humanity in any manner.

The peninsula's loggers, Indian tribes, and fishermen wanted to know whether the new rules were working. They said if the rules were increasing the abundance of salmon, they would accept them; if not, the rules should be changed to allow more fishing and logging. Calhoun asked if I could help him write a request for proposals, including guidelines for the work required and acceptable performance, so organizations would know how to write proposals to do the work and there would be a rational basis for choosing among them.[3]

I gathered a small team: Ken Cummins, one of the world's best freshwater ecologists, who had studied salmon for years; and Matt Sobel, an expert in operations research with considerable experience in environmental matters—both of whom I mentioned earlier. We wrote down a set of things to be accomplished in the proposed study. One seemed obvious: We said that if you wanted to know whether actions you had taken succeeded in increasing the number of salmon, you had to count the salmon.

This was only one of the conditions we set down, but that condition created an uproar! The salmon biologists argued that it wasn't necessary to count salmon, and that in fact it was a bad idea to do so. They got the governor and two senators involved, and the entire thing blew up to huge proportions.

At Calhoun's request, I directed a workshop that brought together the salmon biologists who opposed counting the fish and those who thought it necessary. We asked the biologists who opposed counting salmon why they held that position. We never got an answer that made much sense. The main response was that you would be counting salmon at only one location, whereas salmon lived all over the place and spent much of their life in the ocean. It seemed like arguing that the weather bureau shouldn't measure temperatures at specific points because the atmosphere was all over the place. This was the only time I had encountered or heard of scientists studying a phenomenon arguing against knowing how abundant that phenomenon was. Thinking back over it, I am still perplexed.

Wildlife scientists in east and southern Africa were completely on the other side of this argument. While doing research on elephants, I was fortunate to spend time with Iain Douglas-Hamilton, world-famous for developing ways to identify individual elephants and study their behavior, and was also fortunate to get to know Michael Norton-Griffiths and his colleagues, who were actively involved in careful counts of elephants from light aircraft.[4] They mounted a horizontal rod that stuck out from the fuselage of the plane, outside but visible from the right (copilot's) seat near to where an observer sat. He counted the number of elephants that passed under the rod as the airplane flew at a specific altitude, which the scientists had previously calibrated so they knew the land area covered by the rod as the aircraft moved over the plains, flying straight lines back and forth to cover a large area. They never questioned the value of those counts, which seemed to them, and to me, essential.

That was one of the best-done counts of wildlife; not unique, but not all that common either.

How Many Are There, and How Many More Should There Be?

There are good reasons for counting. First, if you are out hiking in grizzly bear habitat, you want to know how many grizzlies there are, so you have some idea what your chances are of encountering and getting chased by one. Also, the conservation of grizzlies has been a popular environmental concern in the Rocky Mountains, Alaska, and Northwestern Canada. Then, too, the Endangered Species Act requires that an endangered species, such as grizzlies, be restored to their former abundance, as discussed earlier. If we don't count the grizzly bears, how can we know what to restore them to? So it would seem a no-brainer that grizzlies should be counted regularly. It would also seem obvious that conservation scientists would be trying to find out how grizzly bear numbers had changed over time.

Historical accounts, prior to modern scientific monitoring, can be useful. Among the most impressive for excellent observations and insights about land, waters, and ecosystems and species before European settlement of the United States are the journals of Lewis and Clark.[5] When Lewis and Clark saw wildlife, they mainly gave quite general accounts about abundance: "many bison," and so forth. But that wasn't true for grizzly bears, which they began to encounter after they left their first wintering camp with the Mandan Indians in what is now North Dakota.

Grizzlies were the most dangerous wildlife the expedition encountered, animals that made unprovoked attacks on the men.[6] One of those encounters occurred on May 11, 1805, when the expedition was northeast of what is now the Pine Recreation Area near Fort Peck Dam in eastern Montana. A member of the expedition, named Bratton, went for a walk along the Missouri River shore, but soon rushed back, "so much out of breath that it was several minutes before he could tell what had happened," Lewis wrote. Bratton had met and shot a grizzly bear, he told Lewis, but the bear didn't fall; instead, it ran after him for about half a mile, and it was still alive.

Lewis took seven men and trailed the bear about a mile by following its blood in the shrubs and willows near the shore. Finding the bear, they killed it with two shots through the skull. Upon cutting it open, they found that Bratton had shot the bear in the lungs, after which the bear had chased him and then moved in another direction, a total of a mile and a half.

"These bear being so hard to die reather intimedates us all [Lewis's spelling].... The wonderful power of life which these animals possess," the journals continue, "renders them dreadful; their very track in the mud or sand, which we have sometimes found 11 inches long and 7 1/4 wide, exclusive of the talons, is alarming."

This was not their first encounter with a grizzly—that had taken place the previous fall, on October 20, 1804, when they were near Bismarck, North Dakota, about to set up their winter camp. That location, in the Great Plains hundreds of miles east of the Rocky Mountains, considerably extended the eastern range assumed in the twentieth century for this animal (you can find it, for example, in field guides to mammals of North America). Lewis and Clark saw grizzlies during the next spring and into the summer. On approximately 20 days between April 17 and the end of July 1805, they saw these bears—about one encounter or sighting every five days. Most of their sightings were upstream from the confluence of the Yellowstone and Missouri rivers, within what is now Montana. They were especially troubled by them when they were portaging their equipment around the Great Falls of the Missouri River, where the city of that name is today. Their last sighting was near Three Forks, Montana, the headwaters of the Missouri.

Because grizzlies are so big and dangerous, Lewis and Clark recorded the number of bears (usually just one) they saw in each encounter. Reading their accounts, I realized that it was possible to use the journals to estimate the total number of these dangerous animals and learn about their original range and abundance at the beginning of the nineteenth century. Marking where they saw each grizzly on a map of the expedition's route showed that the expedition encountered a total of 37 grizzlies over a distance of approximately 1,000 miles, an average of about four grizzlies per 100 miles traveled. The journal also suggested that the grizzly occurred over some 530,000 square miles Assuming on average that the men could see a half-mile on each side of the river, there would have been about 20,000 bears in 1805.

The area known to have grizzlies today covers 20,000 square miles—just 6% of the presettlement range, based on the journals of Lewis and Clark. Today, 95% of grizzly habitat lies on government land in four states, much of it in four national parks: Glacier, Yellowstone, Grand Teton, and North Cascades. The rare encounter with a grizzly today occurs when someone goes cross-country backpacking in one of these national parks or in adjacent national forests. You are more likely to see them in the Canadian Rockies, although there, too, the chances are low. Chances are greater in Alaska.

Under late-twentieth-century interpretations of the Endangered Species Act, a species can be listed as threatened or endangered if its numbers drop to less than half of the estimated "carrying capacity"—the maximum number of animals that a habitat can support. And the carrying capacity is typically taken to be the estimate of presettlement abundance—the very number calculated from Lewis and Clark's journals.[7] (Note, of course, that this brings us back to the static balance-of-nature assumption about the condition of wildlife prior to disturbance by European people.)

Few modern studies provide estimates to compare with these historical estimates. One was made by the Craighead brothers, two of America's experts on grizzly bears. Theirs was limited to Yellowstone National Park, where they reported an average of 230 grizzlies between 1959 and 1967, an average density of three bears per 100 square miles, similar to the density estimated from the Lewis and Clark journals.[7]

Strangely, with the sole exception of information gathered in Yellowstone, our knowledge of the abundance and density of grizzlies in 1990 was not much better than what someone could have surmised from Lewis and Clark's journals in 1806, when they and their journals returned home to Virginia. If this is what we know about one of the most famous, most dangerous, readily reported, legally threatened, and therefore protected species, whose abundance and whereabouts are of considerable interest to outdoorsmen and government agencies, what could be our knowledge of other species? The answer is, in most cases, much less.

Alternatives to Restoring Populations to Their "Once-Was" Abundance

If it's so hard to count wildlife, and to get counts of them year after year for a long time, is it really necessary to our effort to save endangered species? Once we move our mind-set away from the balance of nature and machine-age nature, with their static carrying capacities and maximum sustainable yields, we can think much more openly, and a whole world of possibilities opens before us.

Several different approaches were tried in the last decades of the twentieth century. The main ones are known as the historical range of variation; a self-sustaining population; and the minimum population size that can be considered "safe" from extinction for a reasonable time.

The historical range of variation acknowledges that there is no single "natural" abundance for any species; instead, there is a range of abundances, all "natural" in the sense that a population expanded and contracted within that range prior to effects of modern civilization.[8] Using this as a basis, a plan to return the grizzlies to their "original" abundances becomes simpler, at least in terms of what we need to know about them. We don't have to discover an exact and perfect number, the idealized carrying capacity. But we still have to know a lot about the history of a species, and believe that the population sizes it reached in the past are going to be sufficient in the future, no matter how different that future environment might be.

There are several problems with the "historical range of variation." Obviously, it isn't easy to reconstruct that history if we have so much trouble finding out an abundance at any single past time. In addition, since the environment has changed, and been changed by us, perhaps the historical range is not applicable in some cases. Sometimes it is just the best we can do at present with current scientific methods.

More-recent programs to restore endangered or threatened animals have begun to focus on the more realistic goal of a self-sustaining population. Apparently, this was the goal for the Fish and Wildlife Service's "Grizzly Bear Recovery Plan." Its objective was "to establish viable, self-sustaining populations in areas where the grizzly bear occurred in 1975." Self-sustaining is a very different idea than presettlement carrying capacity or historical range of variation. It could include a wide range of abundances and, most important, it could be based on what can be learned today about a species' reproductive and death rates and its habitat and ecosystem requirements.[9]

The Grizzly Bear Recovery Plan, so lengthy that it was known by those working with it as "the compendium without endium," lacked an estimate of the total number of grizzlies sufficient to estimate even the more open-ended goal of achieving a self-sustaining population. To repeat, if the compendium was meant to tell how to aid the recovery of grizzly bear populations, how could it do this without some estimate of current and past abundances? Ironically, the best estimate of the number of grizzlies at any time may have been the one dating back to 1806, when Lewis and Clark brought their journals back to the East Coast.

Minimum Population Size

What is the smallest population size that has a minimum risk of becoming extinct? This is a much simpler question that brings us back to the whooping cranes, whom we visited in Chapter 8. We talked about how a small population can have a surprisingly low risk of becoming extinct—subject, of course, to the future variability of its environment, compared with how it was in the past. This is known as the minimum viable population size. Focusing on this, we don't have to know what is the theoretical or real maximum number of a species that could ever be sustained. We don't have to know what its exact past abundance was before the influence of our human ancestors and ourselves, nor even very much about its history.

Instead, we need to know some things about how a population grows and changes over time, turning us toward the kinds of population dynamics we can study in the here and now. We can use that information along with our knowledge of the physiology of individuals, the genetics of populations, and biological evolution. Also helpful will be knowledge about the broader context within which a species persists, what ecologists call the "ecosystem context."

By 1990, scientists had only recently begun to study how small a population could still be reasonably safe from risk. The first estimates were fairly rough, and the earliest ones were based primarily on the genetic variations of domestic animals, mainly cattle, and on the problem known as "genetic drift."[10] This is a term used in biological evolution to refer to what could happen when a very small population of a species becomes isolated. Simply by chance, it is possible that a very limited genetic variation might occur within that population, so that although it could persist at first under the environmental conditions that isolated it, it might not be able to persist if there was any major environmental or biological change (including the arrival of a new competitor, predator, disease, or insect that spread a disease). Based on the narrow genetics of highly inbred domestic animals, a simple 50/500 rule became rather standard: For a short time (say ten years, as it was usually stated), a population should not fall below 50 individuals, and over the long run it should not fall below 500 individuals.[11]

These newer ideas—historical range of variation, self-sustaining populations, and minimum viable populations—are not single goals like carrying capacity or maximum-sustainable-yield population size. They are boundaries. A historical range of variation gives an upper and lower boundary for population size. Self-sustaining populations and minimum viable populations provide a lower boundary. One would try to take actions to keep a population above the lower boundary and below the upper one, even including the occasional introduction of an individual of the same species from another isolated population. And the boundaries need not be treated as hard-and-fast. One could decide not to let a population approach within some percentage of either boundary, just keeping away from it. No actions or forecasts depend on knowing an exact, single number or forcing the population to reach and stay at that number. These are much more realistic approaches, requiring knowledge more likely to be obtained than the carrying capacity or maximum sustainable yield, and never requiring that a population be held to any exact number.[12]

Intriguing also is the observation, discussed earlier, that several species whose numbers dropped below 50 have done well, including the whooping crane and the elephant seal, suggesting that a minimum viable population might be a very small number for some species.

From our perspective today, we can take the 50/500 rule as a beginning approach to a way to think about a minimum population size. Recently, some have attempted to put the estimates on a firmer basis. A year after *Discordant Harmonies* was published, Eric Menges published a chapter in *Genetics and Conservation of Rare Plants* providing an extensive review of what had been thought and attempted in the calculation of minimum viable populations.[13] One study concluded that an average of "5,800 adult animals are needed for a 95% chance of persistence over 40 generations."[14] But unfortunately this analysis was heavily based on the logistic growth curve and therefore not useful in real nature. It is just as likely to mislead as to lead, and (like so many applications of models based on the logistic) is as likely as not to divert attention from the real empirical needs of the sciences that deal with biological nature.

Within the System

The whooping cranes and the California condor provide another kind of insight for conserving and managing natural resources. While the whooping crane is America's tallest bird, the condor has the greatest wingspread, almost five feet. Rare and declining through the nineteenth century and into the mid-twentieth, the condor population dwindled to just above 20 individuals by the 1970s without undergoing the recovery that took place with the crane. Simply being rare did not seem to imply the same fate for the two species. As ecologist Larry Slobodkin (1928–2009) used to put it, "Being rare is different from becoming extinct, as the whooping crane said to the passenger pigeon."[15]

What was the difference between the whooping crane and the condor, so that one could survive on its own while the other seemed about to disappear in the

1980s? It seems to lie primarily in the condition of their habitats and in their supply of life-sustaining resources.

The crane feeds on small animals in wetlands, and such food has been plentiful in its summering and wintering grounds. The crane's food supply and the ecosystems on which this food supply depends have remained intact and self-sustaining. People helped by making both the summering and wintering grounds into protected areas and stopping the hunting of the birds, but otherwise the cranes took care of themselves on their main flyway.

In contrast, the condor's food supply, nesting areas, and other features of its habitat have disappeared or been radically altered. The condor feeds on carrion, primarily of large mammals, and it once survived on wild game in the open savannas of the Sierra foothills and the coastal ranges of southern and central California, and also, some speculate, on marine mammals washed ashore along the central California coast.[16] But much of that game was eliminated with European settlement of central California, replaced by cattle whose carcasses, of course, mainly wound up in meat markets, not often left to rot on the ground for the condors to find. Furthermore, with fire suppression in the twentieth century, much of the condor's habitat changed from open areas where carrion could be seen from afar to dense shrublands in which dead animals are harder to find. Some experts have even suggested that the shrublands made landings and takeoffs difficult for the condors and prevented them from taking food in much of their home range.

From the comparison of the crane and the condor, we recognize that the condition of the habitat is important, actsually more important than simple population numbers. It's better to have a good habitat sustaining a small population than a large population in a poor habitat. Conservation of endangered species is, in this way, understood to depend on the idea of an ecosystem rather than on simple analyses of populations.

Under the old management, the goals of preservationists and the goals of fishermen would have been seen to be in opposition, with a place for only one or the other. One tried to maintain a species either at its carrying capacity to conserve it or at its maximum-sustainable-yield population size for commercial uses. Harvest and biological conservation were always opposed. Under the new management approaches, the debate becomes instead a set of quantitative and specific questions that require scientific study (as well as mediation and discussion among the various parties).

The sea otter of the Pacific Ocean makes a good focus for this discussion, because it has gotten a lot of attention both commercially (for its beautiful fur) and for conservation, to protect it from extinction, because it is a beautiful animal and because of its role within the nearshore marine ecosystems of the Pacific Ocean. As a result, there has been more funding for this species than for many others, and much scientific knowledge is available.

Sea Otters, Sea Urchins, and Abalone

Sea otters feed on shellfish, including sea urchins and abalone, both of which are commercially valuable. Sea urchins feed on kelp, the large brown algae that form undersea "forests." The kelp beds are important habitats and spawning areas for many species. Sea urchins don't eat all the kelp; they graze along the bottom of the beds, feeding on the holdfasts, the bases of the kelp that attach them to the bottom. When a holdfast is eaten through, the kelp floats free and dies.

A study of sea otter populations and their habitats took place on two Aleutian islands of Alaska: Amchitka Island, where the otters were abundant, and Shemya Island, where there were no otters. The high density of sea otters on Amchitka Island was accompanied by abundant kelp beds and few sea urchins, while Shemya Island, lacking sea otters, had abundant sea urchins but little kelp. In an experiment, the removal of sea urchins led to an increase in kelp.[17] Thus sea otters affected the abundance of kelp, but that the effect was indirect. Sea otters do not eat kelp, nor do they protect individual kelp plants from attack by sea urchins. Sea otters eat sea urchins, which reduces the number of sea urchins feeding on the kelp and thus leads to a greater abundance of kelp. Where kelp is abundant, there are many other species using the kelp forests as habitat. Thus the sea otters affect the abundance of kelp and of many other species indirectly.

The way sea otters indirectly affect many other species has led to its designation as a "keystone species" and an "indicator species." Such species, some say, make conservation and management much easier for us. All we have to do is find the keystone or indicator species in an ecosystem and track its abundance. As long as it's doing okay, the argument goes, the rest of the habitat and ecosystem must also be okay.

Intrigued by this idea, several of us undertook a study of a number of species believed to be such indicators. Unfortunately, we discovered that none of them actually indicated what they were supposed to.[18] Either our knowledge of ecological communities and ecosystems is insufficient to determine which are keystone and indicator species, or the concepts just generally don't work. At this time, we don't know. But the idea remains fashionable.

SEA OTTER HISTORY

Hunting of sea otters began at the end of the eighteenth century, when the otters were distributed throughout a large area of the northern Pacific Ocean coasts—from northern Japan, northeastward along Russia and Alaska, and southward along the coast of North America to Morro Hermoso in Baja California.[19] Nobody knows exactly how many there were, but the U.S. Marine Mammal Commission estimates that there were probably between 150,000 and 300,000 sea otters.[20] According to the standard story, a group of shipwrecked Russian sailors survived a winter eating sea otters and keeping warm in sea otter pelts. The furs that they brought back led to the development of the sea otter fur trade and to intense hunting of this species. By the end of the nineteenth century, sea otters were too rare for successful commercial hunting and the species was thought to have gone extinct. Then two small colonies were discovered in U.S. territorial waters, one in the Aleutian Islands of Alaska and the other along the California coast from Monterey Bay south to Point Conception.

Legal protection of the sea otter began in 1911 with an international treaty for the protection of North Pacific fur seals and sea otters signed by the United States, Japan, Great Britain (for Canada), and Russia. Today, the sea otter is protected by two U.S. federal laws—the Marine Mammal Protection Act of 1972 and the U.S. Endangered Species Act of 1973—as well as some international agreements. As explained in Chapter 2, the Marine Mammal Protection Act states that the goal of marine mammals management should be to obtain "an optimum sustainable population,"[21] an idea that can be traced back to the logistic growth curve and its maximum sustainable yields, as used in fisheries and wildlife management. The wording in the law allowed two definitions of an "optimum sustainable population": either the population size that has the greatest production (the logistic maximum-sustainable-yield population size, or half the carrying capacity) or the logistic growth curve's carrying-capacity level. These continue to be the legal basis for management of the sea otter.

By the time the Marine Mammal Protection Act was passed, sea otters had increased from about 50 in California to about 1,000 and recolonized about 200 miles (370 km) of the California coast. They also made a comeback along the coast of Alaska. By the 1980s, the otters numbered between 94,000 and 120,000 in southwestern Alaskan waters,[22] approaching "historic levels," according to the Marine Mammal Commission.[23]

But then, contrary to what was supposed to happen under complete protection, the number of sea otters declined. By 2004 their numbers had dropped to fewer than 48,000 (Fig. 11.2). The Marine Mammal Commission said that "the leading hypothesis to explain the decline, particularly in the central Aleutian Islands, is increased predation by killer whales. . . . Support for this hypothesis includes observations of killer whale attacks on otters, an overall shift in sea otter distribution to areas closer to shore where access by killer whales is more limited, and sustained otter densities in areas inaccessible to killer whales."[24] The commission added that a decline in other prey for killer whales, including some of the great whales, may have shifted their hunt to the otters.

While this seems quite plausible in the natural history of a set of species occupying the same coastal region of a major ocean, it is not something that is included in any of the classic balance-of-nature or twentieth-century machine-age theories about population growth and change, nor in the formal mathematical models that dominated both fisheries and marine mammal management.

Late-twentieth-century controversies over the sea otters concerned this larger role it plays within its marine ecosystem. Conservationists argued that sea otters

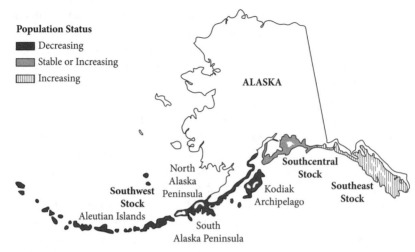

Population Status
- ◼ Decreasing
- ▦ Stable or Increasing
- ▥ Increasing

FIGURE 11.2 How sea otters are doing in Alaska.
(*Source:* Marine Mammal Commission Annual Report 2008.)

are necessary for the persistence of many oceanic species, including a number of economically important ones, that use the kelp forests as breeding grounds or as habitat during parts of their life cycle. But fishermen argued that there were plenty of sea otters to play that role, more than enough, so many, in fact, that the abalone was in danger of declining below a number sufficient for abalone fishermen to make a living. In short, they felt that conservation of the otter had saved the species, which therefore no longer needed complete protection, but could be harvested to some lower, sustainable abundance, and in the meantime had led to policies that were destroying the abalone fishery.

We know enough about the sea otter today to realize two important and perhaps by now obvious things: Its numbers vary despite the most extreme attempts to protect it; and it exists within an ecosystem, responding therefore to a complex set of interacting factors, and influencing, in complex ways, many other species. In the real world of people with different goals and desires, the question is how we can help sea otters to persist, make them readily available for people to watch and enjoy, and at the same time make sure fishermen can harvest a reasonable amount of abalone and other shellfish.

Under the old management approach, the goal was simply to increase the total numbers. But there are other ways to distribute and reduce the total risk, including the establishment of several populations in different locations, more or less isolated from one another and relatively protected from the risk of disease or local environmental catastrophes, such as an oil spill. It is a better bet that sea otters will persist because there are at present two populations, one in Alaska and one in California, than if the total population were concentrated in either location. The greater the number of separate, more or less isolated populations, the greater the chance of survival of this species. From this perspective, recent management has attempted to create a second California population centered along one of the islands in the Santa Barbara Channel, which has led to even more conflict between fishermen and conservationists.

Some things have improved since 1990, especially with the emphasis on monitoring the abundance of the otters. For the California population, which numbers more than 1,000, concerns have extended beyond the population size itself. Scientists and conservationists now actively discuss the extent to which diseases, toxic chemicals, and entanglement in fishing gear might be affecting the otters, but political will lags scientific understanding. As the Marine Mammal Commission's most recent report (2008) notes:

> Lack of funding has been a persistent problem, and in December 2006 the recovery team wrote to the Service's Regional Director emphasizing the importance of an adequate survey program for the stock. The team noted that it had developed a detailed monitoring plan and budget for the next 10 to 15 years but that, to date, limited progress has been made in implementing the plan. At the end of 2008 the Service had not determined what steps would be taken to address this and other research priorities in 2009.[25]

Wildlife in the Twenty-First Century

Today, early in the twenty-first century, wildlife conservation and management still mainly follow the old ways and old ideas. The deep and wonderful knowledge that my fly-fishing friends possessed about wildlife and their habitats and ecosystems still hasn't entered most of the laws, policies, or methods of wildlife conservation and management. This is not just my personal experience; scientists who work for government agencies at various levels tell me they are shocked to discover that their colleagues and sometimes their entire unit approach their tasks with a belief in the classic balance of nature. The result, as always, is a rather haphazard approach to wildlife: many local, specific actions that are intended to do good and some that do; a theoretical basis that confuses and misleads more than it helps. There are some wonderful examples of excellent approaches, such as the U.S. Fish and Wildlife Agency's management of the Aransas National Wildlife Refuge, the wintering grounds for the whooping cranes. Tom Stehn, Whooping Crane Coordinator for the U.S. Fish and Wildlife Service, has been in charge of that management and oversees the continual, excellent monitoring of that population, an example to us all.

But in some ways things are worse, because environmentalism has become so politicized and so much a focus of ideologies, many of them tied to the balance-of-nature idea. This tends to lead to unnecessarily divisive debates, with one side implicitly arguing for maintaining populations at the carrying capacity (don't touch, let nature take its perfect course) and the other for the maximum sustainable yield. Of course there are many thoughtful people who see well beyond these limits, but the debate too often falls victim to this oversimplification.

This chapter and the previous one have explored the extent to which modern rationality, as made explicit in modern science, could help us sustain nature as we believe it should be, as we want it to be. The short answer is that science has so far come up short in achieving these things for us. And in part this is because our pre-scientific, meta-rational beliefs impose themselves on that rationality, taking away from it. I continue to wonder whether we can ever find our way to be at peace with both of our "nature needs"—the ancient inner personal and the modern rational— or whether the Apollonian and Bacchanalian sides of ourselves will never be in harmony.

Many's the time I've stood on a pier at Monterey, California, and elsewhere along the Northern California coast and watched a sea otter floating on her back near shore in the bright sunlit ocean, like a teddy bear princess on her lacy bed of brown kelp. Sometimes she would have a shellfish on her stomach, but even if not, she looked for all the world contented. It is a picture in my mind of nature as peaceful, pleasant, and incredibly simple, where little if anything might go wrong. But as the sea otter's story has unfolded for us, it is clear that the life of this endearing mammal and the persistence of a population of its species depend on a complex set of interrelationships, many things happening at the same time, and some, although linked, happening in a specific progression or at different rates or periodicities. The charming image this single otter portrays shouldn't fool us into believing that simple, formal mathematical equations reducing all this complexity to one or two variables are enough to characterize our understanding or make forecasts about the future of a sea otter population.

The conservation of sequoias, the turn toward controlled fires, and the understanding of the complicated role that sea otters play in their ecosystems and in our hearts give us some hope. Yet by the end of the twentieth century, and still today, most attempts to manage and conserve wildlife remain trapped in the old ways of thinking.

FIGURE 12.1 Visiting the Mauna Loa Observatory in 1980. (Photograph by E. V. Pecan.)

12

The Winds of Mauna Loa

CLIMATE IN A CHANGING WORLD

From the region of an endless summer the eye takes in the domain of an endless winter, where almost perpetual snow crowns the summits of Mauna Kea and Mauna Loa.

—ISABELLA BIRD (1873)[1]

I slowly allowed my eyes to drink in the unusual majesty of the moon. . . . I managed to direct my view homeward, and there in the black, starless sky I could see our marble-sized planet, no bigger than my thumb. . . . In a strange way there was an indescribable feeling of proximity and connection between us and everyone back on Earth. Yet we were physically separated and farther away from home than any two human beings had ever been. The irony was paradoxical, even overwhelming, but I dared not dwell on it for long.[2]

—BUZZ ALDRIN, *THE SECOND MAN TO WALK ON THE MOON*

The Point of this Story

I begin this chapter with my conclusions about climate change and life. Then I explain how I got there.

Scientific observations, both modern records and reconstructions of the past, show us that a complex group of factors affect climate. And as more information has accumulated in the past decades, more questions and doubts arise about the degree to which people are causing climate change. This leads me to be agnostic at present as to whether there is going to be a disastrous human-induced warming. The computer models of climate, on which so many political, ideological, and scientific arguments depend, appeared to be one of our best hopes for insights in the 1980s. But these steady-state models are fundamentally inappropriate conceptually to describe and explain the large-scale dynamics of the atmosphere. There are now more than 20 of these steady-state models, and their forecasts differ greatly, further complicating the situation for us.

The difficulty I have with the current approaches to global environmental sciences is perhaps best illustrated by an experience I had several years after visiting the Mauna Loa Observatory, but still in the 1980s. It was within that turnabout time, the time of awakening awareness of global environmental problems, and I was piloting a small aircraft—a Cessna 182—from Santa Barbara to San Jose, California, to attend a NASA meeting about whether there could be such things as a global ecology and global environmental problems, possibly man-made. I had filed an instrument flight plan, which enabled me to take advantage of the FAA's flight-control system to its greatest extent and, if necessary, fly through clouds and land in instrument-requiring conditions. The weather forecast had been good, with only scattered clouds (meaning clouds would occupy less than 50% of the airspace), and I looked forward to a pleasant flight over California's beautiful coastal mountain ranges.

Once I was in the air, however, it was immediately apparent that the day had grown much cloudier, between broken cloud cover (more than 50% cloud cover) and overcast (100%). I was being routed at 10,000 feet over central California when I saw up ahead a large cloud buildup, the kind on its way to becoming a thunderhead—at least I thought it was only on its way. Then I was flying within that cloud and it was hailing heavily. Within 30 seconds the windshields were iced over, and likely the entire plane was iced over too. Since I couldn't see out, I had to depend completely on the instruments, but I'd had extensive training for this kind of emergency and immediately did as I'd been trained to do. First, turn on the carburetor heat so the carburetor wouldn't ice over and halt the flow of fuel (yes, those planes still had carburetors, not fuel injection). Next, stabilize the plane, making sure that it was flying straight and level, on course and at its assigned altitude, and of course continue to scan and read all the instruments to be certain the aircraft was functioning properly. A minute or two had passed by. Everything seemed stable, even though iced-over.

Then I called the air traffic controller and in my calmest and most professional voice requested a lower altitude. I was told to descend to 8,000 feet, which I did immediately, and there, back in the bright California sunshine, the ice quickly melted away and the plane and I continued to fly safely and normally.

Any similarly trained pilot could have done exactly the same thing, and instrument-rated pilots reading this will probably not be particularly impressed. In a few more minutes I saw another cloud buildup directly ahead, but it reached only 9,000 feet. Not wanting to chance flying within another buildup that looked harmless on the outside, I called the air traffic controller again and requested 10,000 feet. "You were just at that altitude and asked for a lower one," he replied. "Please make up your mind." But he okayed the higher altitude and in bright sun I sailed calmly over that next cloud.

This little story illustrates how one can respond to a potentially dangerous situation safely and calmly. It requires good training, much practice, and an understanding of the systems in which you are involved. The more I understood about how all the systems worked—engine, fuel flow, electrical, navigational—the safer I was. There was no time for fear, or to cross my fingers. There was no time and no need to focus on ideology, mythology, or anything else unrelated to the task at hand.

Without trying to stretch the analogy too far, upon landing safely and making my NASA meeting on time, it seemed to me that the approach used to pilot an airplane—objective, scientific, technically competent, and calm—was the right approach for dealing with potential global environmental problems that were coming to the fore but had not yet reached a condition as treacherous as "iced-over conditions," so to speak. I still believe that this is the correct approach, and am disappointed to see the search for solutions to what many consider the greatest crisis of our time, global warming, focusing on belief, wishful thinking, ideology, opinion without substance, and political orientations.

Not that emotion can't be a part of piloting aircraft. Far from it. As I mentioned much earlier in this book, one of my life heroes, Antoine de Saint-Exupéry, wrote that piloting small aircraft put him in touch with nature in a way unavailable to him or to anyone before the invention of powered aircraft. For him, piloting a small aircraft by himself was strongly emotional, an experience of great beauty and meaning in every sense of the word. Who knows clouds better, in a very human way, than one who has flown an iced-over airplane through their turbulence, winds, and dark gray? Of course we should, and could, approach the management of global environmental problems with connections to our emotions, our sense of beauty. But these should not prevent us from using our rationality, knowledge, and understanding to actually solve a global environmental problem at hand.

Is it natural or unnatural that our species is or may be causing the climate to change? To believe it's unnatural, you have to believe that human beings have a special place among all living things, a special place within the great chain of being, as the ancients put it, and that our special place is an evil one. As Arne Naess (1912–2009), the leading philosopher of the "Deep Ecology" movement, put it: "Mankind during the last nine thousand years has conducted itself like a pioneer invading species" that is "individualistic, aggressive, and hustling. They attempt to exterminate or suppress other species. They discover new ways to live under unfavorable external conditions—admirable!—but they are ultimately self-destructive. They are replaced by other species which are better suited to reestablish and mature the ecosystem."[3]

Is this the view of early-successional species that you, the reader, have arrived at from the rest of this book? Naess has turned the scientific understanding of ecological succession into a morality play. He doesn't view early-successional species as simply one of the ways that Darwinian evolution has expressed itself in vegetation, nor does he acknowledge that ecosystems, as systems, require such species in order to become established. Instead, he makes moral judgments about different ways that species have dealt with evolution and with a changing, dynamic environment.

Taking Earth's Pulse

In the late 1980s, it was clear that the science, conservation, and management of populations and ecosystems could be improved by a major change in worldviews, what scientists refer to as a need for a new paradigm. There were high hopes for a planetary perspective on life, and it seemed possible that this new science could begin free from the ancient myths about nature. But in several major ways that's not what has happened, and this is what I will try to explain in this chapter and the next.

The hope was that we would seek and achieve a general science of the biosphere, meaning a science of the planetary-size system that contains and supports life, and that this would be a science with the standard scientific goal of seeking understanding, not the manipulation of information for political or ideological purposes. It was as if we saw the picture—Earth from the moon, as Buzz Aldrin described it—but we did not hear the music, those dynamics that had made life on our planet possible and persistent for more than 3 billion years.

Global warming has become the major, perhaps for many the sole, focus of concern about the environment from a planetary perspective. Today it is widely accepted that our planet is warming, and most people, including many scientists, believe people are responsible for it. Not all agree on this, however, and global warming has therefore become one of the most contentious environmental issues, extending far beyond science to political and ideological arenas. It reached the point where Nobel Laureate economist Paul Krugman wrote in "Betraying the Planet," an op-ed piece published in 2009 in the *New York Times*, that "as I watched the deniers [those opposing the notion that most global warming is human-induced] make their arguments, I couldn't help thinking that I was watching a form of treason—treason against the planet." Krugman went on to write about "the irresponsibility and immorality of climate-change denial."[4] On the other side, Texas governor Rick Perry, running for president, said in the summer of 2011 that "there are a substantial number of scientists who have manipulated data so that they will have dollars rolling into their projects."[5]

What I Hope to Accomplish in This Chapter and the Next

Some of my scientist friends and colleagues who are experts on climate change and its ecological effects have taken positions on both sides of the debate, to the extent that the discussion sometimes has become nasty, damaging to their careers and to the science itself. I myself have long been interested in this topic, having begun research in 1968 on the possibility that burning fossil fuels might be leading to a global warming. A landmark conference titled *Carbon and the Biosphere* was held in 1973, directed by George Woodwell, who was at that time in charge of ecological research at Brookhaven National Laboratory. With several colleagues, I presented a paper that suggested how a large increase in carbon dioxide might affect forests. My research ever since has included work on possible ecological effects of global warming. Throughout, I have tried to remain objective and independent, focusing on the scientific investigations and their results and trying to keep outside of the worst of the fray.

In this chapter and the next, I would like to take a calmer approach than is typical these days. The purpose of the two chapters is to tell the story of how knowledge and understanding of climate change has advanced in the past several decades, and why this has led to a change in my opinion about possible human-induced climate warming and its potential ecological effects. I also want to clarify what we do and do not know, and how we can improve the way we think about and talk about global environmental problems.

There is a great deal of important scientific research concerning global warming, and the debate about it continues. An article in *Nature* in 2008 put it succinctly: "The climate is changing, and so are aspects of the world's physical and biological systems. It is no easy matter to link cause and effect."[6] In the past 20 years we have made remarkable progress in our ability to reconstruct millions of years of change in temperature and greenhouse gases, using technologies little dreamed of early in the twentieth century. These include the ability to estimate temperatures by measuring ratios of hydrogen isotopes in the ice, to measure carbon dioxide concentrations in ice cores taken from Antarctic and other glaciers, and to use oxygen isotopes from the deep sea to reconstruct temperatures in such places as off the coast of Greenland. But there is still much that we do not understand. My analysis has to do with where the gaps and the faults in approaches lie.

The Basic Questions and Where We Are in Answering Them

The questions that have captured so much attention boil down to these: Is the climate changing, and, if so, how? Is it warming, and, if so, in ways that never happened before? What causes those changes? Are we responsible for them? Will the effects be disastrous for us and for much of the rest of life on Earth?

Whether we are the cause, or one cause, of climate change immerses us immediately in the three ancient questions that have played the bass line, so to speak, of this book:

- *What is nature like undisturbed?* This translates in this context to: What is the climate like when undisturbed by us?
- *What is our effect on nature?* This becomes: Are we causing climate change? If so, is this change bad, not just in a practical sense but in a moral and religious sense? Are we violating the boundaries of our given role on the planet and in the universe by changing the climate? If so, are we therefore evil, and shouldn't all right-thinking, moral people do whatever they can to stop or reverse those changes?

- *What is the effect of nature on us?* This translates to: How does global warming affect the availability of life's necessities for us—water, food, shelter, and, beyond those, our access to the beauty and diversity of life?

These global questions, once raised, involve us not just in the rational questions of what is and how it came about, but in moral, ethical, political, and ideological issues. We met these issues that arise in our mind and heart—Apollo's rationality and Dionysus's rhythms—earlier in the book, in discussions about species and ecosystems. They continue when we consider Earth as a globe. Nature immerses us in them at every one of its levels.

In popular discussions, the terms *climate change* and *global warming* are used interchangeably, but in fact they are distinct. *Climate change* has traditionally referred to any kind of climate change, cooling or warming, from any cause, known, unknown, natural, or human-caused. For example, before global warming became such a (literally) hot topic, astrophysicists and climatologists warned us about a possible "nuclear winter," a deadly cooling of the planet resulting from a thermonuclear war that would throw so much dust into the upper atmosphere that we would suffer a terrible new ice age. *Global warming* is the term (not especially scientific) used to mean present warming of Earth's surface, usually within the context that it is human-caused and bad for us and for all life.[7]

However, the United Nations Framework Convention on Climate Change changed the meaning of *climate change*. Under "Definitions" the convention stated that " 'Climate change' means a change of climate which is attributed directly or indirectly to human activity that alters the composition of the global atmosphere and which is in addition to natural climate variability observed over comparable time periods."[7]

Climate is the average weather, and climatologists use the term *climate change* to mean change in Earth's average climate, expressed as its average temperature (the most important variable for us), average precipitation, and so forth. It is well known that Earth's average climate results from continental and oceanic patterns, regional (within continents and oceans) and local patterns, and that there are "climate zones." A climate zone is a large region, such as the tropics or the Arctic, characterized by a certain kind of climate—certain ranges of temperature, precipitation, and seasonal changes. We know from reconstructions of past climates that these climate zones have moved about during Earth's history. During an ice age, for instance, an Arctic kind of climate moves to lower latitudes. This geographic detail complicates the discussion and the initial questions. It is not merely whether the average temperature of Earth has changed, but what this means for various regions and for the locale where I live.

We learn about climate change in three ways: historical records (from the time that people began keeping them, including today's records); reconstructions of the past through several kinds of scientific indexes that correlate with climate; and computer programs that are models of how Earth's atmosphere works.

Although we prefer to rely on modern scientific observations, these have been done for a short time compared with the climate changes we need to know about. The use of thermometers, barometers, and other instruments to monitor climate began around 1860, and for many decades such monitoring was done only sparsely, in places such as major cities—London, New York—that are not representative of Earth's surface as a whole. Satellite measurements, on which we depend so much today for weather monitoring and forecasting, began at a significant scale only in the 1970s. Satellite measurements of sea ice, for example, began in 1978.

Historical records include recollections in books, newspapers, journal articles, personal journals, travelers' diaries, and farmers' logs. A classic review of these often fascinating historical accounts is in *Times of Feast, Times of Famine: A History of Climate Since the Year 1000.*[9] Although mostly these are recorded as qualitative data, we can sometimes get quantitative information from them, as in dates of wine harvests and small-grain harvests. As another example, a painting of a mountain glacier in Switzerland can be used to determine the elevation to which the glacier had descended by the year it was painted. Then, as the climate cooled further, someone may have written that the same glacier reached farther down the mountain, eventually blocking a river in the valley, which flooded and destroyed a town, whose elevation is also known. Additional important historical information on climate change and human societies and civilization can be found in the books by Hubert H. Lamb, such as *Climate, History and the Modern World.*[10]

How Our Ideas About Life on a Planet Have Changed Since 1990

Is the climate changing? Today the answer is simple: It is changing and has always changed—that's where I began in this book. One of the most accurate reconstructions is from the Vostok Ice Cores, taking us back more than 400,000 years (Fig. 12.2). Answering the other questions about climate change is more difficult and is part of the story of this chapter. Compare this with Figure 1.3 in Chapter 1, which shows 800,000 years of temperature change. Both ice cores show that the climate has always been changing. These recontructions for ice cores, amazing new science, lift us to a global view of life on Earth, past and present.

In my earlier book, *Discordant Harmonies*, I talked about the glory of a planetary view of life, with sunlight displaying the intense tropical colors on Hawaii's Mauna Loa, as the English Victorian traveler Isabella Bird described it in her wonderful 1880 book *Six Months in the Sandwich Islands*, and as I saw it there in the 1980s. At that time, the vision on Mauna Loa seemed a metaphor for the new science of an ecology for the entire Earth's life containing and life support system, which I and others had hoped would develop.[1]

On an October morning in 1982, I traveled the upper slopes of Mauna Loa, Hawaii, driving on a dirt road that climbed more than 11,000 feet above sea level. The trip was much as Isabelle Bird had described it a century earlier: from endless summer along the coast, from dense shade of the lush green tropical

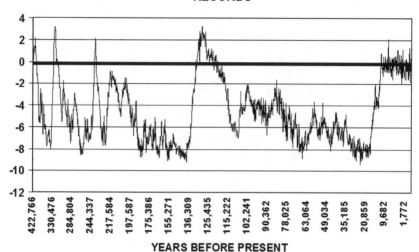

TEMPERATURE DIFFERENCE FROM MODERN RECORDS

YEARS BEFORE PRESENT

FIGURE 12.2 Scientific reconstruction of long-term temperature change, January 2000. (Historical Isotopic Temperature Record from the Vostok Ice Core.) The Vertical axis show temperature change in degrees centigrade. The data available from CDIAC represent a major effort by researchers from France, Russia, and the U.S.A. We ask as a professional courtesy [and as a condition for its use by you] that when you refer to this data set in publications you cite the following papers. J. Jouzel et al. [1987]. Vostok ice core: a continuous isotope temperature record over the last climatic cycle (160,000 years). *Nature, 329,* 403–408; J. Jouzel et al. [1993]. Extending the Vostok ice-core record of palaeoclimate to the penultimate glacial period. *Nature, 364,* 407–412; J. Jouzel et al. [1996]. Climatic interpretation of the recently extended Vostok ice records. *Climate Dynamics, 12,* 513–521; J. R. Petit et al. [1999]. Climate and atmospheric history of the past 420,000 years from the Vostok ice core, Antarctica. *Nature, 399,* 429–436).[8]

forest landscape heavily modified by people, through dry woodlands of 'Ohi'a trees opening to the sun at mid-elevations, and into endless sun-drenched winter seemingly untouched by human hands, a sparse, moonlike landscape of solidified streams of fresh black lava covering swaths of older brown lava.

On this morning, I was struck once again by the brilliant colors, as I had been on every trip from temperate zones to subtropics and tropics. Isabella Bird described it 100 years ago, writing that the coast near Hilo had "glades and dells of dazzling green, bright with cataracts," above which the "snow-capped mountains gleamed. . . . Creation," she wrote, "surely cannot exhibit a more brilliant green than that which clothes windward Hawaii."[11] On the way up, occasional bracken ferns struggling to grow in crevices of older lava were life's last outposts; once they were left behind, life seemed absent, the rich green of tropical vegetation only a distant haze below the thermocline, the hammer and shovel of our own civilization visible only in the dirt road that led upward. Mauna Loa seemed a place relegated to the physical Earth, untouched in one's vision by any life, a place of nature undisturbed. But in this brightness our vision betrays us, for life touches these slopes in the chemistry of the winds.

On this barren slope at 11,500 feet stood a small collection of buildings: observatories painted a light blue, in striking contrast to the gray and black rocks and the intense deep blue of the high-elevation sky, and from which a network of black electric cables and tubes swarmed outward to small towers. I had come to visit a famous observatory (today operated by the National Oceanographic and Atmospheric Administration), which looks not at the stars but at Earth. Since 1957 the concentrations of trace gases in the atmosphere, including carbon dioxide, have been measured at this laboratory.[12] The measurements began as part of the International Geophysical Year (July 1, 1957, to December 31, 1958), and the atmospheric monitoring on Mauna Loa was conceived of by pioneering oceanographer Roger Revelle (1909–1991), at the time director of the Scripps Institution of Oceanography. Revelle persuaded chemist Charles David Keeling (1928–2005) to begin and run that observatory, which he did with great care, creating one of the best and most carefully calibrated global environmental data sets so far.

Curiously, but not unusual in the history of environmental sciences, in the early 1960s the U.S. National Science Foundation temporarily halted support of the measurements, apparently considering them routine and therefore not of scientific interest. Yet the measurements have become one of the foundations of the global-warming debate, widely and continually used by all sides in the discussion. The International Geophysical Year took place during the Cold War, yet managed to maintain scientific cooperation among many nations. (As "Pogo" cartoonist Walt Kelly pointed out, it lasted 18 months, and only governments and scientists could create an Earth year that lasted so long.)

Revelle and Keeling chose Mauna Loa carefully. At 11,500 feet above sea level, the Mauna Loa Observatory stood above local effects of life on the big island, protected from the tropical forests and the tourists by atmospheric conditions. More than 2,000 miles downwind from the nearest continent, Mauna Loa is also distant enough to be protected from major concentrations of continental land plants and animals and major industrial centers and from their combined short-term local variations in the release of gases to the atmosphere. For these reasons, the upper slopes of Mauna Loa form an ideal place to measure the chemistry of our atmosphere. (Release of carbon dioxide from Mauna Loa's volcanoes could be the one artifact that could creep into the measurements.)

From a thin tower, the high-altitude air is pumped down and through a tube into one of the buildings, where instruments scan the moving air with infrared light and measure that light's absorption by carbon dioxide. Within those instruments, light creates information. Alongside are tanks of carefully calibrated gases to ensure that measurements made years apart are consistent with one another. These measurements are a key to life's activity. Like the mouse in Joseph Priestly's jar that you heard about in Chapter 9, all oxygen-breathing organisms give off carbon dioxide as they respire. And like the green plant in another of Priestly's jars, all photosynthetic organisms remove carbon dioxide from the air, converting the carbon to sugar and giving off oxygen as a waste product. Life's activity is summed up by the breathing in and out of carbon dioxide in the atmosphere. On Mauna Loa, the totality of the inhalations and exhalations of all the organisms in the Northern Hemisphere is recorded faithfully, and the observatory has become famous for its monitoring of life on Earth for a generation.

Over the years, the measurements at Mauna Loa have shown two clear patterns (Fig. 12.3): an annual oscillation, with a CO_2 decline in summer followed by an increase in winter, a periodic pattern as regular as the vibrations of a plucked guitar string; and, second, imposed on this rising and falling, a steady annual increase like a rising tone.[13]

The summer decline is the result of photosynthesis on the land in the Northern Hemisphere: the summation of the growing of the 'Ohi'a trees on Mauna Loa's slopes, the spruce and fir at Isle Royale, the oaks and hickories in Hutcheson Forest on the Atlantic coast, the acacia and baobabs on Tsavo's desolate plains, and all the other thousands of species of land plants in the hemisphere. Photosynthesis in the oceans contributes little to these seasonal patterns because most algae are short-lived and lack ways to store organic compounds for long periods.

The increase during the winter is the result of respiration without photosynthesis: the breaths of moose struggling in the snows on Isle Royale; of elephants, with their silent steps, on the Tsavo plains; of the Arctic lynx and rabbit, whose populations are varying wildly year to year; and of billions of people.

Life on the land in its totality touches the slopes of Mauna Loa invisibly, its effects brushed against the black rocks by the winds. Our civilization is part of this invisible touch, reaching the slopes in the form of a continual increase in the concentration of carbon dioxide in the atmosphere, which is a result of burning fossil fuels and clearing land, destroying forests and soils and converting their stored organic carbon to carbon dioxide.

Right or wrong, the proposal that there could be human-induced global warming was a late-twentieth-century novelty whose antecedents lay in the more obscure scientific literature published half a century and more ago. The Swedish chemist Svante Arrhenius (1859–1927) is generally credited with the first formal scientific analysis showing that carbon dioxide in the atmosphere could change Earth's surface temperature. In 1896 he wrote that carbon dioxide is "nearly insensible" to visible light, meaning that it did not absorb this light; that absorption was "chiefly limited to the long-waves part [that we call infrared]. . . . The influence of this absorption is comparatively small on the heat from the sun, but must be of great importance in the transmission of rays from the earth." He wrote that the last ice age could have come about because the concentration of carbon dioxide in the atmosphere was reduced to somewhere between 55% and 62% of its present concentration.[15]

The possibility that modern civilization might be changing Earth's climate first aroused wide discussion among scientists in 1938 with publications by the British engineer Guy S. Callendar (1898–1964), who found that measurements of carbon dioxide concentration in the atmosphere in the twentieth century were

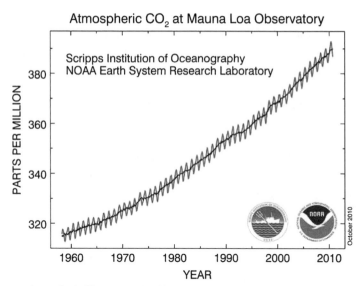

FIGURE 12.3 Atmospheric CO₂ measured at Mauna Loa, Hawaii, Observatory. These observations, begun by Charles Keeling in 1957, have continued since with one short pause. These show two consistent patterns: a steady overall rise in concentration and an annual oscillation representing summer and winter growth changes induced by living things.[14]

higher than those made in the nineteenth century.[16] Callendar suggested that the increase could be accounted for by the amount of carbon dioxide added to the atmosphere from the burning of coal, oil, and natural gas since the beginning of the Industrial Revolution, and he made calculations showing that this was consistent with how much fossil fuel had been burned. He also suggested that the increase might lead to a global warming.

Could it be that the machine age was changing the biosphere? Callendar was attacked by his scientific colleagues for this suggestion, some dismissing the notion simply on the grounds that nineteenth-century scientists could not have done as good a job as scientists in the 1930s, and therefore their measurements were inaccurate. But by 1990—actually, sufficiently demonstrated by the early 1970s, when the measurements extended for more than a decade—Callendar's suggestion of an increase in carbon dioxide in the atmosphere had been confirmed by the measurements at Mauna Loa.[17]

Callendar began a set of controversies that are still unresolved today, although much of the basic science about what changes climate has become so commonplace that it will not be new, as it was in 1990, to most readers. Even so, it is worthwhile to review the basics, familiar to anyone who has watched Al Gore's movie *An Inconvenient Truth*, read his book of the same name, read the many articles, or seen the many TV programs about climate change that are now a standard part of our culture.

Keeling first published his measurements on Mauna Loa in 1973 for 1957 to 1973 (Fig. 12.4). These showed that carbon dioxide concentration in the atmosphere had increased 3.5% in 15 years.[18] The average Earth surface temperature had been on an overall rise since at least 1860 (Fig. 12.5), so at that stage in the development of the science of global ecology, it seemed plausible—actually quite likely—that the increase in the atmosphere's concentration of carbon dioxide might be causing a global warming. This justified legitimate scientific research, both basic and applied. Basic research was justified because the connection between carbon dioxide concentration in the atmosphere and Earth's surface temperature seemed to be a valuable way to gain insight into how Earth's biosphere—its global life-containing and life-supporting system—worked. And the potential great practical effects justified applied research.

At that time, little was known about long-term past climate change compared to what is known today, and the rapid rise in carbon dioxide concentration in the atmosphere seemed unusual, perhaps unique, in Earth's history, as did the increase in average Earth surface temperature. We had none of today's information about how great past changes in temperature were and how fast these changes occurred, and data were similarly lacking about past changes in carbon dioxide concentration in the atmosphere. Although a lot was understood in a general way about the physics of energy exchange between a planet and its sun and the rest of the cosmos, what determined total climate change on a planet that had a great amount of liquid water, life, and plate tectonics was not well known. As an ecologist beginning my career, I believed that global warming was a quite likely result of human activities, and that research about how our atmosphere worked and about the ecological consequences of climate change was important. As I will explain in this chapter, much has changed as a result of our greatly increased knowledge about these topics, which has led me to a different assessment of our global environmental situation.

From the vantage point of Mauna Loa, scientists began to observe our planet the way Priestly watched the mouse in the bell jar, discovering that we influence Earth in much the same way that microbes affect a moose's stomach. As discussed earlier, in 1990 the idea that people could affect the entire Earth in such an invisible way was relatively new, possible only after the technological and scientific revolutions of the previous two centuries. As recently as 1967, such human influence seemed preposterous even to most of those who had thought deeply about nature. For example, in his classic book *Traces on the Rhodian Shore: Nature and Culture in Western Thought from Ancient Times to the End of the Eighteenth Century*, Clarence Glacken dismissed the argument that large, uninhabited areas of Earth, whether forests or oceans, might serve a useful purpose, as suggested by some philosophers: "It is the kind of argument which a defender of the design [argument for the creation of the universe] must be prepared to make," he wrote.[20]

Glacken was saying that those who believe the universe is too special to have come about simply by chance, and therefore must have had a maker—our God—will find themselves making this argument, but nobody else would. No rationalist,

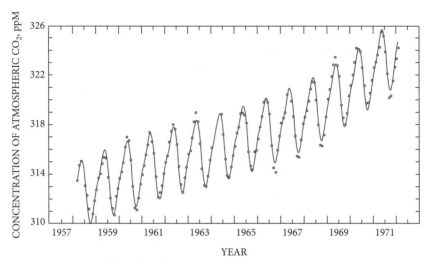

FIGURE 12.4 Charles Keeling's first report of measurements of changes in atmospheric carbon dioxide concentration on Mauna Loa, Hawaii. These confirmed that carbon dioxide was increasing in the atmosphere.[17]

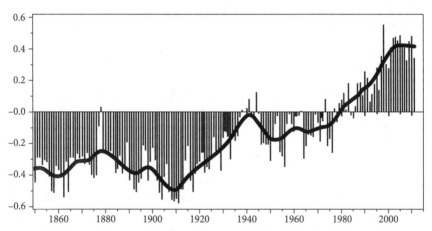

FIGURE 12.5 Changes in Earth's average surface temperature from 1860 to 2011, as reconstructed by the Hadley Meteorological Center, Great Britain. The vertical axis in change in change in degrees centigrade. The heavy line is the average for specific years. The surface cooled between 1940 and the mid-1950s, did not change much for about ten years, then warmed dramatically until 2000. Since then, the temperature has remained within a narrow range.[19]

Glacken infers, would make that argument; it is just too preposterous. And Glacken is no hack writer. His book was "hailed as a monumental work of scholarship and synthesis, bringing together ideas on this vast and universal topic as they never had been before, transcending geography as a discipline but also being recognized as one of the truly great books written by a geographer in this century," according to the University of California, Berkeley, where he had been chairman of the Geography Department. His book was also one of the major influences on me when I wrote *Discordant Harmonies*. In short, even in the 1960s, the decade of modern environmental awakening, the best thinker about the history of ideas about people and nature could not conceive of people having a global environmental effect, and the possibility was rarely mentioned by the public media.

Just 20 years later, in the late 1980s, during the transition that marks the end of the machine age, a headline in a local newspaper—"Restoring Tropical Forests May Ease Warming Trend"—suggested that replanting forests in the tropics might help to stave off a global climate change.[21] In a few decades, we had jumped from dismissing the possibility that any life might affect Earth at a global level to considering a newspaper proposal that we make use of this effect to modify the climate.

Over another 20 years we have jumped again, so that by 2010 people who *didn't* believe that we were changing the climate globally were dismissed as "deniers" and became the butt of derisive jokes on TV's "The Daily Show with Jon Stewart" and PBS's delightful comedy news program "Wait, Wait, Don't Tell Me."

WHAT WE HAVE LEARNED SINCE 1990

We can now return to the basic global-warming questions and put them into today's context, with the great increase in knowledge that has occurred over the past several decades. Because the study of climate, including attempts to reconstruct past climate, is a fast-developing field, you will understand that what I write in this chapter only be can based on scientific knowledge available right now. Of course, this is a controversial topic, and some of you may want to consider any new information that may be available at the time you are reading this. With that said, I turn to the question. Is the climate changing? The answer continues to be simple: It is changing and has always changed. Many studies extend our understanding of climate change far back into the past.

Among the major scientific advances since 1990 is our increasing knowledge of when and how climates changed in the past, and one of the most valuable tools in acquiring this information is the use of ice cores from the Arctic and Antarctic that I have shown. The methods are sophisticated and indirect, and not easy to explain, but the basic idea is simple. As glacial ice forms, it encloses pockets of air that remain sealed inside for millions of years. The carbon dioxide concentration in that air is thus exactly what it was in the atmosphere back then. Based on the decay of radioisotopes in the ice, and on other kinds of observations, scientists can estimate the time that the ice and its air pockets formed.

The temperature of Earth's surface is estimated in a rather indirect and obscure way, referred to as the "oxygen isotope ratio method," and only a few experts are able to do these estimates. Oxygen exists in two isotopes in our atmosphere: a lighter oxygen 16 and a heavier oxygen 18. The speed with which these molecules move is determined by their temperature and also by their weight—the heavier isotope moves more slowly than the lighter one (as anyone who has gained a bit too much weight and then dieted and lost it can relate to). This affects the ratios of the two isotopes when oxygen is combined in a molecule: The warmer it is, the more of the lighter isotope will be moved from one place to another than the heavier one. Scientists who specialize in this correlate the ratios of the two isotopes to what the temperature must have been at the time. These ratios are also frozen in the ice.

In addition to the reconstructions from the Vostok Ice Cores (see Fig. 12.2) taking us back more than 400,000 years, and another taking us back 800,000 years (Fig. 1.3 in Chapter 1), there are perhaps less accurate but far longer reconstructions. These indicate that the climate during the Precambrian era—more than 550 million years ago—averaged a relatively cool 12°C, compared to the modern average of 14.57°C. The Precambrian was the time dominated by bacteria and their relatives, known as prokaryotic organisms, single-celled creatures whose cells are structured differently from ours and other, later-evolved multicellular life-forms. Things warmed up to about 22°C in the Cambrian period, 545 to 495 million years ago, the time of the rise of multicellular creatures and a heavily oxygenated atmosphere. Then the climate got very cool in the Ordovician/Silurian transition, around 443 million years ago. A mass extinction of marine species—about 85% of those species—took place in the late Ordovician. Evidence suggests that what geologists call "a brief glacial interval" caused these extinctions.[22]

The climate warmed again in the Devonian, 417 to 363 million years ago, a time known as the "age of the fishes" because so many species of fish evolved, including ancestors of sharks and of the first amphibians. On land, it was also the time when forests first evolved. Another mass extinction of marine species took place toward the end of the Devonian. It remained comparatively warm during the Carboniferous, 363 to 290 million years ago, the time of the formation of great deposits of coal.

The climate cooled a lot at the end of the Carboniferous, and warmed again in the Triassic, 245 to 208 million years ago, one of geological time periods dominated by dinosaurs. It's been quite a roller-coaster ride.

IS IT GETTING WARMER?

Has Earth's average surface temperature risen in modern times? A warming trend began around 1850 and lasted until the 1940s, when temperatures began to cool again, followed by a leveling off in the 1950s and then a further drop during the 1960s. After that, the average surface temperature rose. No one disagrees that Earth's average surface temperature rose from the mid-1960s until the beginning of the twenty-first century. (Temperature change in the past nine years is more ambiguous; see Fig. 12.5.) According to the National Oceanographic and Atmospheric Administration (NOAA), the global average temperature since 1900 has risen by about 0.8°C (1.5°F), averaging about 0.2°C (0.36°F) per decade in the past 30 years.[23] At the time I am writing this, the past two decades have been the warmest since global temperatures have been monitored.[24] The eight warmest years have occurred since 1997, and the warmest years since direct surface air temperature has been measured were 1998 and 2007. So the answer is yes, it has been getting warmer, on average, since the middle of the nineteenth century.

WARMER THAN EVER BEFORE?

Look again at the reconstructions of Earth's surface temperature in Figure 12.2. There were quite a number of times when temperatures before the twentieth century were warmer than the warmest in modern times (Fig. 12.6). The temperature averaged above 14.57°C, the modern maximum, from approximately 420,000 years ago to 410,000 years ago; 324,000 to 323,000; 239,000 to 237,000; 130,000 to 124,000; and during several shorter spans, around 11,191, 7,555, 4,423, and 2,847 years ago.

FASTER THAN EVER BEFORE?

One of the most worrisome things about climate change and the buildup of carbon dioxide from 1957 to 1990 was the apparent rapidity of both. By 1990 it seemed, at least to us ecologists, that this was unprecedented and therefore something of special concern.

Until recently, it was thought that the climate changed very slowly before the Industrial Revolution—temperatures were believed to change no faster than 2°F (1°C) in 1,000 years. But new studies, summarized in J. D. MacDougall's 2006 book, *Frozen Earth: The Once and Future Story of Ice Ages*, reconstruct climate back 400,000 years from the ice cores obtained in Antarctica and deep ocean waters off Greenland and indicate that there have been many intervals of very rapid temperature change. Some of the most dramatic were changes of 14°F to 24°F (7°C to 12°C) within 50 years.[25] That's a lot faster and greater than anything projected from the global climate models for the immediate future. One rapid cooling occurred in the Northern Hemisphere at the end of the last ice age. Known as "the Younger Dryas event," it began approximately 12,900 years ago and took place over approximately six centuries. It appears to have been caused by large-scale changes in atmospheric and oceanic circulation.[26]

HOW MUCH ICE IS MELTING, AND HOW FAST?

A major concern is whether global warming will lead to a great decline in the volume of water stored as ice. Melting of glacial ice raises the mean sea level, which could "drown" small island nations and increase ocean storm damage to major

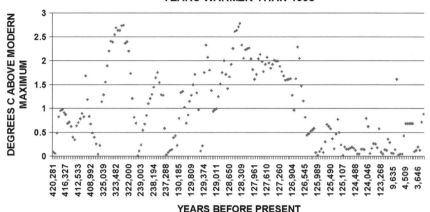

FIGURE 12.6 Years warmer than 1998, the warmest years (reconstructed) records from the Antarctic ice core. In this graph, only those past times are shown when the reconstructed temperature exceeded that of 1998, the warmest modern year to date. The data are the same as in Figure 12.2. (P. Brohan, J. J. Kennedy, I. Harris, S. F. B. Tett, and P. D. Jones [2006]. Uncertainty estimates in regional and global observed temperature changes: a new dataset from 1850. *J. Geophys. Res, 111*, D12106, doi:10.1029/2005JD006548).

cities along coasts. That melting would also change the reflection and reradiation of sunlight reaching the surface, in turn changing the climate. Mountain glaciers are also often major sources of water for lower-elevation ecosystems and for cities, towns, farms, and industry.

Of course, glaciers have been on average melting since the end of the last ice age. The most recent continental glaciation ended about 12,500 years ago with a rapid warming, perhaps as brief as a few decades.[27] This was followed by a short global cooling about 11,500 years ago. At present, many more glaciers in North America, Europe, and other areas are retreating than are advancing. In the Cascades of the Pacific Northwest and the Alps in Switzerland and Italy, retreats are accelerating. For example, all eight glaciers on Mt. Baker in the Northern Cascades of Washington were advancing in 1976. Today all eight are retreating.[28] Some estimates, basically linear extrapolations from present conditions, are that if present trends continue, all glaciers in Glacier National Park in Montana could be gone by 2030 and most glaciers in the European Alps could be gone by the end of the century.

But not all melting of glaciers is due to global warming. For example, the study of a decrease in glacier ice on Mt. Kilimanjaro in Africa shows that melting is not the primary cause of the ice loss. The glaciers of Kilimanjaro formed during the African Humid Period about 11,000 to 4,000 years ago. Although there have been wet periods since then—notably in the nineteenth century, which appears to have led to a secondary increase—conditions have generally been drier.

Since they were first observed in 1912, the glaciers of Kilimanjaro have decreased in area by about 80%. The ice is disappearing not because of warmer temperatures at the top of the mountain, which are almost always below freezing, but because less snowfall is occurring and ice is being depleted by solar radiation and sublimation (ice is transformed from solid state to water vapor without melting). More-arid conditions in the past century led to air that contained less moisture and thus favored sublimation. Much of the ice depletion had occurred by the mid-1950s.[29]

In addition to many glaciers melting back, the Northern Hemisphere's sea ice coverage in September, when the ice is minimum, has declined an average of 10.7% per decade since satellite remote sensing became possible in the 1970s. Recent studies, again basically linear extrapolation, forecast that if present trends continue, the Arctic Ocean might be seasonally ice-free by 2030 to 2050.[30]

On the other hand, the central ice cap on Antarctica has grown during the same time. Satellite measurement from 1992 to 2003 suggests the East Antarctica ice sheet increased in mass by about 50 billion tons per year during that period. Greenland's ice sheets have been one of the major focuses of the debate on whether Earth's glaciers are melting, with some scientific papers claiming these glaciers are melting and others that they are increasing. One of the most thorough and careful analyses concluded that "current Greenland warming is not unprecedented in recent Greenland history. . . . there is a general agreement that the ice sheet is thinning close to its margins and thickening in the ice sheet interior." The paper goes on to state that "a well-documented increase in the ice sheet melt area [occurred] during recent years" but also found "evidence of glacier acceleration at least in some parts of Greenland."[31]

It's clear that the polar regions are complex. Changing patterns of ocean and atmosphere circulation in the Arctic and Antarctic regions influence everything from snowfall to melting of glacial and sea ice and movement of glacial ice.

One of the major concerns in the global-warming debate is the fate of sea ice—whether it is decreasing in a major way that could lead to great changes in the world's climate, oceans, and life, and whether this is a novel change, never having happened in tens of thousands of years. An example of the use of historical records is a recent reconstruction of changes in the southward extent of Arctic sea ice, based on what whalers wrote down in their sailing ships' logbooks each day as they hunted bowhead whales between Alaska and Siberia from 1850 to 1910. These are compared with recent observations of Arctic sea ice from modern powered ships (Fig. 12.7). The comparison of historical and modern ship observations shows that at the end of winter, in May, the southern extent of sea ice in the middle of the nineteenth century was similar to that in the 1970s. In contrast, Arctic sea ice between Alaska and Siberia extended much farther south in the middle of the nineteenth century than in the 1970s.[32]

WHAT CAUSES CLIMATE CHANGE?

Research during the past several decades has added greatly to our understanding of what causes climate change on Earth. As explained earlier, *climate* means the average weather. Weather and climate are the result of basic laws of physics. Any physical object at any instant has a temperature, which changes with the amount of energy received and released. My friend and colleague Harold J. Morowitz (1927–) wrote about this in his 1992 book *The Thermodynamics of Pizza.*[33] He began with a question: Why is it always the cheese in a pizza, not the crust or the tomato sauce, that burns the roof of your mouth? Harold explained the basics of energy exchange between a solid body and its environment, and showed that it was in part the high heat-holding capacity of the water in the cheese that caused it to burn your mouth.

Admittedly, a slice of pizza is simple compared with Earth's surface. In terms of energy exchange, our planet has several "surfaces." There is the top of the atmosphere, which receives energy from the sun and reflects and reradiates energy from Earth's surface and from the lower atmosphere below—a surface that, of course, includes solids and liquid water, further complicating the exchanges.

The top of the atmosphere has its energy budget, as does the surface we walk and swim on below and whose temperature and climate are of most concern to us. Lots of things and processes affect the exchange of energy to and from our planet, and most of these are changing over time, so our planet's rate of energy exchange and consequently its surface temperature are always changing.

Climatologists have estimated that the difference between a twentieth-century climate and that of an ice age represents a lowering of Earth's average temperature by 7° to 13°F. Therefore a 2°F shift would be expected to cause a large change in Earth's surface characteristics.

Fundamental to climate change are changes in the sun's brightness and periodic changes in Earth's orbit around the sun and the angle of its rotation on its axis. Over geologic time, these combine to set up energy variations that allow for cold or warm times. When these conditions combine to make things colder, then the transparency of the atmosphere and the reflective properties of Earth's surface can determine whether an ice age or a warm time (interglacial) will occur.

MILANKOVITCH'S INSIGHTS: A WOBBLING TOP ON A VARYING ORBIT

In the 1920s Milutin Milankovitch (1879–1958), a Serbian engineer, looked at long-term climate records and realized that there were three major climate cycles: one with a duration of about 100,000 years, another of about 40,000 years, and a third of about 20,000 to 40,000 years. He realized these could be explained by the way Earth revolved on its axis and rotated around the sun. Our spinning Earth is like a wobbling top following an elliptical orbit around the sun. Three kinds of changes occur.

May 1850–1869

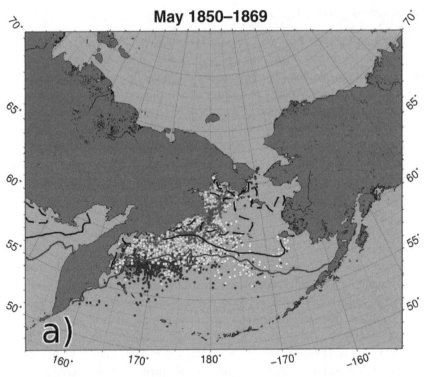

September 1850–1869

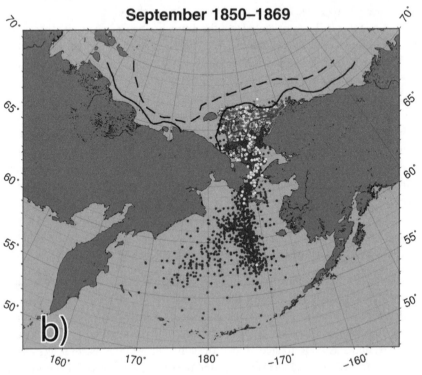

First, the wobble means that Earth is unable to keep its poles at a constant angle in relation to the sun. Right now, the North Pole points to Polaris, the North Star, but this changes as the planet wobbles. The wobble makes a complete cycle in 26,000 years. Second, the tilt of wobble also varies, over a period of 41,000 years. Third, Earth's elliptical orbit around the sun changes. Sometimes it is a more extreme ellipse; at other times it is closer to a circle. This variation occurs over 100,000 years.

The first two cycles alter the amount of sunlight that reaches each latitude of Earth; the first affects the total amount of sunlight reaching Earth. Sometimes the wobble causes the Northern Hemisphere to tilt toward the sun (Northern Hemisphere summertime) when Earth is closest to the sun. At other times, the opposite occurs—the Northern Hemisphere is tipped away from the sun (Northern wintertime) when Earth is closest to the sun. Milankovitch showed that these variations correlated with the major glacial and interglacial periods. They are now called Milankovitch cycles.[34]

While Milankovitch cycles are consistent with the timing of variations in climate, they don't account for all the large-scale climatic variations in the geologic record. It is perhaps best to think of these cycles as setting up some situations in which climates cold enough to produce continental glaciations are possible, and other situations in which climates that cold are unlikely or impossible.

THE GREENHOUSE EFFECT AND THE MUSIC OF THE SPHERES

Visible and infrared light waves consist of a range of frequencies, each wavelength vibrating at one of the frequencies. Think of these as being like sounds emitted from a musical instrument vibrating particles in the air, each tone vibrating at its own frequency—beats per second. Then think of a chemical molecule that those light-wave vibrations reach. The molecule resonates with some of these frequencies and interferes with others. It's something like the way a violin responds to the vibrations of its strings, resonating with some sound wavelengths from the strings and not with others—the light reaching carbon dioxide sort of like a cosmic

FIGURE 12.7 How Arctic sea ice between Alaska and Siberia has changed since the mid-nineteenth century. In these maps, the dots are observations recorded in the logbooks of American sailing ships hunting bowhead whales between 1850. White dots are where the whalers saw sea ice, blue dots only open water. Red lines show the southernmost extent of any ice reported by modern powered ships between 1972 and 1982. Black solid and dashed lines are the 20% and 40% mean ice concentration, respectively, for 1972–1982. May represents the ice extent at the end of winter, September at the end of summer.

(Source of historical and 1970s sea ice observations is Andrew R. Mahoney, John R. Bockstoce, Daniel B. Botkin, Hajo Eicken, and Robert A. Nisbet (2011). Sea Ice Distribution in the Bering and Chukchi Seas: Information from Historical Whaleships' Logbooks and Journals. *Arctic.* 64, (4): 465–477.)

electromagnetic song interacting with many "soundboards" (actually lightboards). Each compound acts like a unique instrument, giving out its own signal. The amount of light allowed to pass through or reflected by a compound is one of the primary ways to determine the chemical composition of atmospheres and enables astronomers to determine the chemical composition of the surfaces of astronomical bodies. It is also used to detect the existence and amount of any specific compound in a sample of gas on Earth.

Until 2008 the scientific consensus, widely disseminated to the public by Al Gore's 2006 movie *An Inconvenient Truth* and by many articles and some books, was that over hundreds of thousands of years carbon dioxide increased and then the average surface temperature increased. Critics pointed out that a careful look at the graphs Gore used showed that in fact the temperature rose first and carbon dioxide followed, making it impossible for carbon dioxide to have caused the warming.

At first this was rejected by most of the scientific community, but the scientific consensus turned around by 2008 when an 800,000-year record from Antarctica and shorter records from Greenland glaciers became available (Fig. 12.8). These confirmed that rises in carbon dioxide lagged temperature rises by perhaps as much as 800 years or more. A year before that, the consensus had already turned around so greatly that an article in *The New Scientist* said: "These lags show that rising CO_2 did not trigger the initial warming at the end of these ice ages—but then, no one claims it did."[35]

It is good science that changes its conclusions as evidence increases and improves. What was unfortunate here was that by 2006 global warming had become so highly politicized that the leading role of carbon dioxide in climate change had to be true in spite of appearances to the contrary—as Cooper, who studied Isle Royale at the beginning of the twentieth century, might have said.

SOLAR CYCLES

The sun goes through cycles, sometimes growing hotter, sometimes cooler. Today, solar intensity is observed directly with telescopes and other instruments. Variations in the sun's intensity in the past can be estimated by the amount of certain radioactive nuclei that are found in glacial ice cores. Over shorter geologic time periods, carbon-14 provides such an index. Over longer time periods, beryllium-10 provides an index of solar activity (This isotope has a very long half-life—half of the amount produced is still around after more than 1.3 million years.) Carbon-14 and beryllium-10 are produced in Earth's upper atmosphere by cosmic rays, and the amount of these rays that reaches the atmosphere is affected by the solar wind—the "wind" of particles emitted by the sun. A hotter sun produces a stronger solar wind, which results in a smaller quantity of carbon-14 and beryllium-10 reaching Earth's surface and getting trapped in glacial ice. These methods indicate that during the Medieval Warm Period, from approximately AD 950 to 1250, the amount of

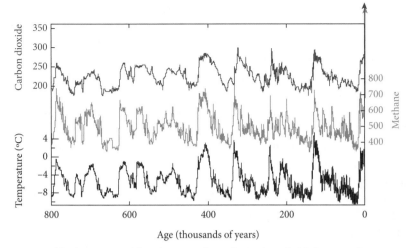

FIGURE 12.8 The longest record from ice cores shows that carbon-dioxide increases lag temperature increases. (E. Brook. [2008]. Windows on the greenhouse. *Nature, 453*, 291-292.)

solar energy reaching Earth was relatively high, and that minimum solar activity occurred during the fourteenth century, coincident with the beginning of the Little Ice Age, showing that variability of sunlight explains part of Earth's climate variability.[36]

Recent astronomical research, much of it since 1990, shows that changes in the sun's energy output correlate quite well with the average surface temperature in the Arctic (Fig. 12.9A), and this correlation is stronger than between Arctic temperature and atmospheric carbon dioxide concentration (Fig. 12.9B). It gets warmer in the Arctic when that great lightbulb in the sky brightens up. Before the scientific age, the sun seemed to change only in its diurnal pattern, day to night, but otherwise seemed constant.

ANCIENT LIFE REMOVED CARBON DIOXIDE FROM EARTH'S ATMOSPHERE

During the ancient history of Earth, the removal of carbon dioxide from Earth's atmosphere by the earliest photosynthetic organisms was like opening a window in a greenhouse; Earth's surface must have cooled. As life evolved on Earth and grew more abundant, it changed Earth's chemistry and the characteristics of its surface, particularly the reflection and absorption of radiant energy. Astronomers tell us there is evidence that the sun became hotter during Earth's early history, so these two processes may have tended to counteract each other. As the sun heated up, the amount of carbon dioxide in the atmosphere declined; as a result, Earth warmed less than a lifeless planet would have. Because of the high concentration of carbon dioxide in the Venusian atmosphere, Venus's surface is much hotter than it would be if the planet supported life.

In addition to producing and taking up greenhouse gases, life affects climate in several other ways. It changes the brightness of the surface—how much of the sun's energy is absorbed and reflected. It also increases and decreases the amount of dust in the atmosphere; changes the roughness of the surface; and evaporates water, which turns into clouds, cooling things. Life decreases the amount of dust in the air by reducing erosion; this warms the surface. Forest and prairie fires add ash particles to the air; this cools the surface. The surface roughness is a relatively minor effect. Winds are slowed down and usually stopped completely at the surface (except with hurricanes and tornadoes, where the winds move things at the surface). This is a frictional loss of energy that changes the way the atmosphere moves over the surface, which can in turn affect climate. Forests make a rougher surface than ice, and slow the winds more. An Earth with more forest and less ice than today would tend to slow the winds more than our present solid Earth does.

Any major change in the abundance and distribution of life on Earth's surface will change the reflection and absorption of visible light and infrared radiant energy and therefore the surface temperature. Some human activities, particularly those of a technological civilization, can cool the climate by introducing into the atmosphere pollutants, such as dust and smoke. Some speculate that even fires lit by prehistoric man could have influenced the climate, cooling things a bit because the fires would have put ash particles into the atmosphere. Those particles reflect sunlight, letting less reach the surface and therefore cooling the surface. And as I've already briefly mentioned, in the 1980s, before global warming became such a hot topic, American astrophysicist Carl Sagan (1934–1996), British astrophysicist Michael Rowan-Robinson,[37] physicist and professor of atmospheric science Richard Turco, and some leading climatologists were warning us about a nuclear winter that could happen because a thermonuclear war would put so much dust into the atmosphere that we would experience a new ice age.[38]

Although the greenhouse effect of carbon dioxide was first proposed in the nineteenth century and then again in the early twentieth, only much later in the twentieth century did scientists come to understand that other small, trace compounds that are in the atmosphere because of life—methane and oxides of nitrogen—could have similar effects on climate. These also absorbed infrared light at wavelengths that atmospheric physicists refer to as being in the "atmospheric window." This window includes those wavelengths in the infrared that are not absorbed by water vapor, and therefore an atmosphere without carbon dioxide and these biologically produced compounds is transparent to these wavelengths. Through this window a large fraction of the heat from Earth is released into space. This loss of heat was essential for life.

At the heart of the argument that we are undergoing human-induced global warming is that the increase in Earth's average surface temperature is caused by, and therefore strongly correlated with, an increased concentration of carbon dioxide in the atmosphere due to the burning of fossil fuels, destruction of land

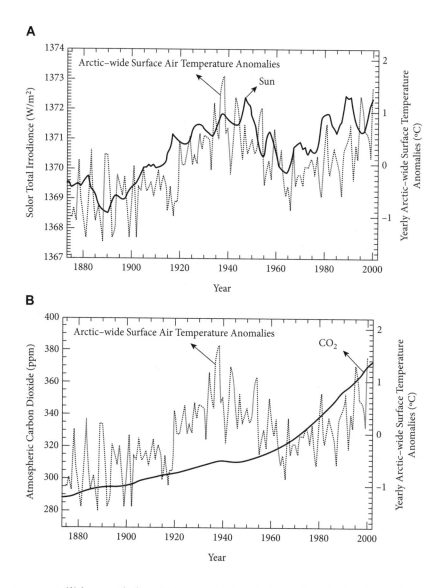

FIGURE 12.9 (A) Average solar intensity (solid lines; technically the "irradiance"). Since 1880, this has varied from year to year, but only by about 2%, from the low point (1890) to the high point (1950 and again in 1990 and 2000). But the sun emits so much energy that this variation appears to affect our climate. For example, Arctic-wide air temperature changes (dotted lines) pretty much parallel changes in sunlight intensity. Temperature is reported as "anomaly," which is the difference between what climatologists call Earth climate's "normal," the temperature from 1960 to 1980 or 1990, and what it is measured at or estimated to be in a specific year. Here that temperature "anomaly" is correlated with the estimated total solar irradiance (top panel) and with the atmospheric carbon dioxide (CO_2) mixing ratio (bottom panel; solid lines) from 1875 to 2000. (B) Change in carbon dioxide concentration and average Arctic surface temperature since 1880. (A, B: W. H. Soon [2005]. Variable solar irradiance as a plausible agent for multidecadal variations in the Arctic-wide surface air temperature record of the past 130 years. *Geophysical Research Letters, 32*, L16712, doi:10.1029/2005GL023429; and W. Soon [2009]. The solar Arctic connection on multidecadal to centennial timescales: Empirical evidence, mechanistic explanation, and testable consequences. *Physical Geography, 30*, 144–184.)

vegetation, and, to some extent, human-induced increases in methane, nitrogen oxides, and chlorofluorohydrocarbons. Because the small, biologically produced compounds absorb infrared light in a portion of the atmospheric spectrum that is otherwise transparent, very small changes in the amounts of these compounds in the atmosphere could have very large effects on Earth's temperature. This was and is a key assumption in the global climate models through the 1980s and until the present. Evidence for changes in the surface temperature and for carbon dioxide and methane concentrations in the atmosphere came primarily from glacier ice cores in Antarctica, whose glaciers are the oldest and provided the oldest records, going back about 800,000 years.

No one doubts that carbon dioxide is a greenhouse gas and that in an atmosphere with no other greenhouse gas, no dust, no water vapor, and a solid planetary surface without liquid water or ice, carbon dioxide concentrations will change the climate. The problem yet unsolved is what combination of factors dominates climate change on a planet that has life, plate tectonics and volcanic activity, and water as vapor, liquid, and ice. But that's a much more subtle question unlikely to have a simple single answer.

It is important that, at least for Arctic temperatures, the correlation between temperature and carbon dioxide is not nearly as good as between temperature and sunlight intensity (compare Figs. 12.9A and 12.9B). In addition, it is just as possible to argue, as some have done with the ice-core data, that it is temperature change that causes the carbon dioxide change. This contrary argument is also based on some other data, but in these cases it is somewhat subtle, making use of information about El Niño events—such as changes in ocean currents, upwellings, and temperatures—that have a periodicity of about seven years. An analysis that includes these variations provides the basis of the argument that temperature causes carbon dioxide change, not the other way around as explained later.

Suffice it to say, without having to take sides in the debate at this point, that the causality is much more complicated than the popular media would have you believe, and that the role of carbon dioxide in actual biospheric temperature changes is still an open question. What we can agree on is that both Earth's surface temperature and atmospheric CO_2 concentration have increased overall since 1880, and that the temperature rose faster between 1975 and 2000 than it had between 1880 and 1975.[39]

As evidence has accumulated, methane has come forward as a more and more important greenhouse gas. The atmospheric concentration of methane more than doubled in the past 200 years. Methane is an important biological product as well as a mineral. It is the primary compound in natural gas. Certain bacteria that can live only in oxygenless atmospheres produce methane and release it. These bacteria live in the guts of termites and in the intestines of ruminant mammals, such as cows, which produce methane as they digest woody plants. The bacteria also live in oxygenless parts of freshwater wetlands, where they decompose vegetation, releasing methane as a decay product. Methane is also released with seepage from oil fields and from methane hydrates. Methane is released naturally from mineral resources and from these biological activities.[40]

Our activities, too, release methane—from landfills, burning of biofuels, production of coal and natural gas, and agriculture, such as raising cattle and cultivating rice. As with carbon dioxide, there are important uncertainties in our understanding of the sources and sinks of methane in the atmosphere. Some analyses suggest that this gas contributes approximately 12% to 20% of the anthropogenic greenhouse effect.

Human activities have contributed other greenhouse gases to the atmosphere: chlorofluorocarbons (CFCs) and nitrogen oxides. CFCs have been the primary refrigerant gases, because they are stable and nontoxic. They are also used as propellants in spray cans. Because they affect the stratospheric ozone layer and also play a role in the greenhouse effect, the United States banned their use as propellants in 1978. In 1987, 24 countries signed the Montreal Protocol to reduce and eventually eliminate production of CFCs and accelerate the development of alternative chemicals. As a result of the treaty, production of CFCs was nearly phased out by 2000. Prior to that, CFCs in the atmosphere were increasing about 5% per year. Some estimates are that approximately 15% to 25% of the anthropogenic greenhouse effect may be related to CFCs.

Nitrous oxide is increasing in the atmosphere and is estimated to contribute as much as 5% of the anthropogenic greenhouse effect.[41] Agricultural fertilizers and burning of fossil fuels add to the concentration of this gas in the atmosphere.

As I have explained, if Earth's average temperature is to remain constant, there must be a balance between the energy received from the sun and the energy lost to space. If the rate of loss decreases while the amount of sunlight that reaches Earth remains constant, the temperature must rise, and as it does, the rate of energy loss by radiation will increase. In this way, a new steady-state temperature would be attained, higher than the previous one. This steady state would persist only if the energy inputs and outputs remained constant, which, as should be clear by now, they never really are for very long.

THE COLOR OF EARTH'S SURFACE AFFECTS CLIMATE AND WEATHER

The color of any surface affects the absorption, reflection, and transmission of light radiation. A completely black surface is both the perfect absorber of that energy (as you would know from wearing black clothing out in bright sun) and the perfect transmitter. An ideally perfectly black body absorbs 100% of the electromagnetic energy (light, infrared, radio waves, cosmic rays, etc.) it receives, and transmits 100% of the energy from the interior to the external environment. An ideally perfect white surface is just the opposite—the best reflector and worst transmitter. An ideally perfectly white body would absorb none of the incoming electromagnetic radiation, reflecting 100% of it and transmitting 0% from the interior to the external environment. (For those of you unfamiliar with this topic in physics, here is the difference between electromagnetic emission and reflectance: A surface of a body receives energy from inside. When that energy reaches the surface, it can either pass through it and be *emitted* by the surface, or blocked and made to stay within the material of the body. In the latter case, the body then heats up. Energy reaching a body from outside can be absorbed or *reflected* by the surface.)

All that we see as colors have absorbing and reflecting properties in between perfect black and perfect white. Thus a very dark rock surface exposed near the North Pole would absorb most of the sunlight it received in the summer, warming the surface and the air passing over it. When a glacier spreads out and covers that rock, it reflects most of the incoming sunlight, cooling both the surface and the air that comes in contact with it. Ice makes ice, so to speak. Bare dark rock leads to the exposure of more bare dark rock.

The color of vegetation also affects the climate and weather in the same way. If vegetation is darker than the soil, it warms the atmosphere. If it is lighter than the soil, it cools the surface.

ROUGHNESS OF EARTH'S SURFACE AFFECTS THE ATMOSPHERE

Above a completely smooth surface, air flows smoothly. Above a rough surface, air becomes turbulent and gives up some of the energy in its motion. This in turn can affect the weather. Forests are a much rougher surface than smooth rock or glaciers, so in this way, also, vegetation affects weather and climate.

THE OCEANS AND CLIMATE CHANGE

The oceans play an important role in climate, because two-thirds of Earth is covered by water, and water has the highest heat-storage capacity of any compound, so a very large amount of heat energy can be stored in the world's oceans. There is a complex, dynamic, and ongoing relationship between the oceans and the atmosphere. If carbon dioxide increases in the atmosphere, it will also increase in the oceans, and over time the oceans can absorb a very large quantity of CO_2. This can cause seawater to become more acidic ($H_2O + CO_2 = H_2CO_3$) as carbonic acid increases.

Ocean currents oscillate, producing warmer or cooler periods of a few years to a decade or so. These natural oscillations result at least in part from the interaction of the much lighter air in the atmosphere and the much denser ocean water. Some scientists attribute the cool winter of 2009–2010 to natural ocean–atmosphere oscillations, and also suggest that these caused a cool year in 1911 that froze Niagara Falls.

EL NIÑO AND CLIMATE

A curious and historically important climate change linked to variations in ocean currents is the Southern Oscillation, known informally as El Niño.[42] From the time of early Spanish settlement of the west coast of South America, people observed a strange event that occurred about every seven years. Usually starting around Christmas (hence the Spanish name "El Niño," referring to the Christ child), the ocean waters would warm up, fishing would become poor, and seabirds would disappear.

Under normal conditions, there are strong vertical, rising currents, called upwellings, off the shore of Peru. These are caused by prevailing winds coming westward off the South American continent. They move the surface water away from the shore and allow cold water to rise from the depths, along with important nutrients that promote the growth of algae (the base of the food chain) and thus produce lots of fish. Seabirds feed on those fish and live in great numbers, nesting on small islands just offshore.

El Niño occurs when those cold upwellings weaken or stop rising altogether. As a result, nutrients decline, algae grow poorly, and so do the fish, which either die, fail to reproduce, or move away. The seabirds, too, either leave or die. Since rainfall follows warm water eastward during El Niño years, there are high rates of precipitation and flooding in Peru, while droughts and fires are common in Australia and Indonesia. Because warm ocean water provides an atmospheric heat source, El Niño changes global atmospheric circulation, and this causes weather changes in regions far removed from the tropical Pacific.[43]

WHAT SCIENTIFIC OBSERVATIONS ARE TELLING US TODAY ABOUT CLIMATE CHANGE

In sum, here is what today's scientific observations and reconstructions of the past tell us. The climate has always been changing. As I said earlier, a warming trend began around 1850 and lasted until the 1940s, when temperatures began to cool again, followed by a leveling off in the 1950s and then a further drop during the 1960s. After that, the average surface temperature rose. No one disagrees that Earth's average surface temperature has been rising since the mid-1960s. But we know from the geologic record that before the nineteenth century, when temperatures began to be recorded, it had at times been as warm, in fact warmer, than it is today. There are a number of rapid climate changes, from warm to cold and from cold to warm, in the geologic record, some seemingly as fast as what modern global climate models ("general circulation models," or GCMs) forecast might happen in the future.

Glacial ice has been melting since the end of the last ice age, although there have been cooler episodes in the past 12,000 years, when glaciers advanced. In the last decades of the twentieth century, it was possible, based on available evidence, to believe that most if not all glaciers were melting back as the climate warmed, and as a result of that warming. But modern observations, including satellite remote-sensing observations begun in 1978 and now giving us a record of more than 30 years, suggest a much more complex pattern. Some glaciers are melting back while others are growing, and this holds for sea ice coverage as well as the remains of continental glaciers and mountain glaciers. Recent research shows that some mountain glaciers are melting not because of the warming—it is still below freezing where those glaciers are—but as a result of very wet episodes in the nineteenth century that provided ample water for the mountain glaciers to increase, and then, in periods of less precipitation, more typical of recent centuries, to be melted by sunlight. Comparing historical ship observations of Arctic sea ice with late-twentieth-century ship records indicates that at the end of Arctic winter, in May, the southern extent of sea ice in the middle of the nineteenth century was similar to that of the 1970s. But in contrast, Arctic sea ice between Alaska and Siberia extended much farther south in the middle of the nineteenth century than in the 1970s.

Greenland's glaciers waxed and waned several times during the twentieth century, sometimes with part of its glaciation increasing while another geographic part was decreasing. The same is taking place in Antarctica, with part of that continent's glaciers shrinking while other parts in other areas are increasing.

My colleague John C. (Craig) George, Wildlife Biologist North Slope Borough, studying marine mammals in Barrel, Alaska, tells me that the sea up there is remarkably open and that when sea ice forms it is young—ice of the year—and therefore thinner, containing less water, than was characteristic in the recent past.

As has happened repeatedly in geologic history, when the climate warms, the amount of water in glaciers decreases; when the climate cools, the amount of water in glaciers increases. This happens today and will continue to happen as the climate continually changes.

Since sea ice and continental glaciers have significant effects on climate, what is happening and what has happened to these in the past are important clues about the direction of climate change. Present trends leave us with the conclusion that, on the one hand, the air temperature near the ground and ice, both on land and sea, change together—warming at the same time or with, geologically speaking, short lags; cooling at the same time or, again, with, geologically speaking, short lags. A simple extrapolation from this idea is that the world's ice must be growing or sinking uniformly, but that contradicts observations and reconstructions. The current pattern of ice change is geographically complex, and a careful scientist would not make simplistic generalizations as if all the ice of all the world were changing in the same way due to the same causes.

Present observational science—modern scientific observations and reconstructions of the past—therefore leaves us with the conclusion that Earth's surface has experienced a warming trend from the 1960s to the beginning of the twenty-first century, and that the overall trend since 1860 has been a general warming. Present scientific observations leave us having to be agnostic as to whether this warming trend will continue for the next decades or centuries.

In the Arctic, where of course much of Earth's ice exists, the strong correlation between short-term changes in solar energy output and short-term temperature change suggests that sunlight intensity may be, as it always has been, one of the fundamental drivers of Earth's surface temperature.

To what extent might people's actions be a significant cause of present climate change, primarily by changing the concentration of greenhouse gases in the atmosphere? Scientists now conclude that carbon dioxide concentration lags temperature change by 800 years. The poor correlation between Arctic surface temperatures and CO_2 concentration suggests that although there is a greenhouse effect from CO_2 and other gases, this is not the cause of current climate change, or at least not the primary driver. On a simpler planet, say Venus or Mars, without life affecting its "biosphere," without any or much liquid water, without plate tectonics, the role of greenhouse gases is much easier to calculate. On our complex planet, this is not the case.

We are left to depend on the output from computer models of climate. Perhaps more important, the implication for our actions is that, as will be better explained in the following chapter, there are many things we can and should be doing that benefit biological diversity, agriculture, fisheries, forests, and wildlife. This should also reduce the rate of and amount of warming, if greenhouse-gas concentrations are in fact a leading cause of today's climate change. We would do well to put our money where it will provide benefits of these kinds at the same time, and avoid putting our money into actions that could be detrimental to today's habitats, agriculture, fisheries, forestry, and biological diversity and are of debatable value in controlling climate.

This is what I have gathered since 1990 from scientific observations. The third approach to understanding climate change is the use of computer models, which we turn to next.

How to Think About Climate Forecasting

[We] face an unimaginable calamity requiring large-scale, preventive
measures to protect human civilization as we know it. . . . The reality of the
danger we are courting has not been changed by the discovery of at least two
mistakes in the thousands of pages of careful scientific work over the last 22
years by the Intergovernmental Panel on Climate Change . . . the scientific
enterprise will never be completely free of mistakes.

—Former Vice-President Al Gore, in a February 28, 2010, *New York Times*
op-ed[44]

One of the amazing societal changes in the past 20 years is the degree to which people have come to accept computer forecasts of what might happen with global warming. Paleoecologist Anthony Barnosky wrote in his 2010 book *Heatstroke: Nature in an Age of Global Warming* that "the nature of climate science is computer models and probability calculations."[45] As if there really could be a science without empiricism, without direct observations, development of observational methodologies—such as satellite remote sensing of sea ice cover and techniques to reconstruct climate and atmospheric chemistry far into the past. And this assertion is from an ecological scientist. No wonder the forecasts from the large-scale climate models are, for many people, a new reality. If the climate models tell us that something is happening or will happen, plenty of people believe it must be true. This, again, is reminiscent of early plant ecologist William Cooper, who, as I mentioned, looked at the forests of Isle Royale and decided that despite appearances to the contrary, the world must be the way we have always believed it to be.

Forecasting has always been difficult and has usually been wrong. People have always been fascinated by predictions but also viewed them with ambivalence: We do and don't want to know what the future holds; we fear the predictions might be true, or might not. Charles Mackay, in his wonderful 1841 book *Extraordinary Popular Delusions and the Madness of Crowds* (still in print), observes that in earlier centuries those who made such forecasts were called necromancers and were grouped with the worst of the alchemists. His description of the prediction of a great flood in London in the sixteenth century gives us pause for reflection.[46]

In 1523, London "swarmed" with "fortune-tellers and astrologers, who were consulted daily by people of every class in society," he writes, and in June of that year several soothsayers "concurred in predicting that, on the 1st day of February 1524, the waters of the Thames would swell to such a height as to overflow the whole city of London and wash away ten thousand houses." Interestingly, "The prophecy met implicit belief" and was repeated "month after month." As a result, many people left London, and by January 1524 "at least twenty thousand persons had quitted the doomed city."

Mackay continues: "Bolton, the prior of St. Bartholomew's, was so alarmed, that he erected, at a very great expense, a sort of fortress at Harrow-on-the-Hill, which he stocked with provisions for two months." Many sought to join him; wagons brought boats there in case the flood made other forms of transportation impossible. "Many wealthy citizens prayed to share his retreat; but the prior, with a prudent forethought, admitted only his personal friends, and those who brought stores of eatables for the blockade." (All punctuation is Mackay's.)

What happened on the fateful day? "The Thames, unmindful of the foolish crowds upon its banks [who had dared to come watch the flood], flowed on quietly as of yore. The tide ebbed at its usual hour, flowed to its usual height, and then ebbed again, just as if twenty astrologers had not pledged their words to the contrary. . . . The obstinate river would not lift its waters to sweep away even one house out of ten thousands."

And what about the soothsayers? "On the morrow," Mackay continues, "it was seriously discussed whether it would not be advisable to duck the false prophets in the river. Luckily for them, they thought of an expedient which allayed the popular fury. They asserted that, by an error (a very slight one) of a little figure, they had fixed the date of this awful inundation a whole century too early. The stars were right after all, and they, erring mortals, were wrong." (There was no such disastrous flood of the Thames a century later either.)

Today, computers seem to have changed our negative view of forecasts. If it comes from a computer and obscure, complex software that most of us cannot understand, we believe it must be true. Looking back with our cell phones in our hands, our fingers on the keyboards of our laptops and iPads, our heads firmly clasped by earphones from our computer-based, digital music player, it's easy for us to dismiss events of that primitive time as mere foolishness. It seems impossible, with our wisdom and vast scientific knowledge, to fall into such traps—our forecasting methods are so much better than sixteenth-century London's!

Modern computer-based weather and climate forecasting began quite cautiously in the late 1980s. Climate modelers who developed the computer programs to forecast climate change readily admitted that the models were crude and not very realistic, but they were the best that could be done with available computers and programming methods. They said our options were to either believe those crude models or believe the opinions of experienced, data-focused scientists. Having done a great deal of computer modeling myself, I appreciated their acknowledgment of the limits of their methods. But today such statements are drowned out by claims on each side: at one extreme, that the forecasts are true; at the other extreme, that nothing in them is worth considering.

Thirty years later, it's time to step back and consider what computer models can tell us. They can tell us three kinds of things. First, they can help us understand a present situation, much as you use a calculator to find out what your total assets are worth. A calculation of Earth's energy budget at any instant is this kind of use of theory and modeling. Such forecasting methods are useful if you want to figure out why the cheese in pizza burns the roof of your mouth, even though you know that soon the pizza will cool off and no part of it will burn you.

The second thing computer models can tell us is the implications of what we know (the facts) and what we assume about a system that interests us, such as a forest or the biosphere. This is the best use, and in reality, no matter what anybody tells us, this is actually what the computer programs do. When a computer works correctly, it simply makes calculations based on what you told it, and shows you, whether you like it or not, the rational implications. This isn't the truth in a real-world sense, but it can be extremely helpful if you understand these results for what they are. Think of it this way: The computer is saying, "If you believe all that stuff you told me, then this is what it implies."

The third use of computer models is to make forecasts, as we are now often told they are doing for us. This is how weather forecasts are made, and we all use them, just as I did on the day I flew into freezing rain in a thunderhead. We know weather forecasts are not always correct, or even close to correct, but they are still helpful in planning a picnic.

When I was on the faculty at Yale University's School of Forestry and Environmental Studies in the 1970s, we were told that scientists at IBM were developing computer programs to forecast the weather. But unfortunately, even with the fastest computers available at that time, forecasting the weather 24 hours in advance took 48 hours of computer time.[47]

It was during the 1970s that the first important computer modeling of greenhouse-gas effects began. The Indian scientist Veerabhadran Ramanathan developed some of the first of these models.[48] It was also during the 1970s that James Hansen directed the development of one of the first of these models at a NASA laboratory. In them, the atmosphere was divided into very large rectangular boxes on the order of 5° latitude and longitude on a side. Every box was in contact with its six neighbors. The computer calculated the exchange of energy and of greenhouse gases for every box at every one of its six sides for every change in time. The exchanges were calculated from mathematical equations that represented in a simplified way the real changes in the real atmosphere. The entire atmosphere was represented this way, which required a large number of boxes and many calculations and took a large amount of computer time.

Some of the boxes were as much as 1,000 kilometers on a side, so the atmosphere over all 48 U.S. contiguous states was represented by just 12 of the boxes, which seems very crude. But according to Spencer R. Weart, a physicist who has written a history of these models, this early NASA climate model "produced a surprisingly realistic simulacrum of atmospheric circulation, including even a jet stream (the real jet stream is often much narrower than a thousand kilometers)."[49]

These models did not actually reproduce a climate change, because these were steady-state models. Instead, the computer program would be run twice, once with a "normal" climate—that is, without any increases in greenhouse gases—and the second time with a "transitory" climate, one with a constant amount of greenhouse gases added. Then the two would be compared and the differences between them would be published as representing the effect of climate change. These tests required huge numbers of calculations. For example, by the late 1980s one of the best known of these models required "50,000 time steps of ten minutes each" to reach an equilibrium—that is, to solve the example. Running the program twice so a comparison could be made required 100,000 time steps, each involving all the exchanges between every neighboring box.[50]

The purpose of an environmental forecast is either to support a decision process or to test a scientific hypothesis. To support a decision process, it must be clear which decisions the forecast expects to improve. In a decision-making process, the problem that we face is deciding what actions are necessary, when they must be taken, and how to rank concerns about the environmental effects of climate change in relation to other pressing environmental problems. To mitigate the effects of global warming, two distinct kinds of actions are needed: long-term actions, such as reducing emissions of greenhouse gases; and short-term ones, such as designing an appropriate nature reserve.

The global-warming debate has a number of ironies, discussed in this chapter and the next. The primary irony is that our attempts to deal with, understand, and respond to global warming—a major change in climate—are using scientific analysis based predominantly on systems fixed over time, in a balance, in a steady state, even though everyone agrees that the climate is always changing and has changed in the past, at many different rates and to many different extremes.

In certain situations, a steady-state analysis is necessary so we can understand what facts tell us. For example, the energy balance of Earth is complex, with the different inputs and outputs, absorptions, reflections, reradiation. To calculate that budget for a specific time, one assumes all is at steady state for an instant, which makes the computations straightforward and readily understood. This is an important use of scientific theory, but different from forecasting. It is a limited use of steady-state mathematics and computer simulations. What is called for is a science of the biosphere that is a science of non-steady-state systems. At this point, it should also be clear to readers of this book that life has been a cause of climate change for several billion years, and in that sense life-caused climate change is "natural." If this is true, and it is, then is it "unnatural" for our species to cause the climate to change?

The earliest of these models considered only the atmosphere, leaving out any effects of the oceans and vegetation. Then oceans were added, and later some of the models began to include effects of Earth's vegetation.

Several of these global climate models existed by the late 1980s. It concerned all scientists involved that each model gave a different forecast, and the scientists took these differences seriously. In the work that I and other ecologists did during that decade, we would use the output from four of these models and see what changes each led to in forests when used with the forest computer model we had developed.

TIPPING POINTS, NORMAL CLIMATES, AND ANOMALIES

In the past decades, discussions about global warming often include the assertion that our atmosphere is about to reach a tipping point. In 2008, James Hansen wrote in a chapter of a book titled *State of the Wild: A Global Portrait of Wildlife, Wildlands, and Oceans* that "our home planet is dangerously near a tipping point at which human-made greenhouse gases reach a level where major climate changes can proceed mostly under their own momentum. Warming will shift climatic zones by intensifying the hydrologic cycle, affecting freshwater availability and human health. We will see repeated coastal tragedies associated with storms and continuously rising sea levels."[51] These statements are all declarative: *is . . . are . . . will be. . . .* There are no reminders that these are forecasts from computer programs that have many problems and limitations. This was in marked contrast to the careful scientific approach of the 1980s, when climatologists told us that the climate models were still crude but were the best we had.

Also in 2008, Jim Hansen and colleagues wrote in an article published in the *Open Atmospheric Science Journal* that the "realization that today's climate is far out of equilibrium with current climate forcings raises the specter of 'tipping points,' the concept that climate can reach a point where, without additional forcing, rapid changes proceed practically out of our control."

A tipping point is an important concept worth thinking about as we wonder what is happening to our climate. Things tip if they can move away from a balance point. You don't think of your car's tire moving away from a balance point as you drive along and the tire rotates. On the other hand, stand a ballpoint pen or your cell phone or a 2×4 board on end and you will find one position—called its "equilibrium point"—in which it will stand there precariously by itself unless a tiny push tips it over. That's the tipping-point idea. But tipping points apply only to systems that have equilibrium. With nature, this means only a balance-of-nature system can tip.

Think about an airplane flying through turbulent clouds. The aircraft is designed to have stable points, and a pilot's goal is to keep the aircraft stable. But airplanes have degrees of stability, and so we can talk about the relative stability of the airplane. Today it is common to believe that for any system, like an ecosystem or an airplane, there are just two conditions: complete stability (at equilibrium and not changing) and chaos (without any order, even without any notion of cause and effect). This was made popular by Edward Lorenz, who discovered that early computers, with relatively crude representations of numbers, could show drastically different results with very small changes in the value of one of those numbers. He envisioned the same thing applying to living systems, so that the fluttering of a butterfly's wings, creating a tiny turbulence in the air, could change the future—change history. If that's true, I can promise you that because some of my airplane landings were rough enough to create a hell of a lot more turbulence than a butterfly could, I may have been a major factor in altering the future. But obviously, that's not how real-world biological and ecological systems work.

In the early days of aviation, there was a lot of experimenting with the relative stability of airplanes. A completely stable airplane, one that could not change from its equilibrium, could not be turned, so it would eventually crash into something (or run out of fuel and then crash into something). A very unstable aircraft is extremely sensitive to any change in the controls, so a small movement of the stick by the pilot would cause huge changes in direction, and the pilot would spend all his time just trying to get the plane to fly straight and level. He would be so distracted by this that he would be quite likely to crash.

Airplanes move in three dimensions: up and down, left to right, and rotating around the long axis. In the early days of aviation, pilots found that each kind of motion seemed easier with a different degree of stability. For example, it might be easier to fly one of those early planes if the rudder control (left and right) was less stable than the rolling (aileron) motion. This was a practical matter, determined by experience (empiricism, we scientists would say), helped by theory. There wasn't and isn't a perfect balance of nature for all airplanes in all situations all the time.

Compared with ecosystems and the biosphere, the small airplane I was guiding through the thunderhead was a very simple system. From this perspective, it's hard to imagine the entire biosphere about to fall off its climate cliff, but that's what the leading climatologists would have you believe. This may be partly due to a lack of understanding of complex systems and partly the result of training in a certain class of mathematical analyses, beginning with algebra and then differential and integral calculus and applying these to digital computers. One easily goes astray, because programming, I can tell you from personal experience, can be so difficult, challenging, and all-encompassing that you quickly forget why you got to the point in a computer program where something wouldn't solve correctly and spend days trying to get the thing to work. A worldview vanishes, the atmosphere of the programming language turns into a thunderhead, and you find yourself working in iced-over conditions, can't see out, and forget what you need to do.

KEEP IT SIMPLE

The desire to make these huge, complex computer programs more and more detailed, to include more and more aspects of the biosphere, and thereby make them more realistic, brings its own problems: the more boxes, the more exchanges of energy and chemicals have to be calculated, with more and more equations. Each of those equations has constants in it, called "parameters," such as the intrinsic rate of growth of a population in simple population models. These have to be estimated, ideally from observations and experiments. But we already know that the fundamental problem with the conservation and management of wildlife is those simple-to-think-of numbers that nobody bothers to get. Expand that by all the features of the biosphere. The result is that the numerical value of many of those parameters is guessed at. It becomes easy, in Mackay's words about the forecasts of the Thames flooding in 1524, "by an error (a very slight one)" to get one of the forecasts wrong.

There is a kind of ecological uncertainty principle: The more you try to explain all the details, the more likely you are to make quantitative errors that lead you astray. The more details you seek to include, the greater the chance of errors that lead you astray. Yet, if you make your model (your theory) too simple, you are likely to miss the very qualities that determine what actually happens.

Even the relationship between two things, such as sunlight and photosynthesis of plants, could be diagramed in several different ways, so that different equations could be used, each with different kinds of parameters. It gives you a headache just thinking about this. It begins to seem easier to guide an airplane all by yourself through a thunderstorm.

The solution is to return to Occam's razor, which has been an underlying principle of the scientific method for centuries. It is the statement made by William of Occam (c. 1288–c. 1348), an English friar and philosopher, who wrote that "entities must not be multiplied beyond necessity [*entia non sunt multiplicanda praeter necessitatem*]." This has been used in modern science to mean that one should continue to add complexity to a theory only to the point where all observations are accounted for, and no more. Also, if there are two explanations that account for all observations, pick the simpler one. A prime example of this is the difference between the Ptolemaic explanation of the motion of the planets and Kepler's explanation. The ancient Ptolemaic one assumed that Earth was at the center of the solar system and that the sun and planets revolved around it. Kepler's analysis, based on observations by the Scandinavian Tyco Brahe (1546–1601), had Earth and the other planets revolving around the sun. (Tyco Brahe was famous during his lifetime for accurate astronomical, including planetary, observations.[52]) At a certain level of detail, both theories give adequate forecasts, but Kepler's had much simpler orbits and was much simpler overall.

In 1979, involved in the work I have described in this chapter, I went to the planetarium in Rochester, New York, and enjoyed the performance. Computing systems were still comparatively primitive—the IBM PC was a year away—and I was curious as to how the marvelous machines were able to model the paths of the planets in our solar system and display them beautifully on the hemispheric ceiling. The people who ran the planetarium told me that their machines used analog computers, comparatively primitive devices. They said it turned out to be easier to program this kind of computer with Ptolemy's equations—the sun and planets circling Earth—than with Kepler's, which required precision not available on the analog computers. At a certain level of detail, both explanations worked. Although with certain kinds of primitive computers it was easier to program Ptolemy's equations, they were in fact more complex than Kepler's, and originally in the early development of modern astronomy, on the basis of simplicity alone, Kepler's was to be chosen. In addition, of course, many other observations since Galileo, including satellite probes to Mars, Venus, and other planets, prove Kepler right.

As I've explained, the earliest computer models of climate looked only at the atmosphere; the oceans and life were simply what we call "boundary conditions," things that never change and are left out of our calculations, or treated as mathematically constant, and involved, if at all, only in terms of conduction of heat energy in and out. This, of course, seemed completely wrong to oceanographers and to us ecologists, since we knew that water and life had great effects on climate.

The oceanographers got interested first in the global climate models, and then in the late 1980s and early 1990s we terrestrial plant ecologists started to get directly involved. One of my graduate students, Jon Bergengren, who came to me with a background in astrophysics, did his Ph.D. thesis adding the world's vegetation to the climate model at the National Center for Atmospheric Research (NCAR) in Boulder, Colorado. At the time, Steve Schneider (1945–2010) was still the director of that center and we worked together, through Jon, to add the vegetation. Jon's world vegetation consisted of 150 "species" or "biomes"—depending on how you wanted to think about them—of vegetation, adding a lot of complexity to the NCAR climate model. And even so, this was a very simplified model of world vegetation, with the vegetation responding to climate and changing, and then those changes affecting climate in a quite simplified way.[53]

During this time, scientists at Lawrence Livermore Laboratory got in touch with me and said that they wanted to enter the global climate-modeling game and asked me to help. They wanted to take things a step further, creating a model that not only involved oceans and vegetation but also considered the dynamic cycling of nitrogen and phosphorus, in addition to the greenhouse-gas carbon-based compounds. There wasn't much money, however, just enough to fund a graduate student, and that funding faded rather quickly, so that model, at least to my knowledge, never got done.

The desire to account for every possible factor pulled us away from Occam's famous advice to seek the simplest explanation consistent with observations. We were instead trying to put in everything we could think of that we believed played a role, even if we couldn't measure it accurately or test it well. And since each new factor (world vegetation or oceans) and each new detail within a factor (a kind of vegetation, or a layer within an ocean) greatly increased the complexity and the need for more equations (shapes of relationships of causes and effects) and parameters (the constants in those equations), the chances that guesswork would lead to very misleading results increased greatly.

In the first decade of our new century, I got to know F. Scott Armstrong, professor at the Wharton School of Business at the University of Pennsylvania and one of the world's experts on forecasting methods in general, which in his career he has applied primarily to business operations. I also got familiar with his landmark 2001 book, *Principles of Forecasting—A Handbook for Researchers and Practitioners*.[54]

Scott said at a meeting where we both spoke that for most of his career he hadn't gotten involved in global warming, that he knew much more about how to forecast whether a particular brand of soap would sell well six months from now than about environment. But people began to ask him to take a look at the forecasting methods in use for global warming—the global climate models and their extensions. He said two important things as far as I was concerned. First, that complex models in general didn't work—they didn't make accurate forecasts; simple models worked best. And second, that there was little evidence that any of the global climate models had been validated (shown to make accurate and realistic forecasts).

In his book, Armstrong lists about 144 criteria for the complete validation of a forecasting model. He said at the meeting that any particular model did not have to meet all those criteria, but it would be nice if a model met some of them. Examining the global climate models, he concluded that these met none of the criteria. In short, they had not been validated in any legitimate scientific way.

I've explored this myself and agree that one can't find the standard kinds of validation for the big climate models. What you do find gets somewhat technical, but it's important. You find what are called curve-fitting exercises, where the person using the computer program adjusts a parameter's value up and down until the output from the program matches observations. This is how you can get a better estimate of a parameter, but it is not an independent test of the model. A more complex kind of curve-fitting involves several factors and their parameters. One of the big unknowns about the atmosphere is the degree to which water that evaporates from the ground stays as vapor (and therefore functions as a greenhouse gas) in the atmosphere, and how much of it condenses into clouds (which reflect visible light back into space and cool the surface). The difficulty of forecasting whether water was going to remain as vapor or condense into clouds is what got me flying into that thunderhead over Southern California. The weather forecasts for that day—the best available anywhere in the world with the best of current technology—didn't get it right.

ALTERNATIVE WAYS OF THINKING ABOUT THE BIOSPHERE

With the irony of complexity, and the assumption that the models were of a steady-state system on my mind, I began to think again about the wilderness at Isle Royale National Park. To summarize the main points of this chapter, I will return with you briefly to that island wilderness. One pleasant June evening years ago, I took a break from ecological research at Isle Royale National Park and went canoeing in a large inlet named Washington Harbor, hoping to see some of the moose that populated this isolated wilderness island in Lake Superior. Upstream, an old cedar arched gracefully over the waters, framing the forest and the deepening sky beyond. The serenity and beauty of the scene rivaled the best of America's landscape painting. For that moment, the remote island wilderness appeared as tranquil as a still life, as permanent in form and structure as brush strokes on canvas at the Louvre.

Soon after I had pushed out from shore, a large bull moose stepped carefully into the cold lake waters and began a slow traverse of the shallows, searching for water irises, lilies, and other water plants that were some of his favorite summer foods. He circled the shallows for 20 minutes, rarely stopping to feed. In this northern wilderness, June was too early for water plants, and as the moose edged his way over to the north shore, he found little to eat. Suddenly, he galloped through the shallows, scrambled out of the inlet, and began kicking vigorously at the shore. He dashed up a short bluff, breathing rapidly, turned, raced down and kicked again where the sand and waters met. Of course I can't say why he acted that way, but it sure looked like he was furious with the harbor for denying him food. Nothing could have contrasted more with the idyllic scenery of that evening than his bizarre, chaotic, and perplexing behavior. But in the almost half-century that I have studied nature's character, I have come to realize that the seeming constancy of the harbor symbolized a false myth about nature. The moose that kicked at the shore—complex, changeable, hard to explain, but intriguing and appealing in its individuality—was closer to the true character of biological nature, with its complex interplays of life and physical environment on our planet.

The big bull moose came to mind as a reminder of the difference between the way much of environmental science has been approached and the way nature actually works. As I've made clear in this chapter, most of the major forecasting tools used in global-warming research, including the global climate models and those used to forecast possible ecological effects of global warming, paint a picture of nature more like a Hudson River School still life than like the moose that kicked at the shore. These forecasting methods assume that nature undisturbed by people is in a steady state, that there is a balance of nature, and warnings that the climate is at a tipping point mean that the system is about to lose its balance.

Since nature has never been constant, there is something fundamentally wrong in most approaches to forecasting what might happen if the climate warms. The paradigm is wrong and has to change. But such fundamental change in human ideas never comes easily, and is often resisted by those whose careers have been based on the old way of thinking. That the general circulation models are such complex computer programs, developed over so many years, makes a fundamental change in the entire way of thinking about climate dynamics and its ecological implications all the more difficult.

We who work in environmental sciences and on global warming need to open ourselves to a much greater variety of ways of thinking about nature. We need to develop forecasting methods appropriate for always-changing, non-steady-state systems where chance—randomness—is inherent.

Among the various things I have tried in my four decades of work on the effects of global warming were a few computer models of the carbon dioxide cycle, small computer programs that used approaches quite different from the standard at the time to the question of what might happen if carbon dioxide were to increase rapidly from human actions. I created a strange little model of small boxes, each representing a biome (biomes, you will recall, are what we ecologists call major ecosystems on Earth, such as tropical forests). These "competed," so to speak, for carbon dioxide in the atmosphere through their photosynthetic organisms, and returned some of that carbon dioxide to the atmosphere as the model's "creatures" respired, or died and decayed.

The results were as strange and surprising to me as the moose who kicked at the shore. The carbon dioxide in the atmosphere didn't just build up over hundreds of years and then slowly decline to the same perfect equilibrium concentration that existed in Earth's atmosphere before the industrial age. No, instead it oscillated strangely, because the biome that had the fastest rate of uptake "outcompeted" the others, pulling the carbon dioxide concentration down so far that the plants and algae in other biomes didn't have enough and died back, giving up their stored carbon dioxide to the atmosphere.

That strange little computer model was at the time just as ephemeral for me as that evening canoe ride at Isle Royale. It got me thinking about how a complicated, intricate, always-changing system could respond to a novel input. The computer, caring even less about me than the bull moose did, simply showed me exactly what the consequences of my assumptions were.

I didn't publish that work, because it was so simple, yet different, and seemed more a personal insight than a definitive forecast. But looking back now at the bull moose and that little computer model, I believe that we have been on the wrong path in our view of the way nature works, and we need a fundamental change in our paradigm. This can come about only in an intellectual atmosphere that is open, free, and wildly experimental. Such an atmosphere would let us accept that natural ecological systems are likely to be full of surprises. And once we open ourselves to those possibilities, perhaps we won't find ourselves caught between defending weak science or lashing out, like that bull moose, at what seems to stand in our way.[55] We also might begin to have a variety of very different kinds of models of the biosphere, rather than a multitude of computer programs that all describe the atmosphere the same way—as a series of boxes connected by differential and integral calculus equations and that attempt to deal with every detail of the biosphere.

With today's programs, to get the global climate models to give results that agree with observations, the programmers could modify the percentage of water vapor that condensed, or could leave that alone and add some effect of people. There was more than one way to skin the cat of climate forecasting, so to speak. The problem was that the results became our new reality, rather than a technical modification of a complex computer program. The output was unvalidated in standard scientific ways.

In Summary . . .

Whether or not we cause climate change, these changes greatly affect us and all life, and we need to take these changes seriously while remaining scientifically objective. That is the purpose of the next chapter. Everyone who has lived on the Pacific Coast of the United States and Canada knows those mornings when the ocean fog rolls in like a gray blanket. Standing at Hendry's Beach in Santa Barbara, below the bluffs where on clear days hang gliders floated, I could see the hills behind the city. The ocean, borrowing its color from the sky, was a deep grayish green. Sounds were deadened on these foggy mornings. Sometimes a dolphin or two would play in the surf not far offshore, but none broke the waves this morning. No longer living in Santa Barbara, but out there visiting in the early years of this new century, I was thinking about where our new science of the biosphere had come and hadn't come. The rationalism that had always seemed so natural and ordinary to me was losing out. Excellent science was being done about this global view of life and its sustaining system, but emotion, politics, and beliefs seemed to be getting the upper hand, blocking a clear view just as the morning fog was doing over Santa Barbara.

The view I had seen and loved from Mauna Loa showed an Earth whose entire atmosphere we are capable of changing and have begun to change. Nature in the largest sense was clearly a system that has varied over time and space at many scales. We have the power to change the biosphere, and we are forced to make choices; nature in the large does not provide a single, simple goal. There are many themes in nature's symphony, each with its own pace and rhythm. We are forced to choose among these themes, some of which we have barely begun to hear and understand.

FIGURE 13.1 Fishing for wild salmon on the Rogue River in Oregon, one of America's most famous salmon rivers and one of the first eight to receive the Wild and Scenic River designation. (Photograph by the author.)

13

Life on a Climate-Changing Planet

Reality must take precedence over public relations, for nature cannot be fooled.

—PHYSICIST RICHARD FEYNMAN IN THE FINAL REPORT ON
THE CHALLENGER DISASTER

Life has had to deal with environmental change, especially climate change, since the beginning of its existence on Earth. Species adjust or go extinct, and both have happened. For life-forms with our kinds of cells—eukaryotic, the kind with distinct organelles—the average existence of a species is about 1 million years, and on average one species goes extinct a year, at least of the species we have named and know including those we know from fossil records.

Organisms adjust to environmental change in three ways, from fastest to slowest: behaviorally, physiologically, and genetically. Ecologist Larry Slobodkin used to demonstrate the first two playfully during a lecture by picking up a piece of chalk and tossing it to one of the students. The student would duck or catch the chalk, which Larry pointed out was the behavioral response, first and fastest, and then within 20 seconds would blush, the physiological adjustment, second fastest. These, he would explain, were not only relatively fast but used little energy in a population. If these failed to make a successful adjustment, a population's genetic makeup could change, with genes transmitted to the next generation that led to characteristics better adjusted to the changed environment, obviously a much slower adjustment.

Individual mobile organisms migrate as an adjustment to climate. Plants and other nonmobile species adjust by having seeds or other propagules that move easily. Wind, water, and animals provide the major transportation. In any population there is a mixture of genetic types and, as Darwin explained a long time ago now, those better adapted to the current climate left more offspring than the less adapted, and over time a population evolved to fit the new climate. But this genetic adjustment took time, and since the climate is always changing, it could be that at any one time a population would be adjusting genetically to a climate that had been present but had passed or was passing. It was and is an eternal dance, populations never quite in perfect harmony with their present environment. If the rate of environmental change is too fast, populations cannot adjust and go extinct. Dealing with environmental change has always been part of being alive.

Early man was part of this dance between life and environment. *Homo erectus*, the first of our kind who left Africa, would likely have migrated as a matter of course. They may not have thought of it as migration in our modern sense; they were going where the environment, including sources of food and water, was better. Environmental change and moving along with it were only natural.

With the beginning of civilization and the construction of buildings that could last a long time, and with investments of time and effort in agricultural fields, as well as the discovery of specific sources of minerals and the building of mines to get them, people's lives changed in ways that led to a desire for constancy. Establishment of property rights and national boundaries (beginning with tribe-established land boundaries) augmented the need and desire for constancy of place and of environment. One can argue that it is our species that most needs and most desires constancy and has therefore formed worldviews that not only require environmental constancy but have turned it into a fundamental belief, a folkway, a series of myths.

The more technologically and legally advanced a civilization, the greater the need and desire for environmental stability, for a balance of nature. Hence, our dilemma. Rather than claim the world is constant except for our sinful interference with it, we need to acknowledge and work out ways to live with environmental change. This can include doing our best to stop or slow that change, as we do in the short term with agricultural irrigation, stabilizing the "precipitation," so to speak. But the harder we work to force environmental constancy onto our surroundings, the more fragile that constancy becomes and the greater the effort and energy it takes. The use of groundwater for crop irrigation illustrates that fragility. Large aquifers that took many thousands of years to develop are being depleted for crop irrigation over comparatively short times—decades or centuries.

A major example of this depletion is the Ogallala aquifer (also called the High Plains aquifer), stretching from South Dakota into Texas. It stores a huge amount of water and is the main water source in the area. Its use began in the 1940s. Today water is removed up to 20 times faster than it is naturally replaced. In southwest Kansas and the panhandle in western Texas, it is said that supplies may last only another decade. Lower Cimarron Springs, famous in the nineteenth century as a watering hole along the Santa Fe Trail, dried up decades ago due to pumping groundwater. Millions of dollars will be needed to find alternative sources.[1]

Achieving short-term stability at the cost of long-term fragility is a trade-off. It makes more sense, in retrospect, that the earliest civilizations, like Egypt and Persia, were established downstream on a river system whose flow varied year to year but was relatively constant compared to much of the rest of the land.

When I give talks about the discordant harmonies of nature and my changing views of global warming, a common response is "Why bother to point this out? Everybody believes in global warming, and doing something about it doesn't hurt anything and can only benefit." But in our real world, the choice to take one action means that other actions are not taken. We are well aware these days of the worldwide limits of capital and cash to do things, and we must choose carefully. There's the rub.

One of the things missing from the global-warming debate is to put the issue of global warming within the set of major environmental problems and then establish priorities based on what can be done, what needs the most immediate action, and what is most important. In addition to possible effects on climate, major ways that human actions are decreasing overall biological diversity, include: (not in an order of priority) habitat destruction; overharvesting of renewable living resources; chemical pollution; removal of groundwater; depletion of mineral resources necessary for life, especially sources of phosphate; and introduction of exotic species that harm other species and are undesirable from our point of view, and just plain inadvertently causing species to become threatened with extinction.

These are here-and-now problems, not based heavily on hypotheses and computer programs that are unvalidated, steady-state, and therefore of questionable use in forecasting. Moreover, sometimes actions that the public is told will help mitigate or reduce global warming create or worsen other environmental problems. For example, in Indonesia, 44 million acres (18 million hectares) of tropical rain forests are said to have been cut down to plant palm trees to produce palm oil, to be used as a biofuel. This is justified as being good for the environment because it is supposed to reduce emissions of greenhouse gases and therefore reduce the rate of global warming. But this habitat destruction further endangers orangutans and Sumatran tigers, already threatened with extinction.[2] While few, if any, knowledgeable environmental organizations will be fooled by the claim that this is going to be environmentally beneficial, the European Union and the government of Malaysia have been considering what to do about the biofuels from these plantations, taking the possibility seriously that using these to fuel cars and trucks in Europe will offset some of the greenhouse gas production by these vehicles and is therefore justified and on the whole environmentally sound.

Singling out global warming from other environmental issues is a one-factor approach, which has been too common in environmental policy decisions. For example, as the Clean Water America Alliance points out, the use of water resources requires considerable energy, but water use and energy use are treated as separate issues most of the time in environmental policy analysis. Because global warming gets so much attention and so much funding, this single-factor approach is a particularly important aspect of this issue's policy analysis.

In many cases, as I hope to show in this chapter, actions to help solve another environmental problem can also be beneficial in reducing undesirable effects of climate change. For example, as I discuss in *Powering the Future: A Scientist's Guide to Energy Independence*, moving away from fossil fuels toward wind and solar energy reduces the human contribution of greenhouse gases to the atmosphere while also reducing habitat destruction (from the mining of fossil fuels) and air, water, and ocean pollution (from the mining, processing, and burning of these fuels), benefiting biodiversity and human health and well-being. The same can be said for a move away from fission-based nuclear power plants, whose toxic substances last up to millions of years (the U.S. government is seeking a warning sign that will keep people away from nuclear waste dumps for 10,000 years). We will return to the question of how to set priorities in the Postscript.

The politization and ideology-driven beliefs about global warming, on both sides of the issue, prevent a calm, rational examination of where actions to mitigate global warming could fit into a set of priorities. Indeed, even making a claim that such prioritizing is possible leads to a change in viewpoints and will likely upset many who believe now that global warming is a present and future reality with disastrous effects. We need to be able to put the discussion within a rational context. Among other aspects of this context, we need, as Thomas Friedman wrote on September 14, 2011, in the *New York Times*, "to start taking steps, as our scientists urge, 'to manage the unavoidable and avoid the unmanageable.'" Not just about climate change, but in establishing an integrated, multifactor approach to our major environmental problems.[3]

Climate Change and Human History

Climate change over the past 18,000 years, during the last period of major continental glaciations, has greatly affected people. Continental glaciation ended about 12,500 years ago with a rapid warming, perhaps as brief as a few decades.[4] This was followed by a global cooling about 12,800 years ago that lasted about 1,000 years. Effects of climate change on civilization and cultures are well known especially for two historic events: the Medieval Warm Period followed by the Little Ice Age.

MEDIEVAL WARMING

The Medieval Warm Period lasted about 300 years, from AD 950 to 1250. During that time, Earth's surface was considerably warmer than what climatologists today call "normal" (meaning the average surface temperature during the past century or some shorter interval, such as 1960–1990). Since weather records were not kept then, we don't have a global picture of what it was like. Available evidence suggests that parts of the world, in particular Western Europe and the Atlantic Ocean, were warmer some of the time than they were in the last decade of the twentieth century.

In Western Europe, it was a time of flourishing culture and activity, as well as expansion of the population; a time when harvests were plentiful, people generally prospered, and many of Europe's grand cathedrals were constructed.[5] There was less sea ice, so the upper waters of the Atlantic Ocean must have been warmer. Viking explorers from Scandinavia traveled widely in the Far North and established settlements in Iceland, Greenland, and even briefly in North America. Near the end of the tenth century, Erik the Red, the famous Viking explorer, arrived at Greenland with his ships and set up settlements that flourished for several hundred years. The settlers were able to raise domestic animals and grow a variety of crops that had never before been cultivated in Greenland. During the same warm period, Polynesian people in the Pacific, taking advantage of winds flowing throughout the Pacific, were able to sail to and colonize islands over vast areas of the Pacific, including Hawaii.[6]

While these cultures prospered in Western Europe and the Pacific during the Medieval Warm Period, other cultures appear to have confronted a more difficult climate. Long, persistent droughts (think human-generational length) appear to have been partially responsible for the collapse of sophisticated cultures in North and Central America. These included the people living near Mono Lake on the eastern side of the Sierra Nevada in California, the Chacoan people in what is today Chaco Canyon in New Mexico, and the Mayan civilization in the Yucatán of southern Mexico and Central America.

We don't know what caused the Medieval Warm Period, and the details about it are obscured by insufficient climate data to help us estimate temperatures during that period. We do know that it was warm, and we can't blame that on the burning of fossil fuels. Clearly more than one factor caused that warming.

The Little Ice Age

The Little Ice Age followed the Medieval Warm Period. It lasted several hundred years, from approximately mid-1400 to 1700. In northern Europe, summers were wetter, both summers and winters were colder; and snow cover persisted longer. With longer snow cover and later springs, grain yields were low and people had to slaughter dairy cattle when their hay ran out. Other crops failed. In the Swiss Alps, some mountain glaciers advanced to fill valleys, flood valley streams, and destroy villages. The population was devastated by the Black Plague, whose effects may have been exacerbated by poor nutrition as a result of crop failures and by the damp and cold that reached out across Europe and even to Iceland by about 1400. Wine harvests were poor, and vineyard cultivation shifted southward. People starved.[7]

In Norway and in the Alps, mountain glaciers expanded downslope, riding over villages, blocking rivers, covering farms. Towns were abandoned in England and Scandinavia. The 1690s were among the worst years. In 1690 a wet summer and a severe winter ruined harvests and caused widespread famine in Europe.[8]

The cold climate was widespread beyond Europe. In the Kiangsi province of China, the centuries-old cultivation of oranges was abandoned in the late 1600s after a number of years of frequent frost. In Ethiopia, Portuguese travelers saw the snow line on the mountains coming down lower than they had before. In east Africa, mountain glaciers on Mt. Kenya and Kilimanjaro advanced downslope.[9]

Travel and trade became difficult in the Far North. Viking colonies in North America were abandoned, and those in Greenland declined greatly.

Indirect evidence suggests that the same bad weather occurred in eastern North America. So when you try to put yourself in the place of the colonizers, you have to imagine their experiences as harsher in terms of weather than those of our century. Winters had longer periods of harsh weather; frosts tended to occur later in the spring and earlier in the fall than is the average of the twentieth century, so growing seasons were shorter. It was a hard time to become a successful farmer in a new land with new crops, harder than if we tried to do the same today with our generally milder climate.

The experiences of Rev. John Williams of Deerfield, Massachusetts, recounted in T. J. Stile's book *In Their Own Words: The Colonizers: Early European Settlers and the Shaping of North America*, illustrate the tough weather. Captured by Indians in early 1704, he was forced to travel on foot through thick snow and cross Lake Champlain when it was covered with ice in March.[10]

Samuel de Champlain's experiences also reveal the harshness of the climate. In 1604, he explored the coast of New England, then returned to the island of St. Croix above modern-day Maine in early September. There, he wrote, "the snows began on the 6th of October." By December, "the cold was sharp, more severe than in France, and of much longer duration," while the snow in the nearby mountains "was from three to four feet deep up to the end of the month of April."

If the colonization had taken place a few centuries earlier—before the fourteenth century—the climate would have been much milder and therefore settlement in some ways easier.

This leaves us with a major problem: How can we live with environmental change in physical, cultural, and personal terms? Salmon can help us think this through.

HOW SALMON HAVE PERSISTED THROUGH
ENVIRONMENTAL CHANGES OF MANY KINDS

During the past half century, ecologists and other environmental scientists have been asked to participate to a greater and greater extent in setting policy, which has brought them into politics and the political processes, and has forced them to work out a role for themselves in society. Among those most involved in such processes have been Dr. Lee Talbot of George Mason University; Robert Lackey, a fisheries scientist of the University of Oregon;[11] and Sidney Holt, one of the originators of the International Whaling Commission. This happened to me as well, as my involvement with salmon conservation shows.

One summer day in 1993, I was eating lunch in a small café in Gold Beach, Oregon, the town at the mouth of the Rogue River. The Rogue is one of the most famous salmon-fishing streams in the Pacific Northwest, and Gold Beach is the major jumping-off place for sportsmen who come to fish for salmon on the Rogue. That day, I could see several open boats on the river, each with a guide sitting in the stern by the outboard motor, talking to a man at the prow, who was trying to cast carefully and accurately into the rough, rolling waters where surf and stream met. Looking toward the ocean, I saw bobbing heads of sea lions, also fishing for salmon.

Now and again a jet boat filled with tourists splashed away from a dock and headed upstream, on its way to Paradise Lodge, a resort reachable only by boat or on foot, within the legally designated Wild and Scenic River stretch of the Rogue (Fig. 13.1).

The Rogue was one of the first eight rivers given that legal status, which was established by national law in 1968. Congress determined that "certain selected rivers of the Nation which, with their immediate environments, possess outstandingly remarkable scenic, recreational, geologic, fish and wildlife, historic, cultural or other similar values, shall be preserved in free-flowing condition, and that they and their immediate environments shall be protected for the benefit and enjoyment of present and future generations." Laws like the Wild and Scenic Rivers Act, the Wilderness Act of 1964, the Marine Mammal Protection Act of 1972, and the Endangered Species Act of 1973 gave us great encouragement: Not only was the environment becoming a popular public issue, but we were doing something about it, taking steps to conserve and improve nature around us.

The Rogue certainly meets the Wild and Scenic Rivers Act's criteria. It begins as a small outflow from the volcanic rocks that contain and create Crater Lake, in beautiful rugged country that a group of us visited in a winter's snow. It flows across the fertile Willamette Valley through the small city of Grants Pass—the starting point for those who are planning to canoe, kayak, or raft down the river—and then flows westward through the coast range to the Pacific Ocean. Where I sat that day was where now and again one of the more adventurous tourists arrived at the dock after a several-day trip from Grants Pass through the wild and scenic stretch of the river. I was with two men involved with a small organization to help conserve salmon, and with the five members of a scientific panel I had put together for a study the state of Oregon had asked me to do. As explained in Chapter 11, the goal of that study was to investigate, from an objective scientific basis, the relative effects of forest practices on salmon. Right after lunch I was going to direct one of a series of public meetings. I had decided that the project should involve more than scientists from outside the state, and should include people who lived in Oregon and depended on salmon for their livelihood, or were just interested in the conservation and management of those fish and their habitats. Thus the work was an exercise in the application of science within a democracy.

Given what I had learned in the year before and written in my book *Discordant Harmonies*, I knew that data about salmon and their habitats would likely be less than ideal. One purpose of the public meetings was that we hoped to get valuable insights from the people at large for our study. The meetings had certain ground rules: Anybody who wished to could make a statement, as long as they sent in a written request in advance saying what they would talk about. (Having to write something down was a way, we hoped, to keep out any crazies who might turn up and want to talk.)

Ken Cummins, one of the members of the scientific panel and, as noted earlier, an expert on freshwater ecology and on salmon and fly-fishing, turned to the

Oregonians and said, "I understand that you are pretty upset with the way the government is handling the salmon issues."

"Yes," one of them replied. "When the government told us they could manage salmon, we thought that meant we could all manage to *have* salmon."

That's the real-world dilemma that faces us when it comes to doing our best to help nature. Aside from our theories, beliefs, prejudices, and opinions, what can we actually *do* to make things better, or at least keep them from getting worse? Salmon epitomize the challenge, because they confront environmental constraints at every level of their existence, from local to global—from the size of gravel in a short reach of a stream where an adult seeks to lay eggs, to large-scale ocean circulation patterns and the advances and retreats of glaciers over hundreds of thousands of years.

As explained in the opening of Chapter 12, in the 1980s I had hoped that the newly emerging science of the biosphere could develop free from the ancient myths about nature, that it would allow a fresh start, not only for global problems like global warming but also for the conservation and management of our living resources—fish, wildlife, endangered species of all kinds. The opportunity to lead a major study about salmon for the state of Oregon seemed a wonderful chance to help develop these new approaches.

Here's what happened. As I said in Chapter 11, salmon conservation in the last decades of the twentieth century was motivated by the conviction that these fish required old-growth forests along the streams where the young are hatched and later return as adults to lay eggs. That was the justification for the rule established during the Clinton administration to stop all logging near a salmon stream, out to a distance equal to twice the height of a tree that typically grew along the streamside, also mentioned in Chapter 11.

After lunch that day in Gold Beach, I went to the public meeting in town, mainly meeting with commercial salmon fishermen and salmon-fishing guides. During the meeting, a small, wiry man with a patch over one eye stood up and introduced himself as Jim Welter, who had lived his entire life in Brookings, Oregon, as a fisherman. He said, "I don't know much about science, but it just makes sense that if these salmon are born and reared in freshwater streams and spend about a year there, and then go to the ocean and return when they're three or four, the amount of water flowing in the stream where they were born ought to make a big difference in how many survive and return."

It made a lot of sense to me, and it was refreshing to hear something constructive, especially when I had only recently learned that the Bonneville Power Administration, which built and ran the big dams on the Columbia and Snake rivers, had spent $2.5 billion on salmon research and restoration and, according to one of their top executives who spoke to me, those dollars hadn't yielded a single sign of improvement in the salmon. How could a big agency spend that much money and have absolutely nothing to show for it? I found out, but that's the subject of another story, another time.

Jim Welter did more than provide us with a little wisdom based on years of experience. In my career working on natural living resources, I'd come across people who did provide that kind of insight, almost always interesting, but Jim took it several steps further. He went to the state of Oregon's Department of Fish and Game and got the data for the counts of salmon crossing a dam on the Rogue and the Umpqua rivers. Of the more than 20 rivers that flowed to the Pacific Ocean in Oregon south of the Columbia River, these were the only two where the state actually counted salmon.

Discovering that the state didn't know how many salmon it had in most of its rivers was pretty disconcerting. I'd been hired to tell them what was happening to salmon and why, and this required basic information about changes in salmon numbers over time—information that did not exist, I had only recently discovered, for most of the rivers.

Jim Welter went next to the U.S. Geological Survey and got the data for stream flow for each year on those rivers for the time that salmon had been counted. This was a remarkable step, especially because, to my knowledge, no agency of the state or federal government had done this comparison. Even more remarkable was that Jim had gotten a friend who knew a little about science to help him graph the two kinds of data. He brought in a huge hand-drawn graph (this was in the days before PowerPoint, and anyway, Jim wouldn't have used that). A nonscientist actually doing an analysis of data. Once again, no government agency had gone this far.

Sure enough, as Jim pointed out, if there was a high-water year, then four years later a lot of salmon swam upstream. If there was a low-water year, then four years later few salmon returned. Jim provided the first important insight into what might be a major factor influencing salmon abundance.

We were so impressed with Jim's suggestion and his graph that we contracted with Ben Stout (1924–2007), a forester and statistician, to do a formal statistical analysis of these two data sets. And sure enough, it turned out that you could account for 80% of the variation in salmon abundance from water flow alone, and you could thereby forecast pretty well four years in advance whether or not there would be a good salmon year. Since the methods in use at the time set the catch sometimes a few months before the fishing season opened, and didn't give the fishermen much chance to prepare, this seemed a remarkable advance. We wrote it up as a scientific paper and proposed it to the state and to salmon fisheries scientists.

In the years since, once in a while I call Jim and ask how he's doing. Sometimes he asks, "Them government fellows ever listen to what you told them?" And I have to admit that they hadn't. Another time Jim said on the phone, "If only we weren't so greedy, everything would be all right."

Although Jim wasn't trained as a scientist, he was a natural at it. Gathering data, looking at it, thinking about it, graphing it, and coming up with insights: That was just good science. And sad to say, we had seen little like it, certainly not from the large staff of the Bonneville Power Administration. But as I said, that's another story. If you want to hear about why BPA and other scientists did not think to plot water flow against salmon returns, write me and I'll set that story down. Of course, their science may have improved since then.

Jim Welter represents one kind of person we desperately need to help with our environmental problems: a good observer invested in natural resources without any ideological bones to pick, open to new ideas, willing to look at primary data in a fresh way, to construct graphs, and not jump to conclusions. When I think about acting locally to help nature, I think about Jim Welter, who had more foresight with his one eye than many government employees do with two.

That water flow alone could forecast so well about salmon is all the more surprising considering the life cycle of these fish. The streams of the Pacific Northwest of North America, where six species of salmon live, are subjected to several different kinds of large-scale disruptions. The largest and least frequent of these are the glacial and interglacial ages, coming and going at intervals of tens of thousands and hundreds of thousands of years. More frequent but still occasional are the failures of steep bedrock walls (called headwalls) at the upper reaches of the streams, greatly altering stream habitats. Forest fires occur even more frequently. It's as if there were a rhythm and discordant harmony to these disturbances, with ice ages coming and going as the bass notes (those with the lowest frequency of occurrence), headwall failures the baritone, and forest fires the tenor, ecological succession a kind of alto, and predation by sea lions a kind of soprano. Ecologists refer to these multiple processes at different scales of time and space, all interacting and affecting life at the same time, as a hierarchical structure, but we can just think of it as the complex discordant harmony of nature.

The standard story about salmon is that they are amazingly adapted, through biological evolution, to always find their way back not only to their natal stream but to exactly the part of that stream where they were spawned. But studies show that about 15% of salmon "make a mistake," returning to the "wrong" stream, not the one where they were born. This has been viewed as a failure of the salmon's internal guidance system. But consider what would happen if salmon succeeded in returning to their natal stream 100% of the time.

Streams go in and out of use as habitats for salmon, which are fish of northern waters, some not far south of the Arctic ice edge. Sockeye are especially characteristic of these cold northern waters. They are not found south of Oregon and are not traditionally abundant even there. In contrast, king (Chinook) salmon are still found as far south as Santa Barbara, California, where they spawn in streams even today. With the advances of glaciers during an ice age, salmon had to move to more southerly streams as the northernmost streams became covered by ice, and streams just south of the ice verged on going out of business as far as salmon were concerned. Then, when an interglacial time began and their southernmost streams no longer provided suitable habitat, salmon had to move to more northerly streams again.

The streams that flow into the Pacific Ocean on the West Coast of North America mostly originate in very steep terrain, the edge of rugged mountains, like those the Rogue River flows through and rafters love to float down. Water and ice erode the bedrock, and on very steep slopes near the headwaters large sections of a rock wall occasionally give way, filling the stream bed with rocky debris of many sizes. When this happens, the stream is unusable by salmon. Over time, however, the flowing stream gradually carries the debris downstream and sorts it by size, moving the largest sizes where the stream flows fastest, dropping the smallest in quiet pools. Eventually large beds of pebbles just the right size for salmon spawning accumulate in some reaches of the stream, and the stream becomes prime salmon habitat. But water erosion continues, and eventually the pebbles of just the right size for salmon are transported down to the ocean. Once again the stream goes out of service for the salmon until another headwall failure starts the process anew.

At even shorter intervals, forest fires create large openings along the streams, and ecological succession begins. Alders are among the important early-successional trees along these streams. These have nitrogen-fixing bacteria living symbiotically in their roots. The trees provide the bacteria with food and habitat; the bacteria provide the tree with a necessary nutrient. As a result, alder leaves are nitrogen-rich. When they fall into a stream, they fertilize the water, making the stream more productive of the small invertebrates that the little salmon eat. If the forest is never allowed to clear, alders disappear and the stream can become too infertile to support many salmon.

It's best to view the six salmon species as living in a kind of dance over the decades, centuries, thousands, tens of thousands, and hundreds of thousands of years, moving among many streams as some go into business for salmon and some go out of business. Against this complicated pattern, salmon have evolved, adapted, and continue to persist.

Salmon and Global Warming

If there is to be a global warming, what might happen to salmon? If the warming continued at a rate that salmon had experienced in past times, and they had persisted through those, then, assuming other aspects of the stream habitats remained suitable, each species of salmon would migrate north as they have in the past. Streams and rivers now famous for salmon fishing might no longer be, but those that are home to few if any salmon might become prime habitats for them.

Three things might prevent this. The warming could happen so quickly that even the 15% who make a "mistake" and go to the wrong streams can't create migrations fast enough, and populations might become extinct. Or streams and rivers might be so altered by other events, including human activities, that too few provide adequate habitat and a species thus becomes vulnerable to extinction. Or the warming might be so great that no stream, even those farthest north, could support salmon, and all wild salmon would go extinct.

How likely are these events that could extinguish the salmon of the Pacific Northwest? That's the hub of a debate among scientists. In the 1980s, available evidence suggested that before the arrival of humans, climate changed rather slowly compared to what was being forecast to happen within the fossil-fuel-burning modern biosphere. But that changed in the twenty-first century, when scientists were surprised to discover that some warmings had happened very quickly in the past, perhaps in only a few decades, as mentioned in the previous chapter. To tell you about this, we will leave the salmon for a while as they swim upstream and downstream on the Rogue River, but we'll come back to them and their possible fate before the chapter ends.

A Warming in a Forest

In the 1980s and early 1990s, my experience with the scientific community led me to believe that research to develop global climate models followed standard scientific methods. At that time, these models predicted that if we doubled the concentration of carbon dioxide in the atmosphere—an increase that was expected to occur in approximately the next century from the burning of fossil fuels—then Earth's average surface temperature would rise about 4° to 8°F. The effects were forecast to be much more pronounced at higher latitudes than at lower ones, more pronounced in the Boundary Waters and at Isle Royale than in Gold Beach, and even less in Tsavo. In Minnesota and Michigan the temperature might climb 6° to 12°F.[12]

Between 1988 and 1992, Congress provided a budget for the Environmental Protection Agency to support research on possible effects of global warming, and for those four years I and other forest ecologists used the computer model my two associates and I had developed, and variations of it, to see what the results from existing climate models implied for forests and their endangered species.

We obtained projections into the twenty-first century from several climate models for monthly temperature and rainfall at several locations in the Midwest. One of them was Virginia, Minnesota, not far across Lake Superior from Isle Royale and the weather station nearest to the Boundary Waters Canoe Area. (If you want something to put that town on the map, Virginia, Minnesota, is where Bob Dylan grew up.) According to the climate models of that time, the soils of the Boundary Waters and Isle Royale would become warmer and drier.

Using these projections, the forest model projected that if the expected climate warming occurred, the spruce and fir forests of the Boundary Waters and Isle Royale would be replaced by sugar maple and red maple, which grow to some extent in these areas but had been more characteristic of areas to the south since European settlement of North America. The intensity and kind of effects would depend on soil type and soil water conditions. In the Boundary Waters Canoe Area and the adjacent Superior National Forest, habitats well suited to balsam fir in 1990, especially lowland, poorly drained wetlands, were projected to change to predominantly sugar maple, but the change would be gradual until 2005, when balsam fir would likely have begun to show a major and rapid decline in areas where global warming would begin to make the soils much drier.[13]

Thus the boreal forest would be replaced by northern hardwoods, and the habitat of the moose would change dramatically; the waters at Isle Royale that in 1990 reflected a gliding canoe and a moose and boreal forest trees along the shore would mirror a different image. Even more striking, the projections suggested that these effects might be readily noticeable in as little as 30 years—by 2020—just around the corner for us in the second decade of the twenty-first century, but in the 1980s seeming rather far off, to be seen by the next human generation.

To the south in southern Michigan, the effects were projected to be even more dramatic: Forests that were transitional, between northern hardwoods (typically dominated by sugar maple) and oak forests to the south, would be converted to oak woodlands or savannas, or even to treeless prairies, which, before they were plowed and converted to farmland, were much farther west.

Clearly, climate change could have led to major dislocations in the forest industry in the Great Lakes states. In the northernmost areas, the timber industry had focused on a certain complement of species, primarily softwoods—northern pines, firs, and spruces, and some boreal forest early-successional species, such as aspen, whose wood was used to produce paper pulp and construction materials. As I mentioned, the forest model projected that these would change to hardwoods such as oaks and maples, which are useful for furniture and decorative purposes and have much longer rotation times (harvesting is less frequent) than the softwoods. Thus there might be a major shift in the character of the forest industry, a shift whose costs to that industry would need to be evaluated. The shift would require different equipment and lead to sales to new markets.

How likely does this seem today for the forests stretching from northern Minnesota to southern Michigan? Unfortunately, I couldn't find a sufficiently detailed forest inventory in Minnesota for the years since 1990 that would tell us whether this change is occurring there. However, a careful inventory of forests in neighboring Michigan was published in 2009.[14] In that state, there was only a 1% decline in the amount of timber between 2000 and 2004 (the most recent year for which information is available), and the amount of harvested softwoods—the conifers that the computer projections indicated would decline the most—remained essentially constant during those years. We can't tell one way or the other from these data whether there had been any effect of global warming on the forests by 2004, and, besides, the computer model projects that noticeable changes would be just beginning during that time.

In 1990 a major environmental concern was the possible global warming effects on forests, but since then radical changes have taken place in the forest industry and in the public perception of forests. Most of the major national timber companies, such as International Paper, have sold off most of their land, primarily to timber investment-management companies (TIMOs, a kind of real estate investment trust) and to conservation organizations. For example, in 1994 about 80% of the forestland in Maine was owned by industrial forest corporations, but by 2000 most of it, 60%, had been purchased by TIMOs. (Most of the rest of that 80% of Maine's forestland was sold to conservation and environmental organizations.[15]) A considerable amount of the wood used in the United States now comes from logging done in foreign nations by U.S. timber corporations.

Prior to the big sell-off of forestlands, American forest companies had harvested the timber, planned how to do it, and made products from it. They employed professional foresters, and there was an assumption within the forest industry that the profession of forestry and the science on which it was based played an important role in improving harvests and maintaining the land. Although their practices were often heavily criticized by environmental groups, both sides believed in sound management of forests, and in the 1980s and 1990s the two sides made many attempts to work together to improve forest ecosystem sustainability.

In contrast, TIMOs are primarily financial investors who view forestland as an opportunity to profit by buying and selling timber. It is unclear how much sound forestry will be practiced on TIMO-owned land, but there is less emphasis on professional forestry and forest science,[16] and far fewer professional foresters have been employed. The danger is that forestland viewed only as a commodity will be harvested and abandoned once the resource is used. If this happens, it will be the exact opposite of what most people involved in forestry, both in the industry and in conservation groups, hoped for and thought was possible throughout the twentieth century.

We hear much less about the vigor of timber industries and forest production than we did in the twentieth century. The big sell-off of forestland by industrial forestry companies to TIMOs happened without much publicity and remains relatively little known to the public. Public concerns about U.S. forestland have turned much more to concerns about global warming, which seems to have overwhelmed much of the previous focus on forest ecosystem conservation in the Great Lakes states.

Global Warming and Kirtland's Warbler

As part of my research on forecasting global warming's effects on forests, I wondered about the plight of Kirtland's warbler (which I wrote about in Chapter 4) and its very special kind of habitat. Naturalists and ornithologists who had studied this warbler were convinced that it would nest only in jack pines, and only where these trees grew on a very coarse sand in southern Michigan. There was solid natural-history knowledge behind this belief. The warbler had never nested anywhere else, and it builds its nest near the ground, typically on the lowest branch, where the eggs might get damp if the soil below was boggy or just not very well drained. It seemed plausible, therefore, that the warbler would seek to nest only where there were the coarsest of coarse sands, which would drain fastest and on average be the driest.

In 1991, we used our forest model to project what might happen to the jack pines where the warbler nested. The model projected that the warbler would likely be in danger of extinction by around 2010 because its habitat was going to go ker-flooey. Jack pines would be unable to thrive in the temperatures and precipitation that global climate models forecast for southern Michigan, the only place and only trees where this lovely warbler nested. The computer told us these declines should be measurable even in 1991, the year we made the forecast.[17] Our forecasts, by the way, weren't unknown; on the contrary, they attracted a fair amount of attention in major newspapers, both in the United States and abroad (Fig. 13.2).

We suggested that measurements of jack-pine growth be started in the species' habitat to verify the forecasts and to see whether global warming's potential effects on the diversity of life were actually occurring. It seemed to be a good test of whether global warming was happening as forecast and whether its ecological effects were occurring as our forest model projected they would. It also just seemed a good thing to do to help conserve this endangered bird.

People could have started going to southern Michigan to check out our forecasts 20 years ago. Nobody did. I tried to get funding to do this, contacting a number of government agencies and private foundations, including one that had supported my previous research on this species. Nobody was interested.

Even today, amid the furor over global warming, no one seems to remember our forecasts and nobody is rushing out to the forests of Michigan to test whether global warming is having the effects we predicted in 1991. This seems odd, since computer models are the basis for the current alarm about global warming. What could explain the lack of interest in comparing a past forecast to today's forecasts? Perhaps our earlier forecast was based on methods inferior to today's? Indeed, some of my colleagues, less enthusiastic than I about my computer modeling, including the forest model, criticized our approach when it was first published.

But wait a minute. Given the usual progress of science, won't future methods always be better than past methods? If this is why nobody is paying attention to our earlier forecast, then, ironically, no decades-long ecological projections would ever be taken seriously, or even remembered when the time came due, for that matter.

What this suggested to me was that the primary uses of, and interest in, such forecasts were political, not scientific—that scientists as well as politicians were using them for political and ideological purposes to influence public behavior. The question was not whether the forecasts were scientifically valid, but how much impetus they could provide to move society today for their proponents' purposes.[18] This may have been a misjudgment on my part, and in the first few years of the twenty-first century I wondered if I was correct, and then something happened that deepened my concerns.

FIGURE 13.2 Our forecasts about the possible fate of Kirtland's warbler, endangered by global warming. These forecasts attracted some media attention, although they are little known today.

Forecasting How Many Species Might Go Extinct Under Global Warming

In 2004, a major newspaper asked me to review and comment on a paper titled "Extinction Risk from Climate Change" that was about to be published in the prestigious scientific journal *Nature*. The authors were Professor Chris D. Thomas of the Centre for Biodiversity and Conservation, University of Leeds, Great Britain, and 18 of his colleagues, including Alison Cameron from the Royal Society for the Protection of Birds.[19] (Major newspapers routinely ask a scientist to review a scientific paper that is likely to generate considerable public attention, and I had done this fairly often for a variety of newspapers.) This paper seemed right on target for my concerns, and I readily agreed to review it, thinking it must be good since it was going to appear in *Nature*. It made a striking statement up front in the abstract: "We predict," the authors wrote, "on the basis of mid-range climate-warming scenarios for 2050, that 15–37% of species in our sample of regions and taxa will be 'committed to extinction.'" That grabbed my attention.

But I was quickly disappointed in their analysis. In my view, the paper used unacceptable data and inappropriate theory to arrive at unjustified conclusions. Explaining this to you requires getting deeper into the science than I had wished, but here goes.

The theory the authors used is known as the "species-area curve." Suppose you want to know how many species live in a certain area or ecosystem—say, your backyard or a park that you like to visit. As should be clear by now, it isn't easy to find every species and know for certain that you have done so, even in a small area. But there's a good way to estimate what the likely total number would be. Here's what you do: You divide the area up into squares and go to them at random (perhaps using a roll of dice to determine which one you'll visit next). In each square, you count all the species you find that you haven't seen in the area before. Obviously, in the first square you visit, every species will be "new" and will be listed. But in the second square, you're likely to see at least one if not most of the species you saw in the first square. And in the third square, chances are you'll see even a greater percentage of species you saw in the first two squares. The more squares you visit, the fewer "new" species you find. After you finish, you plot the number of species seen in total against the number of squares visited.

Scientists did this for the vegetation in the Konza Prairie Biological Station in Kansas, one of the few remaining unplowed and therefore original tallgrass prairies still in existence. Its biological diversity, especially the number of species, is of interest to many naturalists, conservationists, and ecologists because the tallgrass prairie was very diverse, especially for natural vegetation in temperate climate zones (Fig. 13.3). The result is a curve that slopes upward but gets shallower and shallower as you visit more and more squares. You can see that the curve will eventually level off—there will be a maximum number of species you will be able to count, because eventually you will have seen every species in the area. But you save yourself considerable effort by using your graph, because you can see at how many species the curve will completely level off. And that's your best estimate of the total number of species where you are looking.

The species-area curve is an information-gathering curve, not a cause-and-effect curve. But Thomas and his colleagues used it as if it were a curve of causality. They wrote that if you assumed that the total area available to a set of species was going to decline due to global warming, then you could estimate how many species would

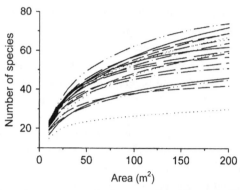

FIGURE 13.3 A set of species-area curves for the vegetation at Konza Prairie Biological Station, a 13.5-square-mile area of native tallgrass prairie in the Flint Hills of northeastern Kansas, land that was never plowed and is therefore among the few remaining original tallgrass prairie sites in the United States. Each curve shows counts of plant species in one of the watersheds. As the area examined increases, the number of new species found (species not originally seen in that watershed) increases toward a maximum. The curve is a way to estimate how many species are there without having to measure every square foot in the watershed.[20]

go extinct simply by walking backward down the species area curve. That makes a curious and false assumption: that area causes species. The authors also assumed that the area available to any species would decrease due to global warming, not just move to different geographic locations (and perhaps even increase for some). So the theory used did not justify the conclusions.

Furthermore, the authors used crude estimates lacking any statistical validity of the total areas of major kinds of ecosystems (biomes) as though they were actual, not estimated. For example, they estimated that boreal forests would shrink by just a bit less than 4%. But since the 1970s I had been concerned about how poorly we were able to estimate the area of each major kind of vegetation and the amount of organic matter and carbon that each stored. I knew that the estimates in use were quite rough and could not stand up to scientific scrutiny. For quite a number of years I tried to persuade the major scientific agencies to fund research that would improve these estimates, but nobody was interested. That must seem odd today, with so much discussion about carbon sequestering by forests—that is, using forests to remove carbon dioxide from our globally warmed atmosphere. Many estimates can be found today as to how much carbon could be removed this way, but few are based on scientifically valid information, even now.

Finally, the Mellon Foundation, which at the time liked to fund valuable research that was not finding funding elsewhere, gave me a grant for a project whose goal was to make the first statistically valid estimate of the area and biomass of any major vegetation type. I chose the boreal forests of North America for several reasons. First, because I was familiar with them and, more important, there were relatively few tree species in them, which would make our study simpler. Second, the terrain they occupied was, on the whole, relatively flat, also making our study easier. Third, this was one of the most extensive vegetation types—occupying about 2 million square miles in the United States and Canada, equal to more than 50% of the total land area of the United States—and one of the most commercially important. Since the global climate models forecast that effects of global warming would be more pronounced at high latitudes, boreal forests should also respond to global warming more quickly than, say, tropical forests.

My colleagues and I began by examining where existing estimates of biomass of the boreal forest of North America had been done. We mapped those and realized that most were within an easy drive of a university campus, although the boreal forest extended far north in Canada to extremely remote areas. Scientists who had studied these forests were not looking at global effects, but were focusing on a wide variety of local things, and it didn't matter to them that the places they went were not, in total, statistically representative of the entire boreal forest.

Then I went to the map and imagery library at the University of California, Santa Barbara, which had one of the best map collections in the world. I found 23 published maps of the boreal forest, but most were just copies of one of the others. Only three maps were distinct in their origins: two from Canada and one from Russia. Using some computer mapping methods that were sophisticated for the 1980s, we "rubber-stretched" and reconfigured the maps on the computer so they could be overlaid on each other with what mapmakers call the same "geographic projection" and could be compared.

We discovered that the area all of them agreed was boreal forest was half the sum of all the areas that at least one of them called boreal forest. In other words, in 1980, botanists and plant geographers didn't know the area covered by the boreal forest to better than a factor of two. You could say that boreal forests covered twice as much area as the next mapmaker said they did, and the maps available would let both of you be correct in terms of the information available at the time. If we didn't know the area that was truly boreal forest within a factor of two, then a change in 4% would be meaningless.

NOBODY'S EVER BEEN THERE BEFORE

How do you estimate the total area and amount of organic matter in something as huge as the boreal forest of North America? We did it in a way similar to how political polls are taken: We did a stratified random sample, on the ground. This required that teams go all over vast areas of North America, from Minnesota to Maine and from there up to the northern limits of the boreal forests, as far north as Hudson Bay in Canada and into the eastern area of Alaska and adjacent Canada. There were two teams, each with seven people (graduate students and postdocs). Each team traveled in its own Chevrolet Carryall (and sometimes also on trains, buses, aircraft, all-terrain vehicles, and on foot). Wherever they stopped to make measurements, they would divide into two field crews of three, while the seventh person took notes and kept things organized.

Two of these seven-person teams traveled across the boreal forests to sites we had selected using the best maps available and a computer to select specific locations at random. In those pre-GPS days, with relatively primitive laptops and before widely available cell phones, we gave each team a pack of 3 × 5 cards with specific directions to each site.

You can imagine what kinds of adventures these students had, turning up near somebody's home or at a government forest or government-owned or non-government-owned nature preserve in some remote place in the Far North and announcing, "We're from the University of California and we've come here to count and measure your trees!"

"Huh?" we could imagine people thinking.

The teams did get to some very remote areas. One team took the Canadian Railroad to Churchill, Manitoba, at the southern shore of Hudson Bay, some 2,000 air miles from their starting point in Santa Barbara, to seek out a few of those randomly chosen small locations, which turned out to be on the lands of the Dene First Nation Aboriginal people of Canada, who lived there. The crew members visited with them and explained what they had come to do. The leaders told them that the place to go was several days' hike through wetlands. The team set off gamely with loaded packs, but, perhaps pitying them or simply curious, several First Nation people followed along and loaned them an all-terrain vehicle.

Another site our computer picked out at random was in northern Canada, east of Alaska and on the eastern shore of a remote lake, reachable only by float plane from the nearest town. The team leader told me that on the flight over he asked the pilot what the shore and forest would be like where they were going to land. The pilot said he didn't know, hadn't been there. "Come to think of it," he added, "*nobody's* ever been there."

Such is the fate of earnest students of nature. But all team members survived and returned healthy.

My experience in forests over the decades had persuaded me that forests were just plain intrinsically highly variable—that if you hiked into two forested stands (as the foresters call them) and thought they looked identical, they could differ by as much as 20% in anything you could measure about them. That wasn't a rigorous scientific conclusion but a bit of natural-history experience. As a result, I set the goal of our study to get the first statistically valid estimate of the biomass of the North American boreal forest (and therefore the first statistically valid estimate of the biomass and carbon content of any major vegetation type anywhere in the world) as having a statistical error of 20%. That is to say, I planned and hoped that the number of sites visited at random would yield an average value whose statistical estimate would be plus or minus 20%.

Our results gave a 23% statistical error, and I was happy.[21] But this made me even more skeptical of the conclusions in Thomas and his colleagues' paper. If we couldn't know the biomass any better than plus or minus 23%, then what hope did we have that anything we measured about this forest, or any major vegetation type, would be known with better than the 4% accuracy that their paper required? Not much.

Here's another thing to add to the mixture: Best available estimates indicate that the world's forestland increased 23% between 1990 and 2000. If you are really concerned about forests, you will be delighted to know that forestland expanded so much that, according to Thomas and his colleagues' analysis, forests are much more diverse now than they were. However, forestland declined between 1980 and 1990 by about that same 23%. That estimates of forestland are this volatile gives us comfort on the one hand, because, if true, then forests recover very fast. On the other hand, it may suggest that the estimates themselves, and not the actual forest area, are volatile. If so, the entire discussion might be completely academic and without any real-world merit. In short, the area of forest may in fact be about the same today as it was in 1980—the estimates are just very weak.

Another problem with the analysis by Thomas et al. is that it ignores any other causes of change in area covered by major vegetation types and assumes that the only change in the environment is global warming. But consider this: Agriculture occupies 38% of Earth's land area (excluding Antarctica, which has no farms or woody vegetation). World food prices rose almost 40% in 2007, in part because of increasing agricultural lands devoted to biofuels. In Indonesia and Malaysia, tropical rain forests have been cleared rapidly in the twenty-first century to plant coconut palm oil trees. An article in the *New York Times* in 2007 stated that in Indonesia "the deforestation speed is 2.8 million hectares a year" out of a total of 91 million hectares. Interest in palm oil has increased for its use in biofuels, which some argue will help us deal with global warming.[22]

The demand for food is likely to increase greatly in the next years and decades because of the rising standard of living in China and India and the growing populations in the 14 nations that have the largest percentage of undernourished people. The U.N. Food and Agricultural Organization (FAO) predicts a world food crisis, and 36 nations were already facing such a crisis in 2010. By August 2011 a surge in U.S. commodity prices was "due to China and India importing more wheat, corn and other foodstuffs to meet the evolving dietary needs of their burgeoning middle class. Both wheat and corn prices were up sharply. Chinese buying of American corn was "at a record high," according to the online Small Cap Network.[23] And in 2011 "for a second year, China has bought significant volumes of corn . . . they will still top 975,000 metric tons (76.9 million bushels) and make China the number eight U.S. corn market," according to the U.S. Grain Council.[24]

People of the world need a great increase in food production, which can come about by improving production per acre and by using more land to grow food. The world's food supply is greatly influenced by what are politely called "social disruptions," such as wars and the rather arbitrary actions of certain kinds of governments. In Africa since 1960, these have included 20 major wars and more than 100 coups. Recent and current wars in the Middle East can be added here.[25] The pressure for more agricultural land for both food and fuel, coupled with such social disruptions, seems capable of changing the area of any vegetation type by quite a lot, more than, say, 4%.

All of this implies that if, because of your concern about extinction of species, you want to focus on the decrease in land area occupied by major vegetation types, you had better focus on land conversion for agriculture at least to the same extent, and probably to a much greater extent, than on the hypothetical effect of global warming. The bottom line here is that you don't even have to decide whether you believe global warming is likely or unlikely. All you have to do is consider the important factors that might cause a major increase in extinction rates, and rank the possible effects of global warming with all the others. This will lead you to realize that it's a good idea to consider more than one possible cause, and that global warming may not be at the top of the list.

AN ANCIENT CONUNDRUM

Soon after Thomas et al.'s paper was published and attracted a lot of attention, Danish plant physiologist Henrik Saxe got in touch with me to discuss what we might do to better estimate the possible effects of global warming on extinctions and on biodiversity in general. We organized a workshop, purposefully inviting scientists from a variety of disciplines and with varied points of view about global warming, plant geography, biological diversity, and forecasting methods. We wanted the results of the workshop to be as objective and scientifically sound as possible, as free as could be from anyone's ideological, political, or moral philosophy, and we believed that the best results would come from a wide-ranging discussion.

Henrik and I and 17 colleagues who had attended the workshop published a scientific paper in *BioScience* in which we analyzed the methods available to forecast what global warming might do to biodiversity.[26] We pointed out what we called a "Quaternary conundrum" (the Quaternary is the period that includes everything from 2.5 million years ago to the present on Earth). We explained the conundrum: "While current empirical and theoretical ecological results suggest that many species could be at risk from global warming, during the recent ice ages surprisingly few species became extinct, few in South America, few in most of the world." We noted that "in North America, for example, only one tree species is known to have gone extinct," and that "large extinctions were reported mainly for tree species in northern Europe and for large mammals (over 44 kg) in the Northern Hemisphere."[27]

We hoped that this open, objective discussion and the paper that resulted would open up a free-ranging reconsideration of what might happen to the diversity of life under global warming. But instead, in its 2007 annual report, the Intergovernmental Panel on Climate Change wrote: "Approximately 20 to 30% of plant and animal species assessed so far are likely to be at increased risk of extinction if increases in global average temperature exceed 1.5 to 2.5°C (medium confidence)."[28] They essentially restated, although a little more cautiously, what Thomas and his colleagues had concluded. And in 2008, James Hansen wrote: "The implications are profound, and the only resolution is for humans to move to a fundamentally different energy pathway within a decade. Otherwise, it will be too late for one-third of the world's animal and plant species."[29]

What was presented in a highly speculative, and in my view seriously flawed, scientific paper in 2004 had, despite facts to the contrary, become part of our new reality by 2008. There is no hesitancy in Hansen's statement. It "will be too late," he writes, contradicting the fossil record for the past 2.5 million years and ignoring the modern methodological problems that led to his assertion.

The low extinction rates during climate changes over the past 2.5 million years may be partially because of biological evolution—adaptation to change. Since Darwin, we know that evolutionary adaptation to a changing environment is a plus, a necessity for life to persist—like the salmon making the "mistake" of going to the "wrong" stream and thereby surviving in a changing environment.

This appears to be happening in response to recent climate change. Two of the longest studies of animals and plants in Great Britain show that the two species that have been studied the longest are adjusting to recent and rapid climate change. One is a 47-year study of a bird known as the great tit, scientifically as *Parus major*. This study, one of the longest for any bird species, shows that these birds are responding behaviorally to rapid climate change. It's a little of "the early bird gets the worm." A species of caterpillar that is one of the main foods of the bird during egg laying has been emerging earlier as the climate has warmed. In response, females of this bird species are laying their eggs an average of two weeks earlier. Both birds and caterpillars are doing okay.

The second study, one of the longest experiments on how vegetation responds to temperature and rainfall, shows that long-lived small grasses and sedges are "highly resistant to climate change."[30] As a result, the authors report, changes in temperature and rainfall regimes during the past 13 years "have had little effect on vegetation structure and physiognomy."

These studies demonstrate something that ecologists have long known and that has been one focus of their studies: Individuals, populations, and species have evolved with, are adapted to, and respond to environmental change, including climate change, both behaviorally and physiologically. Even more profound is evidence of rapid biological evolution in response to climate change. Rapid genetic adaptation to climate has already been documented for a few wild organisms for which long-term studies of field populations have been conducted.[31] Invasive species, also, have evolved at surprisingly rapid rates since their arrival in a new habitat in the twentieth century.[32]

Yet the scientific literature on global warming tends to fall into a peculiar biological contradiction, illustrated by a book published in 2009 titled *Heatstroke: Nature in an Age of Global Warming*, by paleoecologist Anthony D. Barnosky.[33]

Barnosky discusses observed changes in behavior and distribution of species, which he believes to be responses to global warming. He views these only negatively, as a clear indication of global warming's terrible effects, forcing species to undergo change, forcing nature away from a steady state. He considers such change a threat to species and a potential disappointment to those of us who seek what we saw as children and won't find it—or at least not where we first saw it. But the conclusion should be the opposite: If species did not and could not change, they would become extinct. Present observations of adjustments to climate should be viewed as a positive for the conservation of biological diversity.

This is not to take away from the problems that clearly confront some species, especially long-lived animals that typically have low reproductive rates. For them, if the climate changes too fast, there won't be time for evolutionary adaptation. Black guillemots (also called sea pigeons) illustrate the concerns some scientists have about global warming's effects on some species. These birds nest on Cooper Island, Alaska, where their abundance has declined. It appears that this is a result of warmer temperatures in the 1990s, which caused the sea ice to recede farther from Cooper Island each spring. This recession was occurring too early in the year for the immature chicks to survive on their own. Adults feed on Arctic cod, which they catch under the sea ice and then carry back to the nest to feed their chicks. The distance from feeding grounds to nest must be less than about 18 miles, but in recent years the ice in the spring has been receding as much as 155 miles from the island before the chicks are able to leave the nests. As a result, the black guillemots on the island lost an important source of food.

The future of black guillemots on Cooper Island depends on future springtime weather. Too warm and the birds may disappear. Too cold and there may be too few snow-free days for breeding, and in this case, too, they might disappear. This may be a situation where the adjustment can't happen, or can't happen within the time necessary for the species to persist.

In sum, the rapid warming that is occurring now is a potential threat to species that are unable to migrate or quickly adapt. However, like so much about life, this is not an all-or-nothing situation: Any environmental change will benefit some species and threaten others. Recent research shows that the climate did change rapidly at times during the past 2.5 million years, and it appears that few species overall went extinct simply because of that. For example, the recent findings show that the most recent continental glaciation ended about 12,500 years ago with a rapid warming, perhaps brief as a few decades or 100 years. This should encourage us about the future of biological diversity under rapid climate change.

It would make decisions a lot simpler if life *were* an all-or-nothing proposition, and that kind of simplicity is one of the appeals of the way global warming is presented—as an unqualified disaster that requires all our attention and energy. In the real world, however, the choices are complex. We find ourselves in a situation where we may have to choose which species we should try hard to save and which ones we will not be able to help. Some say that even thinking about such choices is immoral, that it is only our arrogance that leads us to think we could play such a godlike role. But the reality is much humbler. We have come to know ourselves as the greatest thinking and tool-making creature on the planet, the only one that can make such changes and such decisions. Of course, it's not as if we wish to be in this situation—indeed, it is terribly burdensome—but we must try to do our best, for our own species and for the others that surround us and claim this planet as their home. From this perspective, we find ourselves forced to see how well we can forecast the effects of climate change—and other major environmental change—on life around us.

HOW WE FORECAST RESULTS OF GLOBAL WARMING

One of the most common ways that ecological effects of global warming are forecast is by using what are called "bioclimatic-envelope models," a forecasting method that has been applied to vegetation, mammals, birds, reptiles, amphibians, and butterflies. These models begin with a map of the current known distribution of a species overlaid on a map of the current climate, expressed usually as just a temperature range in which the species occurs. Then anticipated climate changes under global warming are drawn, so to speak, on a map, and the species is then forecast to be found wherever the temperatures approximate those where the species occurs today.

This seems quite sensible, and it has its uses. However, it assumes a steady-state relationship in several ways. The climate alone is assumed to be in a steady state. More specifically, as discussed in Chapter 12, the "normal" climate, as the climate modelers refer to it, is taken to be the climate between 1960 and 1980 or 1990. The present geographic distribution of the species is assumed to be in a steady state with the present climate—there is no possibility that the species is in the midst of changing its range in response to past climate conditions or other factors. Returning to our salmon in the Pacific Northwest, this approach to forecasting would assume that all salmon species and all of their constituent populations were in an exact balance with the present climate, which in turn could not have been changing at any rate that could have affected salmon in recent years.

In addition, these models consider only the *primary* present geographic distribution of species. They would not consider nor acknowledge that Costa Rican parrots, introduced as pets by people but allowed to go wild, are very much at home nesting and reproducing in San Francisco. If you viewed Earth's surface only in terms of one of these models, and therefore perceived your forecast only within that filter, the picture in Figure 13.4 would be impossible. You would have to deny its existence, relying, in a very *Plato's Republic* way, on your idealized world of that model, while the mere Platonic sort of shadow world, the one we live in, would not be allowed to exist. Thus these forecasting models are likely to overestimate extinctions even when they realistically suggest changes in the ranges of many species.

But this can't be true for our salmon. They are always, continuously, adjusting to change, not only changes in air temperature (over freshwater and ocean) but also

FIGURE 13.4 An impossibility if you use certain global warming forecasting methods: Costa Rican parrots at home in San Francisco. (Photograph by the author.)

forest fires; headwall failures; competition with hatchery fish and effects from the genetic characteristics of hatchery fish; changes in farming, forestry, and fisheries practices; changes in the abundances of their prey, which include many species of ocean fish; and their natural predators, like the sea lions bobbing near the mouth of the Rogue River. None of these are allowed in the bioclimate models; they are tone-deaf to anything but a simple representation of climate. The complex harmony that continuously involves salmon in an ecological and evolutionary dance is canceled out, as if we are listening to nature with a pair of Bose noise-reducing earphones, represented in this case by the steady-state global climate and bioclimatic-envelope computer models.

The models not only bring us back to the balance of nature but also use the mathematics of the machine age, differential and integral calculus. There is nothing inherently wrong with using that mathematics, but it does impose a particular worldview on its users, serving as its own blinders.

When Is a Forecast True or Reliable?

Forecasts are always chancy, as the putative 1524 flood of the Thames River illustrated in Chapter 12. What increases our trust in forecasts, our conviction that they must be accurate? The most important factor is what scientists call "validation," which means demonstrating that a model can begin with known prior conditions and predict events that are known to have occurred afterward.

You will recall that in Chapter 12 I noted that the global climate models have been little, if ever, validated. The ecological models that project possible effects of global warming on biological diversity have also been little validated. After all, this is a difficult thing to do, and the scientists involved have considered the problem carefully and done the best they can with existing data, which are often sparse. Some of the scientifically best attempts include tests of the models using current data to reproduce past observations,[34] or use past data to reproduce the present.[35] Others have, appropriately, used data from one geographic region to develop and calibrate a model and then used data from another distinct geographic area to test the model.[36] Among the most valuable tests are those by Miguel B. Araújo and his colleagues, which evaluate the robustness of forecasts against small changes in quantitative values.[37] This is a topic of active research, with its own scientific journals, such as *Global Change Biology* and *Global Ecology and Biogeography*.[38]

Where Has This Journey Taken Us?

Where has this journey taken us and where does it leave us? It began with the emergence of a new global ecology, a science about life on a planet, ours, and my hope that this science would develop free of the ancient myths about nature, free of the constraints of machine-age perceptions of life, and become a strong and powerful science with theory arising from and attuned to the phenomena that were its concern. That this science would break free of the constraints that had held back the other levels of ecological inquiry.

It should be apparent by this point in our discussion that ecology was a rather immature science in the 1960s. It was not exactly a new science, since its formal origins can be traced back at least to 1838 and its informal origins are lost in long past natural-history observations, but immature in that there was a weak connection between theory and observation. Its theory was mostly borrowed from the simpler mathematics of the machine age, viewing species and ecosystems like clockwork mechanisms or early steam engines. Growing up from the curiosity of naturalists, ecology's observations were also, on the whole, immature, often not very sound scientifically, done by serious and devoted people little trained in the formalisms of the scientific method.

A lot was being asked of this yet-to-be science in the wake of World War II, This period saw the invention of the atomic bomb and its potential environmental effects; the introduction of tens of thousands of man-made organic chemicals into soils, rivers, lakes, and oceans; the rapid increase in the numbers of people; and the incredible explosion of technology, with its uses of new kinds of resources; the breakthrough in travel wrought by the jet airplane, making possible the human-induced migration of many species to new continents, which was creating myriad problems—human, animal, and plant diseases; pest animals and plants, from the zebra mussel to killer bees to kudzu; genetically engineered crops and domestic animals, and the increasing ability to control inheritance among a wild variety of species.

For the salmon on the Rogue River, twentieth-century human-led challenges included the use of toxic organic compounds on the farms within the river's watersheds; growing numbers of dairy cattle wading into small tributaries; expansion of towns and industries, especially in the Willamette Valley through which the Rogue passed; advances in logging machinery so that more-remote and larger forest stands could be cut; a rapid increase in international fishing by many nations in the Pacific Basin, affecting the abundances of salmon's prey, competitors, and predators. And on top of all of this, the need exploded to consider the possible global effects of climate change.

The dilemma is how creativity and innovation, so badly needed to advance the environmental sciences, can be promoted, urged onward. The modern jet airliner is an interesting example of the progress of technology and the role of innovation and creativity. The challenge to ecology and related environmental sciences was much like what an aeronautic engineer attributed to the development of modern aircraft. According to William Cook (one of the engineers at Boeing who helped develop the first successful jet passenger airplane, the Boeing 707), the development of airplanes, from the Wright Brothers until the Boeing 707, took place with "meager theories" that "were often erroneous." The desire and need to develop aircraft outpaced development of the theory of fluid dynamics and aeronautics in general. "Intuition played a major role in the creative process. As new ideas were developed and tested, new theories were developed to explain the successes and provide a key for optimizing performance even further," Cook has written. Moreover, the individuals who did key experiments were viewed as "eccentric" and "at first they received no help from the academic or political arenas."[39] Academic institutions, supposed to be Western civilization's center of creativity, were not the source of inspiration for the key developments of the modern airplane, he suggests.

The development of aircraft had one advantage over the development of ecology. The possibility of people learning to make a flying machine was largely dismissed by academicians, while ecologists have sought more and more academic legitimacy, which has often carried them along with conventional ways of thinking.

In this wilderness of challenges, it is no wonder that no clear path emerges. We have our GPS's to find our way across the terrain, but no compass to guide us to understanding and knowledge of nature. Add to this the deep involvement human beings have always had with nature, an I/Thou relationship, profoundly affecting and affected by our religions, our cultures, our folkways, emotions.

Yes, this science has continued to be the captive of the machine-age view of nature combined with this ancient balance-of nature mythology. Jim Welter, confronting the realties of salmon, rivers, forests, and people throughout his life-time and, ironically, little exposed to modern science that had been captured by the old ways of thinking, was freer to focus on nature's reality. It is what we must do.

It's easy to understand why global warming attracts so much attention. It's a great story and a simple one, like those disaster movies—*Towering Inferno, Titanic, War of the Worlds*—only it may be real. In 2006, it led to one of those movies, *The Day After Tomorrow*, a fictionalized description of what might happen with runaway global warming. That same year saw former vice president Al Gore's movie *An Inconvenient Truth*.

I write this in the summer of 2011, when people from the Midwest to Long Island, NY, and from Maine to Georgia are suffering a major heat wave. On July 22, according to the *New York Times*, the temperature in New York City reached 104°F, 2° short of the record, and in nearby Newark, NJ, it reached 108°, an all-time record. "More than 1,400 record-high temperatures have been broken or tied around the country in July alone," the *New York Times* also reported that day. The National Oceanographic and Atmospheric Administration stated that "the com-bined global land and ocean average surface temperature for June 2011 was the seventh warmest on record."[40]

It would be easy to conclude that human-induced global warming was here, was real, and that its existence could not be questioned. That wasn't the experi-ence of people on the West Coast from California to Washington. "Usually by August, most of the snow on Mount Rainier, the sleeping volcanic giant here, has long since melted. The meadows of wildflowers are abloom, and hikers galore are tramping along the trails," according to Katharine Q. Seelye, writing in the *New York Times* on August 13, 2011. "But this year, temperatures have been colder than usual, keeping record mounds of old snow lying around. This has discouraged everyone, from the most rigorous climbers to backpackers, hikers and Sunday drivers."[41] The summer in California's wine country was so cold that one grower, Fritz Maytag, wrote, "We have had the coolest summer I can ever remember, the second in a row, and the grapes are so far behind I think we may have a true wine disaster on our hands here. The mountains are the worst, we are almost always several weeks behind the valley, but they must be scared this year too."[42]

Climate change just isn't simple. An article in the *New Scientist* stated that "though climate change models predict extended droughts and periods of intense rainfall for the end of the 21st century, they don't explain the current droughts, says Martin Hoerling at the US National Oceanic and Atmospheric Administration. 'A lot of these extreme conditions are natural variations of the climate. Extremes happen, heat waves happen, heavy rains happen,' he says."

That article goes on to say that much of the drought in the southern United States and heavy rains in the north of the nation are, according to Michael Hayes, director of the National Drought Mitigation Center at the University of Nebraska-Lincoln, "a result of La Niña."[43]

La Niña and El Niño are periodic variations in currents in the Pacific Ocean. They occur approximately every seven years. Under La Niña, trade winds blow west across the tropical Pacific, pushing the surface water so that it piles up in the Western Pacific—the sea surface can be as much as 0.5 m higher at Indonesia than at Peru.

During El Niño, in contrast, trade winds weaken and may even reverse, so that the westward-moving equatorial ocean current weakens or reverses. The eastern equatorial Pacific Ocean becomes unusually warm. The rise in temperature of sea surface waters off the South American coast inhibits the upwelling of nutrient-rich cold water from deeper levels; the upwelling normally supports a diverse marine ecosystem and major fisheries. El Niño became famous first because of periodic fish population declines off Peru. Birds that fed on anchovies and their relatives either left the area or died, and commercial fisheries suffered. Many fish-eating birds nested on islands off the Peruvian coast, and their droppings over centuries created one of the world's major supplies of phosphate fertilizers. The periodic decline in these deposits, averaging seven years, demonstrated the long and persistent history of El Niño.

Because rainfall follows warm water eastward during El Niño years, there are high rates of precipitation and flooding in Peru, while droughts and fires are commonly observed in Australia and Indonesia. Because warm ocean water provides an atmospheric heat source, El Niño changes global atmospheric circulation, which causes changes in weather in regions that are far removed from the tropical Pacific.[44]

During El Niño, the normal conditions of equatorial upwelling of deep oceanic waters in the eastern Pacific are diminished or eliminated. Upwelling releases carbon dioxide to the atmosphere as carbon-dioxide-rich deep water reaches the surface. El Niño events thus reduce the amount carbon dioxide released from the ocean, influencing the global carbon dioxide cycle. Some researchers have suggested there are strong relationships between El Niño events and changes in the sea ice cover around Antarctica.

In summary, the majority of scientific analysis, policy planning, and actions today that are supposed to help life on Earth deal with global warming suffer from the same balance-of-nature folkway that has plagued twentieth-century environmental work. An integrated, multifactor approach that deals with the real dynamics of ecosystems, species, and populations is necessary if we are going to mitigate the ecological and environmental effects of rapid climate change.

The real world of nature is rarely so simple or flashy, although its charismatic animals can make it seem so on TV nature programs. To really understand it, and figure out the best we might do to help, takes a lot of time and effort—like seven graduate students and postdocs hiking for days through bogs just south of Hudson Bay; like trying to find ways to count salmon on streams and rivers without dams. It takes patience as well, like Jim Welter going to one government office after another to get water-flow data and counts of salmon on the Rogue, then having a friend help draw graphs. It takes thinking about mathematics in ways that few of us choose to do.

And in the end it comes down to whether the rationality that has been at the heart and spirit of science and invention in modern times will come to our aid at this time, when so many people are concerned about the environment. We can only hope it will be so.

FIGURE 14.1 Men at work on the Venetian Canal. (Photograph by the author.)

14

The Moon in the Nautilus Shell

NATURE IN THE TWENTY-FIRST CENTURY

As the plan holds, life is poured copiously throughout a Universe, engendering the things and weaving variety into their being.

—PLOTINUS (THIRD CENTURY CE)[1]

[The newly designed solar-powered car is a] meaningful synthesis of biology and technology, bringing man toward an accommodation with dwindling resources.

—PAUL MACCREADY (1987)[2]

Science itself is now the only field through which the dimension of mythology can be again revealed.

—JOSEPH CAMPBELL IN OCCIDENTAL MYTHOLOGY: THE MASKS OF GOD[3]

Dawn Flight

Leonardo da Vinci Airport lies on the mainland north of Venice along the shore of the lagoon, where it is most conveniently reached from the city by water. I had arranged to leave Venice on an early-morning flight and departed from St. Mark's Square before dawn by water taxi. Before daylight, the city was as much sound as light as we traveled slowly along small side canals where, in the dark, the wavelets of the lagoon and the wake of the boat brushed against the buildings, the sound of city and sea brought together. As dawn broke, we were free of the main islands, speeding into the open waters of the lagoon, where the fresh wind spoke not of the city but of shorebirds and salt grasses.

The airplane took off in full daylight, and from within one of the magnificent creations of the twentieth century, the jet aircraft, we viewed the classic city of history, the city within the lagoon, for one last time. Quickly below us passed St. Mark's Square; the church of Santa Maria della Salute, supported by its 1 million saplings driven long ago into the shifting muds of the lagoon; and the Grand Canal with the Rialto Bridge, site of so many famous paintings of Venice in the hazy light of the Adriatic coast. Wave patterns showed that the water was in motion, and the **321** wake of boats traveling from the city to the mainland showed that they too moved, but all else, in the few seconds it took the city to pass below, appeared stationary. A quick impression, and the scene was gone.

From this vantage, all seemed clear, constant, and beautiful: The pollution of the waters and the crumbling decay of the statues in the acidic air were invisible; nature and the city appeared briefly as one. As in a painting, the view from 1,000 feet was of a graceful still life without a hint of the changes that had occurred over the centuries, wrought by man and by nature. Below was a city founded when people believed that Earth was a fellow creature and nature the product of a divine order. Before the rise of modern science, in the age of Venice as a great city of history, nature was viewed as a whole: The universe, the physical Earth, and life were one, or each appeared to be merely different expressions of the same truths, of the same balance, order, and harmony.

The roar of the engines caught my attention, and I thought about the many systems that made this wonderful machine glide smoothly over the Earth with re-liability, stability, and precision. The steady roar reassured me that the engines were spinning at a constant speed under the control of intricate monitoring systems. Up front, the pilots relied on solid-state radio-navigation and communication systems, each tuned to a specific frequency, and on guidance systems made possible by gyro-scopes spinning steadily. Now I was within a machine with all the qualities of order, harmony, and balance that had once been attributed to the nature I saw below.

Nature and technology have always been part of my own experience. All of the places I have been and described in this book were reached by modern machine, and the equipment that we camped with and hiked with and measured nature with were all products of late-twentieth-century technology. Piloting a small airplane over the landscape and walking through the wilderness had each in their own ways brought me in touch with nature, and I have been uncomfortable over the years with the division between the views of my engineer friends and my conservationist friends, when they seem to see these two parts of my life in oppo-sition. From an aircraft, from within a machine, we are in touch with nature; we see the view below, and we experience the winds and weather. The machine shows us nature in ways we could not see it otherwise. No wonder that modern scientific perspectives on the environment have emphasized the mechanical image; sciences depend on machines for observation.

As the airplane departed from the Adriatic, my mind meandered from thoughts of the shallow European sea to those of the far-off Pacific Ocean and one of its humblest and most obscure creatures, the chambered nautilus (*Nautilus pompilius Linnaeus*), which lives far from Venice in the southwestern Pacific. Although the shell of the chambered nautilus decorates many a coffee table, alive it is a cryptic creature with nocturnal habits, living in the depths of the ocean, as much as 1,000 feet below the surface, and rarely seen alive by human beings.[4]

The nautilus lives only in the outermost chamber of its shell, periodically pull-ing itself forward to the outside of its shell and depositing a wall behind it. As it grows, the nautilus requires a larger protective shield, and the chambers increase in size and the shell develops in a most interesting way, coiling into the convoluted shape of a logarithmic spiral, following a simple but elegant mathematical formula. As its shell grows, the chambered nautilus records two different rhythms of the solar system. Along the opening of the outer chamber of the shell, small deposits of calcium carbonate are laid down in groups of three to five, which are separated from adjacent groups by a ridge known as the "growth line." There are an average of 30 growth lines per chamber, one for each day in the lunar cycle, suggesting that a new chamber is put down each lunar month and a new growth line each day. This implies that the chambered nautilus contains in its shell two clocks: one timed to the sun, the other to the moon. These are relative clocks, marking the number of days within a lunar month.

The chambered nautilus is an ancient form of life whose oldest fossil ancestors have been dated at 420 million years before the present, and chambered nautilus fossils have been found representing most of the geologic periods from that time to the present. Strangely, the number of growth lines per chamber has increased over time. The oldest fossil shells have only nine growth lines per chamber, compared with the 30 of modern shells, suggesting that the lunar month has grown longer and that the moon used to revolve faster around Earth than it does now. This in turn implies that the moon must have been closer to Earth, since the closer a satellite is to a planet, the faster it must revolve to remain in orbit. The "clocks" in the nautilus shell indicate that the revolution of the moon 420 million years ago took only nine days.[5]

The timing of the chambered nautilus's clocks, corroborated by other evidence, has been accurate enough to allow two geophysicists, P. G. K. Kahn and S. M. Pompea, to infer that there have been three major stages in Earth–moon history. In the first period, liquid water was not present on Earth's surface, and the distance between Earth and the moon widened very slowly. Astronomers tell us that the moon's recession from Earth would have been caused by a loss of energy from friction, and that the friction of the tides against the land would have caused the most rapid loss. In the second stage, the oceans appeared to cover all of Earth, or so much that there was still relatively little loss of energy from the friction of the tides, and the moon continued to recede slowly. Finally, when the continents emerged about 600 million years ago and the water began to pound against the shores, the frictional loss of energy from tidal action increased, and the moon receded more rapidly.[6]

This is indirect evidence of when the continents emerged from the oceans. The lines on fossil nautilus shells tell us about the history of the motions of Earth and the moon, about the history of Earth, and about the origin of the oceans and continents, so important in the history of life on Earth. Thus in the chambered nautilus, the solar system, the physical Earth, and life on Earth are linked.

Similarly, our perception of nature could return to a unity, with a great difference from past pre-scientific ideas. The moon in the nautilus shell is not a fable; it is an insight based on sophisticated modern methods of scientific observation, open to tests of validity and accuracy and to disproof. As with the scientific analysis of the fossil nautilus shells, science can take our ideas to a new stage, where the separations between the organic and the machine, and between the cosmos and Earth—distinctions that have dominated ideas since the rise of modern science—are disintegrating.

In the Mirror of Nature, We See Ourselves

The answers to the old questions—What is the character of nature undisturbed? What is the influence of nature on human beings? What is the influence of human beings on nature?—can no longer be viewed as distinct from one another. Life and the environment are one thing, not two; and people, like all life, are immersed in the one system. When we influence nature, we influence ourselves; when we change nature, we change ourselves. Nature interests us not merely as a scientific curiosity but as a subject that pervades philosophy, theology, aesthetics, and psychology. There are deep reasons that we desire a balance and harmony in the structure of the biological world and that we seek to find that structural balance, just as our ancestors desired and sought that kind of balance in the physical world.

Clearly, to abandon a belief in the constancy of undisturbed nature is psychologically uncomfortable. As long as we could believe that nature undisturbed was constant, we had a simple standard against which to judge our actions, a reflection from a windless pond in which our place was both apparent and fixed, providing us with a comforting sense of continuity and permanence. Abandoning these beliefs leaves us in an extreme existential position: We are like small boats without anchors in a sea of time; how we long for safe harbor on a shore.

The change in our perception of nature and the new answers to ancient questions about nature arise from new observations and new ways of thinking that even now seem radical. I wrote in 1990 that the transition taking place will continue to affect us deeply for decades in ways that may not be obvious. By and large, the effects are not obviously a part of our consciousness in the second decade of the twenty-first century, but they nevertheless strike at the very root of how we see ourselves. We have clouded our perception of nature with false images; and as long as we continue to do that, we will cloud our perception of ourselves, cripple our ability to manage natural resources, and choose the wrong approaches to dealing with global environmental concerns. The way to achieve a harmony with nature is first to break free of old metaphors and embrace new ones so that we can lift the veils that prevent us from accepting what we observe, and then to make use of technology to study life and life-supporting systems as they are. A harmony between ourselves and nature depends on—indeed, requires—modern technological tools to teach us about Earth and to help us manage wisely what we have inadvertently begun to unravel.

Once we realize that we are part of a living system that is global in scale and is produced and in some ways controlled by life, and once we accept the intrinsic qualities of organic systems—with their ambiguities, variabilities, and complexities—we can feel ourselves a part of the world in a way that our nineteenth-century ancestors could not, but our ancestors before them did. We can leave behind the metaphors of the machine, which are so uncomfortable psychologically because they separate us from nature and are so un-lifelike and therefore so different from ourselves, and we can arrive, with the best information available to us in our time, at a new, organic view of Earth. In this new view, we will see ourselves as part of a living and changing system whose changes we can accept, use, and control to make Earth a comfortable home for each of us and for all of us collectively in our civilizations.

The machine-age view provided simple and immediate answers to the classic questions about the relationship between human beings and nature. Nature knew best, and nature undisturbed was constant; individuals, depending on which of the interpretations of nature they chose, had a certain fixed relationship to their surroundings.

From the new perspective, nature does not provide simple answers. People are forced to choose the kind of environment they want, and a "desirable" environment may be one that people have altered, at least in some vicinities some of the time.

An awareness of the power of civilization to change and destroy the biological world has grown since the nineteenth century. We recognize that civilization has had a tremendous impact on nature, and it is tempting to agree with George Perkins Marsh that the absence of structural balance in the biological world is always, or almost always, the result of human activity, that "man is everywhere a disturbing agent. Wherever he plants his foot, the harmonies of nature are turned to discords."[7] But today we understand, in spite of our wishes, that nature moves and changes and involves risks and uncertainties, and that our judgments of our own actions must be made against this moving image.

There are ranges within which life can persist, and changes that living systems must undergo in order to persist. We can change structural aspects of life within the acceptable ranges. Changes that are necessary to the continuation of life we must allow to occur, or at huge cost replace what otherwise would have been achieved. We can engineer nature at nature's rates and in nature's ways; we must be wary when we engineer nature at an unnatural rate and in novel ways. To conserve well is to engineer within the rules of natural changes, patterns, and ambiguities; to engineer well is to conserve, to maintain the dynamics of the living systems.

The answer to the question about what the human role in nature should be depends on time, culture, technologies, and peoples. There is no simple, universal answer for all peoples, cultures, and times, but in all cases the answer must be very much influenced by the fact that we are changing nature at all levels, from the local to the global; that we have the power to mold nature into what we want it to be or to destroy it completely; and that we know we have that power. This leads us to a very different kind of answer from those of the Greek and Roman philosophers, their intellectual descendants in the Middle Ages and Renaissance, or the people of the early and middle industrial-mechanical age.

Now that we understand that we are changing the environment at a global level, we must accept responsibility for the changes our actions have wrought and try to minimize these effects and slow down the rates of change as much as possible. First and foremost, we have to use science correctly to estimate, to the best of our objective abilities, which environmental changes we have wrought and which we have not. This requires not only information and understanding but also the political will and social and economic means and policies to accomplish what we need and desire. Recent events suggest that political will among us citizens is ample but volatile: The majority of us are easily manipulated by fast-talking politicians, and therefore easily misguided, especially when the old myths and folkways dominate. It is uncomfortable for us that the new perspective does not give the same simple answers to all questions, but requires that our management be specific and that the answers to questions be dependent on the particular qualities of our goals and the actions open to us. Knowing what to do in each case requires considerable information—surveys, monitoring, knowledge, and understanding—that we as a society have been reluctant to seek. Perhaps we have been too much like those people Peter Kalm met in eighteenth-century America, who believed that the study of nature was "a mere trifle, and the pastime of fools,"[8] or just another of many public issues that come and go and are to be manipulated for political and ideological goals.

A new awareness of biological nature is coming and is inevitable, and it can easily be misused. If we persist in arguing that what is natural is constant and what is constant is good, then those of us who value wilderness for its intrinsic characteristics or believe that the biosphere must be maintained within certain bounds will have lost our ability to live in harmony with nature as it really is. If we don't understand the true nature of populations, biological communities, and ecosystems, how can we expect to husband them wisely? When we had less power, we could live with myths. But today, as Joseph Campbell recognized, "Science itself is now the only field through which the dimension of mythology can be again revealed."[9]

The task that I am encouraging the reader to join in continues the one begun by George Perkins Marsh, a task that acknowledges the great destructive powers of human civilization but assumes optimistically that we will begin to make prudent choices in our dealings with nature. The message of this book is consistent with the ethical outlook of Paul Sears, who wrote that "nature is not to be conquered save on her own terms."[10] I have tried simply to give a modern view of "her" terms. It is also consistent with the land ethic of Aldo Leopold: "Conservation is a state of harmony between men and land."[11] We have not abandoned that belief or Leopold's ethic, but have redefined *harmony*. To achieve that new harmony, we must understand the character of nature undisturbed, the discordant harmony that is the topic of this book.

The proper response to environmental problems we have created with our technology is not to abandon civilization or modern technology, an approach favored by some, perhaps especially those who have most suffered the destructive effects of human actions against the natural world, or who cling to the belief that everything natural (i.e., nonhuman) is desirable and good. We have negatively altered nature with our technology, but must now depend on technology to see us through to solutions. A new understanding of the biological world is necessary for us to learn how to live within the discordant harmonies of our biological surroundings, so that they function not only to promote the continuation of life but also to benefit ourselves: our aesthetics, morality, philosophies, and material needs. We need not only new knowledge but also new metaphors, which are arising from an amalgamation of the organic metaphor with a new technological metaphor evolving from the old machine idea that we have been accustomed to using for the past 200 years.

In this book, we have journeyed through nature and natural history as we understand it today, and we find that at the end we have come, in a sense, full circle; our ideas have evolved from organic to mechanical and now return to a new linkage, connecting life and technology in metaphor and in fact.

As the plane banked and climbed toward the Alps, the central issue of the relationship between ourselves and nature seemed to come into focus. Could that most magnificent machine of the twentieth century, the airplane, serve as the proper model for the system of nature visible below?

Machines can help us see nature, but they alone are not the proper model, the right metaphor for nature. We have things backward. We use an engineering metaphor and imagine that Earth is a machine when it is not, but we do not take an engineering approach to nature; we do not borrow the cleverness and the skills of the engineer, which is what we must do. We talk about the spaceship Earth, but who is monitoring the dials and turning the knobs? No one; there are few dials to watch, only occasional alarms from people peering out the window, who call to us that they see species disappearing, an ozone hole in the upper atmosphere, the climate changing, the coasts of all the world polluted. But because we have never created a system of monitoring our environment or devised an understanding of nature's strange ecological systems, we are still like the passengers in the cabin who think they smell smoke or, misunderstanding how a plane flies, mistake light turbulence for trouble. We need to instrument the cockpit of the biosphere and let up the window shade so that we begin to observe nature as it is, not as we imagine it to be.

Postscript: A Guide to Action

People ask me what the changes discussed in this book imply for the actions we should take in the future. Every natural ecological system passes through many states, all of which are "natural" in the traditional meaning of the word. Because nature is made up of such dynamic systems, which we did not invent and only partially understand, we can only partially control nature. The larger the ecological system, the poorer our power to change it. Therefore, choosing what to do is not a search for the single "true" condition of nature. Rather, it is a design problem. To the extent we can control nature, we can see it as a variety of opportunities, much like a landscape architect looks at a site that will become a park and considers its range of possibilities, rather than as a system for which there can be only one true condition. No doubt the assertion that nature poses a problem of design, not of truth, will affront people who view environmentalism as the pursuit of a single morally just and right nature. But I assert it is the reality, a conclusion I have reached by my examination of nature for more than 40 years.

As an example of the design approach, consider water rights in Montana and Colorado. Many people choose to visit or live in these states because of the excellent outdoor activities they offer, including those on their rivers and streams. But the two states have very different laws governing who owns streams and rivers. In Montana, the water in a stream or river belongs to all the people who live in the state. They own these in common. Any Montana resident can use any part of any of these. In Colorado, the water flowing past a privately owned parcel of land is owned by the adjacent landowner. He can exclude any or all others—the non-owners— from using, even passing by in or on, the waters in that reach of the stream or river. Some believe the water commons (Montana's choice) is the right approach; others believe the private-property approach (Colorado's choice) is correct. We could argue with no more knowledge than we have at present as to which should be the general rule for all states. But better, we can consider these two state water-right laws just different designs, and we monitor those waters over time to see which in fact works better for the stream, its life, and people.

This analysis and the rest of the discussion in this book put us within nature, but in a unique role because, like it or not, desire it or not, we are the only species with the technological knowledge and abilities and the capacity to reason about these and take actions based on our choices. We are stuck with our role as pilots of the biosphere, a strange kind of pilot who can alter only sometimes, to some extent, the trajectory of our life-supporting system. We can only some of the time, in some situations, change nature's path.

This postscript is not meant to be a definitive treatment of all the actions for all environmental issues, but merely an overview of some of the practical implications of the new perspective that I have called for. In *Discordant Harmonies*, I trod very lightly here, because my goal was simply to point out problems with the old ideas and old ways of thinking about nature and the environment. But times have changed; as we have seen throughout this book, people argue vehemently over environmental issues, some even calling for violent action, like burning down the homes of those who disagree with them.[1] I can no longer tread lightly. There are many specific and general things to change and improve, and I will do my best to set them down here.

Before beginning, I want to make clear that many excellent and devoted scientists are working in environmental sciences, and they are often frustrated by the lack of resources available to them, as well as by fashionable notions, influence of politicians and pundits, and ways that their science seems stuck in old ways and seems to keep repeating itself. For example, Professor Tom Lovejoy III, one of the world's leaders in the conservation of biological diversity and the application of ecology in general, expressed his frustrations with the lack of forward movement in biological conservation in a speech to the American Institute of Biological Sciences in 1988 titled "Will unexpectedly the top blow off?".[2] Others have discussed their problems and needs and have urged me to write about them. Due to space limitations, I can mention only a few of the excellent programs and projects as examples of what we can and should do. I hope that colleagues whose programs I have omitted will understand this, and that the ones whose programs will be discussed here will understand that I am trying to be of help and be constructive, not merely critical and negative. I hope they agree with the general points and don't take them as personal criticisms. Lord knows how my colleagues struggle to understand, as scientists, what has been so difficult, elusive, and profound for our species as long as we have walked on this planet.

Citizen Science

This postscript makes a big push for citizen science. By this I mean voluntary participation in periodic counts and measurements of the abundances of populations, the diversity of habitats, and some straightforward chemistry, such as monitoring the concentration of nitrogen, phosphorus, radionuclides, and toxic artificial organic compounds in one's local environment, as part of larger study so each contribution can be integrated into a scientifically and statistically sound sample. This will enhance the data that we have about nature around us and involve citizens directly in natural history. As more and more people become urban residents, there is a tendency to have less and less contact with nature and become a passive recipient of assertions by politicians and others not so much to benefit nature or the rest of society but to serve their own purposes. Citizen science can reverse this, turning more of us into naturalists who are better able to evaluate what politicians tell us about our environment.

The Important Environmental Issues

Since nature doesn't provide us with a simple answer in a single context, we have to clarify our goals. We can't just leave nature alone and get out of the way, hoping this will solve all environmental problems. We need to decide which are the important environmental issues and then select from those the ones that are solvable with current knowledge and methods. Some would say there is only one important environmental issue: global warming. Others, with whom I agree, believe there are a number of crucial environmental issues, some of which are tractable today.

The list of important issues also changes when we include people within nature, as I have done in this book. Let's take as the three overriding goals for us and nature (1) the persistence of the great diversity of life on Earth; (2) a sustained population of human beings; and (3) the continuation of human civilization, democracy, and human creativity. From this perspective, as a start in what has to become a much larger discussion, I can list the following among the leading tractable environmental issues of our time: finding enough energy for people with the fewest negative environmental effects; ensuring water for terrestrial life, including ourselves; ensuring sufficient phosphorus and other essential minerals required by vegetation (including agriculture); greatly reducing large-scale habitat destruction; controlling invasive species; directly assisting the most endangered species that matter to us and play important roles in their ecosystems; reducing the spread of manufactured toxic substances, including radioisotopes; revising our ways of setting harvest amounts for fish and forests to likely sustainable levels.

These are just examples of what appear tractable. We must also think about sobering issues that have been intractable and continue to be major threats to life, including ours: the continuing rapid growth of our own population and the threat of nuclear war, other wars, and other large-scale social disruptions that will increase starvation and malnutrition and lead millions of people to do whatever they can to survive, no matter how detrimental to what we call the natural world. Of course, another very widespread war would dampen, if not completely halt, attempts to preserve large natural areas and save many threatened and endangered species. Recall, for example, my story about returning to the Philippines in 1986 while doing work at the East-West Center in Honolulu, Hawaii. We were trying to find out what ecological science provided to improve land management in Southeast Asia. We talked with fishermen about their destructive practices, including dynamiting coral reefs to catch fish. The fishermen were well aware that the practice was unsustainable, but said they had to live today and let tomorrow take care of itself. (An economist will tell you that the fishermen accepted a very large discount factor so that the future value of fish declined rapidly compared to its present value.) If we hope for a future of wonderful biological diversity, we must do what we can to avoid situations that force people to make such decisions.[3]

INVASIVE SPECIES

Invasive species will likely be a particularly thorny problem. In 2005 a tiny fruit fly that carries and disperses a bacterial disease of citrus plants arrived in the United States from China. This disease, known as "citrus greening" or Chinese *huanglongbing* (yellow dragon disease), had been extending its range and had reached India, many African countries, and Brazil. Wherever the fly and the bacteria have gone, citrus crops have failed. The bacteria interfere with the flow of organic compounds in the phloem (the living part of the plant's bark). The fruit fly larvae suck juices from the tree, inadvertently injecting the bacteria. Winds blow the adult flies from one tree to another, making control of the fly difficult. According to the U.S. Department of Agriculture, citrus greening is the most severe new threat to citrus plants in the United States and might end commercial orange production in Florida. In 2010, the U.S. Department of Agriculture's Animal and Plant Health Inspection Service (APHIS) issued an interim rule announcing a plant quarantine in Florida and several other states and territories to stop the spread of citrus greening. Citrus fruit can leave the state only if the USDA has issued a Federal Certificate.[4]

Modern air travel and large container ships moving freight around the world make it very difficult to control such invasions, especially by insects and bacterial and viral diseases (Table 15.1). It's difficult to pin down the level of threat these potential invasions imply for agriculture, horticulture, and natural ecosystems. These are not so readily open to steady-state computer simulations, or at least they haven't been, and they are also not so easily painted in the large as the kind of threat that politicians tell us is coming from global warming. The fruit fly that causes citrus greening is, of course, just one of many invasive species that have caused major problems in recent decades. It seems quite likely that others will create similar problems in the near future.

We should include in the invasive-species problem the new hobby of informal genetic engineering, where people make use of amazing new knowledge of genetic inheritance to try out new life-forms. Much environmental damage in the industrial age happened because people of goodwill brought exotic species to new habitats. Apparently they believed that good intentions ensured good results. There are many famous examples of this mistake, such as the intentional introduction of gypsy moths in 1869 in Medford, Massachusetts, by Leopold Trouvelot, who thought these insects could be the beginning of silk production in the United States. The story told is that he believed the country was safe from problems with

TABLE 15.1

U. S. Department of Agriculture Interceptions of Exotic Plant Pests in 2007

Major Sources and Events	Number	Percent
Airport, Individual	42,003	61%
Express Carrier	6	0.01%
Inspection Station	2,763	4.02%
Land Border	14,394	20.94%
Maritime	4,518	6.57%
Pre-departure	4,869	7.08%
Rail	16	0.02%
USPS Mail	184	0.27%
Total	68,753	

These are the numbers of known plant pests that have been found on or with passengers arriving in the United State or in packages shipped separately. Air travel overwhelms all other ways of bringing exotic plant pests into the United States, and detection is extremely difficult.

(Courtesy of Alan K. Dowdy, PhD, USDA, APHIS, Plant Protection and Quarantine: Emergency and Domestic Programs.)

the insect because prevailing winds, which are from the west, would merely blow any escaped moths out to sea. Hobbyist genetic engineering could have analogous undesired and unexpected effects.

RESPONDING TO HABITAT DESTRUCTION

Habitat destruction also ranks high as a major environmental problem in the here and now, affecting biological diversity, the quality of human life, the sustainability of human populations, and the continuation of civilization. In spite of much talk, much less is taking place than seems necessary to avoid habitat destruction, but there are encouraging actions. Constructive efforts include the development of new national parks in many nations as a result of work by major environmental organizations such as the Nature Conservancy, the World Wildlife Fund, and Conservation International. An especially interesting recent activity to protect habitats is the leasing of land-use rights. This approach, known as "conservation concession," was initiated by Conservation International and has since become the major work of "Save Your World Foundation" under the direction of Richard Rice. For example, one project Rice established in Guyana, the "Upper Essequibo Conservation Concession," leased timber rights to protect 200,000 acres of pristine rain forest in Guyana for just $0.15 per acre per year, an agreement to last 30 years with the potential to be renewed.[5] An additional benefit of this approach, consistent with one of the themes of this book, is that indigenous peoples can continue to live within the protected forest, in contrast to the requirements of some national parks.

A similar approach has been used in the Predator Compensation Fund (PCF), co-founded by Tom Hill and Richard Bonham. This project is done in cooperation with the Maasai on their land in East Africa. Traditionally, a proof of manhood by a Maasai warrior was the killing of a lion in single combat with it, and lions were also being poisoned so they would not be able to kill Maasai cattle. The PCF agreed to pay the Maasai for cattle killed by lions if the Maasai agreed to stop killing lions. According to Rice, the agreement "has virtually stopped the killing of predators across more than 1 million acres."[6]

Both the establishment of national parks and the leasing of land-use rights are valuable actions, but they differ from each other in several important ways. In general, national parks allow people as temporary visitors, but not as residents, they separate people from nature. Leasing of land-use rights involves people within nature and is a much faster and less expensive method.

SOME THOUGHTS ABOUT GLOBAL ENVIRONMENTAL PROBLEMS

We face three kinds of global environmental problems as a result of human activities: catastrophic, acute, and chronic. The catastrophic include nuclear war and even the possibility of a nuclear winter, which physicists and climatologists warned us about in the early 1980s before global warming became popular.[7] Acute global problems are sudden and short-term, lasting perhaps a year, such as the emission of a toxin carried from its place of origin or the accidental release of an undesirable strain of bacteria. Chronic global environmental problems develop over a long time and typically result from a slow environmental change. Acid rain, depletion of the ozone layer, and the greenhouse effect are major examples of chronic global problems. Most of our current global environmental problems are chronic.

We can divide our goals for the biosphere into two stages: a transition stage and a long-term stage. The transition stage gets us from where we are now to where we would like to be. The long-term stage is a biosphere that we consider healthy, meaning able to persist in a broad set of states that would sustain a high level of biological diversity, along with human cultures and civilization. That is to say, a healthy biosphere would vary within a set of conditions acceptable to us and necessary for the persistence of life.

For example, the rate and magnitude of variation would be small enough for the rate of extinction to return to its precivilization level. Soil erosion would be reduced to a range of levels that could be replenished by human actions that are within our capabilities. Freshwater would again become a sustainable resource. Pollutants that do not decay (such as arsenic and radioisotopes) would be eliminated or reduced to vanishing levels, and the concentrations of other pollutants would be lowered. The abundance of renewable natural resources, especially forests and fisheries, would vary within acceptable limits, subject to management actions within our capabilities, both technically and economically, and we would have constrained our energy use to a sustainable level. The human population would cease to increase overall, but would vary from place to place, increasing in one while decreasing in another, and even undergo some total decreases over time; it would remain below an upper bound.

It's important to note that I'm not assuming that these goals would be, or would have to be, reached by top-down command-and-control methods. As made clear by recent events, environmental issues continue to attract worldwide interest, and people feel very strongly about them. What I'm writing about here are the goals, leaving the societal means open.

Given the current state of scientific knowledge about climate change, and the uncertainties about our role in causing a global warming, current responses would be to our best advantage where the actions we take might lessen what is believed to be both possible undesired climate change and one of the other here-and-now problems, such as habitat destruction.

Counting: Obtaining Necessary Information

Some of the things that need to be done to improve the science of ecology and its associated sciences are straightforward. One, discussed earlier, is a lot more counting. You'd be amazed by how often the essential things about an environmental problem have not been measured and nobody has been interested in them. There are wonderful data about the whooping crane, which, as I mentioned, has been counted in total and in number of young of the year and adults every year since 1938. We have learned a lot from this marvelous data set, and we can learn a lot more. You will remember that we used it to obtain the first mathematically valid estimate of the probability of extinction for any species (at least to my knowledge it was the first). The database is so good that I asked Tom Stehn, the Whooping Crane Coordinator at Aransas National Wildlife Refuge of the U.S. Fish and Wildlife Service and the person in charge of these counts, I asked whether there were similar data for other bird species. He replied, "The whooping crane is unique!"

That's great for the cranes, but sad for the other avian species that people love so much. Among songbirds, Kirtland's warbler, the first songbird for which a complete census was ever done, has been counted annually since 1951 and does pretty well. But there are very few bird species for which we have such complete data, even though birdwatching is one of America's major hobbies. The U.S. Fish and Wildlife Service estimates that almost 19 million Americans go on bird-watching trips away from home, and 40 million do bird-watching at home.[8] By comparison, only about 24 million Americans play tennis.[9] During the Audubon Society's annual Christmas bird count, volunteers go out to see and hear (birds often are identified only by their calls) which species are present in their locale that day. It has become a kind of contest to see which locations have the greatest diversity (Santa Barbara, California, has been near the top for many years, I'm glad to say). Audubon and the Cornell Lab of Ornithology also sponsor an annual Great Backyard

Bird Count (GBBC) in February. Anybody can participate, and the rules are straight forward.

This kind of citizen science is a start and could be important both to increase people's contact with nature and natural history and to expand our knowledge of birds. It could relatively easily be modified to be a scientific and statistically valid sample.[10] What if we had 70 years of counts of adults and young of the year for ten species rather than one? That would help us a lot in just thinking about the dynamics of populations and might also put some meat into discussions about whether climate change is affecting species.

You would think that salmon, the focus of so much environmental concern, would be counted widely. When I was directing the study of salmon for the state of Oregon, a vice president of the Bonneville Power Administration (BPA) told me that by that time in the early 1990s BPA had spent $2.5 billion on salmon research and conservation without any indication of improvement of the salmon or their habitats on the Columbia and Snake rivers. Salmon are counted where there are large dams with fish "ladders," often quite elaborate constructed pathways to allow the fish to migrate up and down the river when they would otherwise be blocked by the dam. At Bonneville Dam, the first of the BPA dams, made famous in Woody Guthrie's "Roll on, Columbia," there are windows and automatic cameras so that, in theory, every fish that moves upstream gets counted. Tourists can stand at these windows and watch the salmon swim by in the turbulent waters.

But south of the Columbia River in Oregon and California, 26 rivers flow into the Pacific Ocean, and salmon have been counted on only two, the Rogue, which we visited in Chapter 13, and the Umpqua.[11] It's a heavy burden on already overworked government scientists to add to these counts, but it's an opportunity for citizen science. When I gave a talk at Long Beach, Washington, in the 1990s, the people told me that there was a lot of concern about how the salmon were faring on one of the tributaries of the Columbia that flowed south in Washington, but residents had no luck getting any government agency to monitor the salmon. So they took it on themselves. They strung a large net across the stream, then persuaded high-school students to go out and stand in the stream, pick up a fish, add it to the number passing upstream, and put it on the upstream side of the net. It didn't take any high tech.

Some new technology and some that's been available for several decades can help with counting fish. Side-scanning sonar can be used to distinguish individual species. All hatchery fish are tagged—a tiny metal rod with a numerical code on it, like a bar code on a package you buy, is inserted in each fish in a way that is harmless. It's trickier and perhaps not advisable to do that for wild salmon—you certainly wouldn't want to try to tag every one, but you might tag a statistically valid subsample.

Forests, too, generally lack adequate counts, which is perhaps even more surprising than the lack of bird-population estimates given the economic and environmental importance of forests and how deeply people have come to love or fear them. In the twentieth century, the U.S. Forest Service claimed that it maintained a series of permanent plots, 30 × 30 feet (just over 9 × 9 meters) square. The species and diameter of every tree on each square was supposed to be measured every ten years. Since the 1970s, when we first developed the computer model of forests, I've sought data of this kind to validate the model. Every time I gave a talk about that model to a group of foresters and explained that the main thing we lacked was data to validate the forecasts, someone who worked for the U.S. Forest Service would stand up during the question period and tell me about the permanent plots and insist that they had been remeasured every decade and that the data were easily available. I always thanked them and asked for the name of someone I could contact to get those data. Then I would call that person, who would tell me to call another person, who would tell me to call a third. Around the fifteenth person, I would be told to call the first person who spoke up at the first meeting.

I've some of the data from these permanent plots, but I've never found a single one that was measured more than once using the same measuring methods so the results could be compared from one decade to another. Often the monitoring was just ignored and not done. Sometimes, measuring fashions changed and the measurements were done differently each time, so there was no way to calibrate what was seen in one decade with what was seen in the next.

In his monitoring of carbon dioxide on Mauna Loa, Charles Keeling provided an excellent example of how to monitor carefully, so measurements at one time could be compared with another. The trick with CO_2 monitoring is that air with known concentrations of CO_2 is needed to calibrate the instruments. But these large tanks of carefully calibrated gas, if used routinely, run out in a number of years. So Keeling used some tanks only to check and confirm the calibration of others, and those special tanks were used so rarely that they lasted decades. Now *that* was careful monitoring. Good common sense too.

When I was following the trails of Lewis and Clark and writing about it, I got interested in how much the chemistry of the Missouri and Columbia rivers may have changed with the advent of modern artificial chemicals, including pesticides. The Missouri drains one sixth of the land area of the lower 48 states, including much of the prime agricultural land. I was surprised to find that no one monitored the concentrations of the most heavily used pesticides in the Missouri River itself. EPA, charged with doing such things, had set up a national system to monitor 60 major tributaries of the great rivers of the United States. The only one that flowed into the Missouri was the South Platte. You would think we'd want to know the concentrations of dieldrin and aldrin and the other major pesticides in the Missouri's main channels, but in practice, in reality, we don't know them. I can understand why EPA, with a limited budget and seeking a statistically sound sample of the rivers of America, would choose to monitor a larger sample of smaller rivers—this makes sense when your goal is a national average. But still you would think, with all our talk about environment, we would want this information for the Missouri River itself and for our nation's other great rivers.

These are just a few examples of straightforward counting that we are missing. We haven't even touched on the also important, frequently essential, more subtle and difficult factors and processes. It makes you wonder, when it comes to some major environmental problems, if we really are involved in an environmental science.

The National Science Foundation's Long-term Ecological Research Program, begun in 1980, has been a pioneer in correcting some of the deficiencies. The original idea was to begin monitoring the same factors at an integrated set of locations representing many kinds of ecosystems throughout the United States. Today it includes 26 sites. The program has diverged somewhat from the original goal because scientists at each site saw projects that interested them that were rather specific, and as a result the ability to compare changes among the sites has not been as successful as originally hoped.

At a much more specific level, monitoring of animal populations, especially of the charismatic large mammals, developed during the second half of the twentieth century, especially in the study of animal behavior, but extending to the ways that behavior affects reproduction and therefore population dynamics. As I mentioned earlier, Iain Douglas-Hamilton is a pioneer in this, discovering that individual African elephants could be identified from cuts, scars, and other markings on their ears and tusks. Roger Searle Payne, who along with Scott McVay discovered the elaborate songs of the humpback whales, was a pioneer in identifying individual whales from unique markings on their fins, which meant that their social behavior could be studied as never before.

A major program still in the planning stages but scheduled to begin operation in 2016 has the goal of correcting some of these gaps in our data. It is the National Science Foundation's National Ecological Observatory Network (NEON), set up as an independent nonprofit corporation whose goal is to "collect data across the United States on the impacts of climate change, land use change and invasive species on natural resources and biodiversity." Many universities are members of NEON.[12] Because the goals are good, I hope this program will succeed, but there are reasons to be concerned about it, too technical for this book. Viewed from a design perspective, it will be a test of whether a large federally funded and centrally run organization can provide essential data, independent of the politics and ideologies that can plague such entities, especially because they are so large that they leave little money and few places for alternative approaches. It is also a test of whether a top-down, command-and-control approach can be successful in America's democracy.

As long as we thought that nature undisturbed sought the single steady-state condition that was right and desirable, we could believe that we didn't need all that knowledge; since the ship of nature was self-guiding, we didn't have to read the dials. With the new perspective, we must have specific knowledge because policies must be specific. Knowledge about nature is essential if we are to achieve a new harmony between ourselves and our environment. We believe we are changing nature, but we cannot know that we are and cannot know how much we are unless we have baseline surveys of the present status of those aspects of interest to us and continue to monitor their status over time. The lack of basic information is one of the reasons we have so much controversy and disagreement about possible effects of global warming.

In summary, in most areas we lack even the most basic information about the condition of nature. We do not have many primary numbers: the number of species, the abundances of populations, and the amount of organic matter by geographic region and for the whole Earth. We need the ecological equivalents of the United States Geological Survey, institutions whose purpose is to describe and monitor the status of ecological systems, just as the Geological Survey creates maps of the geological terrain.

Not only is monitoring of the state of nature required, but scientific research must be an integral part of the management of nature. This is known as *adaptive management*. We lack not only information about the state of nature but also an adequate understanding of how ecological systems function, and we must continue to improve this understanding. Under the old perspective, detailed understanding wasn't necessary, because we thought nature knew best and that our understanding was irrelevant, that we needed only an appreciation of nature, not an understanding of it. In fact, our research at Isle Royale was at times viewed by the park's management as a nuisance that interfered with the park's real function of simply providing recreation for visitors, and some conservation organizations have considered scientific research inhumane and unnecessary. As I have pointed out in this book, some federal programs have failed to obtain data necessary to solve specific problems or provide basic information for many problems. There is, of course, an economic aspect to both the design approach and the scientifically empirical approach. In brief, what is an effective way to organize and fund the long-term collection of necessary data.

Today, we no longer have the luxury of believing that we can live in harmony with the environment without knowledge and understanding of natural systems, including the economics of those systems.

Our Way of Thinking About and Making Forecasts About Nature

In 2008 the editors of *Nature* published an essay titled "The Next Big Climate Challenge: Governments Should Work Together to Build the Supercomputers Needed for Future Predictions that Can Capture the Detail Required to Inform Policy."[13]

This is the nub of the problem that faces us: whether ever more detailed and complex computer models are the path to better understanding of climate and environment in general, or whether simpler is best. The article went on to say that "few scientific creations have had greater impact on public opinion and policy than computer models of Earth's climate." Indeed, that seems to be the case, but is that science or politics? "These models, which unanimously show a rising tide of red as temperatures climb worldwide, have been key over the past decade in forging the scientific and political consensus that global warming is a grave danger," the essay continued. *Nature*'s editors argue that the world needs "simulations good enough to guide hard decisions," and that "today's modelling efforts . . . are not up to that job."

What the world needs, the article said, are better and faster computers to run bigger and much more detailed climate models. The cost, which "might easily top a billion dollars," would be too much for any single nation, so it should become an international effort, the editors said. This would lead to "profound changes" in the community of climate scientists, and would "pull climate modelling into the world of 'big science' alongside space telescopes and particle accelerators—a transformation that would require new, and possibly disruptive, institutional arrangements."

That puts the debate about scientific theory for climate change about as clearly and extremely as can be. Among other assertions, it takes the position that the more details in a computer simulation, the more accurate and realistic its forecasts will be. But this has not been the experience of those who have studied the validation and utility of forecasting methods. In fact, the opposite is generally true: the more complex, the less useful.

Lumping the development of huge climate simulations with space telescopes and particle accelerators, as the essay in *Nature* does, makes a fundamental scientific mistake. Space telescopes and particle accelerators are techniques to gather data. The realization that such data could be useful and important to a science comes about in part from formal theory, but these devices are not creating the theory themselves. Scientists' desire to have such remarkable observational methods arises also from what we have learned since Galileo (and what good observers of nature have known since time immemorial for our species): that just plain observing nature leads to new knowledge, which leads to new insights, which gets people thinking about new theory. Yogi Berra had it right when he said, "You can observe a lot just by watching."

In contrast, the huge climate models are the theory itself, and there is little evidence, and some contradictory evidence, that this is a helpful approach. Developing these huge simulations requires an army of programmers, typically graduate students, who become fascinated with the game of computer programming itself. It's a fun game, as I can testify from my own experience. But to a cynic the effort could take on the image of thousands of technician-serfs building a computer-program Great Pyramid, rather than a path to a James Clerk Maxwell, Max Planck, Albert Einstein, etc., etc. series of wonderful insights about how nature works, arising from careful thinking over a long time and occasional leaps of imagination, unencumbered by such alternative tasks as the details of a computer program.

Perhaps there is a widely different approach, hinted at in Chapter 12's little story about the moose that kicked at Isle Royale's shore: the possibility that strange but imaginative models, based on clear thinking about the phenomena, might offer an alternative.

Our study of nature needs a strong theoretical foundation that arises from the phenomena themselves. Instead, most theory in these sciences has been borrowed from nineteenth-century and early-twentieth-century theory for steady-state machines. Ecology needs a brilliant young mathematical mind that comes into the field with no prior academic education in the standard theories but is acquainted with the available data and is taught natural history by one of the great naturalists, like Jim Welter or Murray Buell, or like many others it has been my privilege to accompany on field trips and research projects and learn from. There is a kind of complexity here that doesn't fit well with past conventions. There is too little use of the mathematics of stochastic processes and game theory, and some of the game-theory attempts I've seen applied to ecological phenomena have none of the essential characteristics of nature's phenomena.

Too much theory in ecology and related sciences has been forced on the subject from outside rather than arising from its own characteristics. It is instructive to remember how Volterra, one of the early scientists to apply differential equation mathematics to ecology, came to do this. As I described in Chapter 3, his son-in-law was involved with commercial fishing and told him about population cycles of two fish species, one the predator of the other. The catch of these varied in a regular way over time, rising and falling, but the predator's rises and falls were pretty much out of phase with the prey's. When the prey's catch reached a low, the catch of the predator reached a high, and vice versa.

If you are familiar with Newton's calculus that explains mechanics, and with James Clerk Maxwell's equations and wonderful analysis of electricity and magnetism, it would seem quite obvious that you could "solve" this fishery problem from the simplest equations for the vibrations of two out-of-phase waves. It's a nifty idea as long as you don't know any of the real natural history, the many factors that actually influence the abundances. Volterra's work was important and fundamental to the development of the science of nature in his time and in the early years of the science. As with Alfred Lotka, the fault wasn't with Volterra but with those who have blindly followed his work as if it were true in spite of many observations to the contrary.

We need a way to teach natural history directly without the overload of misplaced mathematics but with a link to theory in terms of testable hypotheses. I'm not sure what the best mechanism for this is, but I have envisioned a gifted young mathematician with a pencil and pad of paper sitting at one end of a log and Henry David Thoreau at the other, at the edge of the swamp by the edge of town, and hiking through the woods on Thoreau's daily four-hour walks, stopping frequently to consider how seeds are transported by wind and animals, measuring how deep a pond is, and considering the difference between wilderness and wildness. It may be that our culture is entering an era when nobody really wants this; that we are moving away from rationalism, and this will never happen. Of course today, that student would be taking notes on some electronic computing device, an iPad or a future equivalent.

Occasionally graduate students with a background in mathematics, physics, or astronomy, but with little or no natural-history field experience, would come to work on a degree with me. I would get them out into a forest with an experienced naturalist, like Tom Siccama of the Yale School of Forestry and Environmental Studies. Tom had done a Ph.D. thesis that required him to climb to the top of Camel's Hump Mountain in Vermont every day for five years, winters and summers, Sundays and holidays, and make measurements along the way. (Camel's Hump Mountain, reaching 4,083 feet, is Vermont's third-highest mountain and, as part of the Green Mountains, one of the oldest on Earth.)

One of the most brilliant students who ever worked with me, Al Doolittle, came with an impressive mathematics and computer background and an incredible talent with all machines and computers. Al and Tom Siccama and I went on a few hikes in New Hampshire's White Mountains. Al ran and programmed computers and was doing a thesis on forest photosynthesis and growth. It required that he drive a small truck and trailer to the edge of a forest and set up instruments whose readings were recorded on one of the first practical minicomputers, a Digital Equipment Corporation PDP-8E, housed in the trailer. One day a thunderstorm came up and lightning struck the trailer and blew out the computer. Al pulled out electrical measuring instruments and some tools, diagnosed the problem, drove to Radio Shack, bought a $2.85 part, and fixed the computer's hardware. He is the only person I know who, with no formal training in computer repair, could fix a computer's hardware. As far as I was concerned, all Al needed to make a unique contribution to ecology was to become a reasonably good naturalist. But it was not to be—he began to make money programming early CAT scan machines for the Yale Medical School and didn't become a naturalist.

It may interest you to know how alien computers were in those days. Every year Yale University had an amateur music contest in which students would play a musical instrument and winners were chosen. We learned that our computer put out FM radio frequencies and that if you programmed it to go through a simple calculation loop that took just the right amount of time and tuned a radio to the right frequency, the computer would "play" a pure tone. Then you could write a more complex program that made the computer move from one frequency to another. Al took the trouble to write a program that made the computer play Bach's "Jesu, Joy of Man's Desiring," and for fun we entered the computer in the music contest. This wasn't easy to do, as the computer was about as large as a small bookcase and just as heavy, so it was a job getting it to the contest. We thought the other students and faculty would be amused, but they were insulted and angry and made us take the thing away. Writing computer programs that were meant to imitate nature must have cast up similar feelings among many of my ecological colleagues, but I believe I realized that too late.

Some Implications for Environmental Laws in an Ever-Changing Nature

As discussed earlier, the major U.S. environmental laws that focus on species—the Marine Mammal Protection Act of 1972 and the Endangered Species Act of 1973—require and, if successful, lead to a steady state. These laws have had a great beneficial effect. This book is not an appropriate place to get into a definitive discussion of whether and how such laws should change. That is a much larger and complex discussion, extending to major historical, legal, social, and cultural factors that are beyond the scope of this book and of my knowledge. Rather, my purpose is to discuss how policies that result from these laws could be more effective today if they could take into account the non–steady-state character of nature.

PERSISTENCE AND RECURRENCE

A question that naturally arises from the discussion so far is: If classical fixed stability—steady state—is no longer an appropriate concept, what can replace it? In 1975 Matt Sobel, an expert on the mathematics of stochastic (random) processes, and I attempted to provide an answer.[14] We proposed two concepts, which we called *persistence* and *recurrence*, to replace *carrying capacity, optimum sustainable population, maximum* and *optimum sustainable yields.*

Persistence is a pretty straightforward idea. To say that a population (or species, ecosystem, or some factor of interest to us about any of these) is persistent means that it varies only within certain finite boundaries and remains within those boundaries. This is persistence within known bounds of some measure of interest. *Recurrence* means that if a specific condition (such as a certain population size) occurred in the past, it will occur again in the future. These ideas will be familiar to those who work in the mathematics of stochastic processes. For others, they are most easily understood through pictures.

Persistence is illustrated in Figure 15.1, which shows the catch of salmon on the Columbia River from soon after the Civil War, when large-scale commercial fishing began (as distinct from the catch of salmon by Native Americans), through 1990. From 1880 to 1950, the catch varied greatly but stayed above 15 million pounds and below 50 million pounds. After 1955, the annual catch declined. It continued to vary from year to year, but stayed below 15 million pounds. From a management perspective, one could say that the persistence between 15 and 50 was desirable and perhaps even "natural," and give it its own name. (We named it mathematically as Φ_1 (Phi-1), but you can use more ordinary terms if you wish.)

The catch after 1955 was also variable and persisted within certain bounds (our Φ_2 persistence), but the bounds were completely different from the previous. With persistence, there is no need to assume or demand a balance of nature, and management policies are clear. The manager could say Φ_1 is a goal of management, meaning that as long as the catch remains within this range, it's okay. Therefore Φ_2 is not a management goal and some regulations could be established to limit harvests when the catch was dropping toward that range. This seems quite commonsensical and, as an added plus, it doesn't require a constancy of nature or any particular kind of stability of the natural phenomenon.

The concept of recurrence may be somewhat less familiar. It means that a population (or whatever you are interested in) reaches a certain condition, such as a specific population size, and then reaches that condition again. We can talk about "infinite recurrence," meaning that once the condition is achieved, it will occur over and over again forever, but the population or other system does not have to stay in that condition all the time—it might revisit it very infrequently. This idea can be used mathematically for systems that have inherent randomness. Once again, the concept makes no requirement of a steady state or any kind of stability, only that there are some recurrent states.

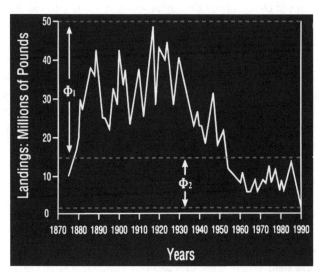

FIGURE 15.1 Catch of salmon on the Columbia River from 1875 to 1990. The catch illustrates the idea of persistence within bounds. From 1880 to 1950, the catch varied greatly but stayed above 15 million pounds and below 50 million pounds. After 1955, the annual catch declined. It continued to vary from year to year, but stayed below 15 million pounds. From a management perspective, one could say that the persistence between 15 and 50 was desirable and perhaps even "natural." The catch after 1955 was also variable and persisted within certain bounds, but the bounds were completely different from the previous. A manager could set the second range of persistence as undesirable. There is no need to assume or demand a balance of nature, and management policies are clear.

Right now, the world's human population does not appear to be recurrent, because it continues to increase. But it's quite possible that some global disaster or something much less, like widespread drought and crop failure, could bring a decline in the world's human population. In that case, specific population sizes would occur a second time. The important question for us would be whether a population size once held by our human population will recur after a decline, or whether we are in for a serious and permanent decline in the abundances of our species.

Persistence and recurrence free us in our ways of thinking about nature, but still allow us to compare different policies and approaches to conservation and management of our surroundings. Those involved with environmental law will see that these two concepts could be powerful in improving some of these laws and policies, but that goes beyond the scope of this book.

A common confusion that arises when the non–steady-state idea is proposed as a basis for policy and action is that once we admit that some kinds of change are natural, we might be forced to accept any and all kinds of change as natural. The discussion throughout this book should have countered that fear as completely irrelevant and unnecessary. What is helpful to us is that natural changes can become our policy guides. Because species have evolved and adapted to specific kinds and rates of change, imposing those kinds and rates should, in general, be benign and permissible, and in many situations desirable, preferable to a policy that seeks to achieve a steady state.

In our laws and policies, we have tended to set down regulations for the cleanliness of air and water, in large part because of their importance to life. But we must recognize that the real focus of our efforts is to maintain life. We are often directed away from this focus by the way policies and laws are formulated; clean-air and clean-water standards designed originally to protect life become, under policies and practices of large bureaucracies, ends in themselves.

We must distinguish between merely the persistence of some kinds of life—say, just bacteria and their prokaryotic relatives—and the maintenance of a biosphere that is desirable to human beings. I have never heard anyone specifically call for a return to an all-prokaryotic (bacteria and their relatives) world, and I doubt that is what anyone would mean. But the constraints on such a world are much less stringent than those for a world that includes animals, plants, algae, and fungi, a world that, in one structural form or another, is what people mean, by "real" nature and "getting back to nature." Accepting the second kind of world as a goal, we must make a further distinction: whether we want the world environment to remain constant or nearly constant with respect to some past condition that we consider desirable, or whether any condition that continues to support a large number of people at a reasonable level of well-being is acceptable.

To sustain the biosphere to meet conditions for the well-being of people, we will have to slow down the rates of changes that we are creating in the biosphere. Oddly, as I suggested earlier people, with our civilizations and cultures, require a more constant world than does most of the rest of nature, which we have treated as if it were the set of systems that required constancy. For our civilizations, temperature and rainfall patterns have to remain within ranges much narrower than those of the past 2 million years. However, this may not be possible or desirable from a very long-term perspective on the survival of life on Earth. (For example, taking a contrary extreme, one could argue for the beneficial effects of continental glaciation, such as restoring the fertility of the soils.)

Insight and Consensus: Two Goals at Variance

We hear the term *scientific consensus* commonly these days, as a goal and as a test of whether we should believe that something is true. The term means that there is general agreement among scientists in a particular discipline that something has been proved. A majority of scientists, or an outspoken minority, agree on something and claim that it must be true because they all believe it is and say it is. But everyone trained in the formal scientific method knows that a vote by a majority of any group is not a scientific proof. Let's pause briefly to review the basics of the scientific method, if for no other reason than it has been very successful in advancing knowledge, technology, and as a result medicine and health.

The historical roots of science lie in the ancient civilizations of Babylonia and Egypt, where observations of the environment were carried out primarily for practical reasons, such as planting crops, or for religious reasons, such as using the positions of the planets and stars to predict human events. But these precursors did not distinguish between science and technology, nor between science and religion. Those distinctions arose first among the ancient Greek philosophers.

The formal scientific method was first described explicitly by Francis Bacon in 1620 as "inductive knowledge."[15] Bacon laid out a procedure showing how science differed from formal logic and from other kinds of knowledge. But since then scientists and philosophers have given much consideration to what is the essence of science. Sometimes scientists make great leaps of insight and imagination, as with Sir Alexander Fleming's discovery of penicillin in 1928. It's a familiar story to scientists, but perhaps bears repeating here. He was studying the pus-producing bacterium *Staphylococcus aureus*. A laboratory culture of these was accidentally contaminated by the common bread mold, the green fungus *Penicillium notatum*, and Fleming noticed that the bacteria did not grow in areas of the culture where the fungus grew. He isolated the mold, grew it in a fluid medium, and found that it produced a substance that killed many disease-causing bacteria.

No doubt many others had seen this mold on bread, and perhaps some even noticed that other strange growths did not overlap with it. But it took Fleming's knowledge and observational ability for this piece of "luck" to occur. This kind of serendipity is a hallmark of many scientific breakthroughs. What Bacon described was a methodical approach, but the scientific inquiry also includes such sudden insights.[16]

Scientists agree these days that the key to the scientific method is disprovability. If you can explain a way that the statement you make can be disproved, then the statement is called a scientific one. If you cannot state such a disproof, the statement is outside of science. An assertion arrived at by agreement among a group of scientists cannot be disproved, and therefore is not a scientific statement.

It's easy to understand why government agencies, politicians, and anybody without formal training in science seeks scientific consensus. What a government agency and a well-meaning politician are looking for is an assertion they can rely on, even if they don't understand how that assertion was arrived at. The simplest way to achieve this, the most direct path, is to find a group of scientists who agree on something. It is all the better if those scientists have special kinds of approval from their colleagues, like winning a Nobel Prize or being elected to the American National Academy of Sciences. At least you can be sure that those scientists have done something that caught the eye of their colleagues and elicited their favorable opinion.

But we also know that consensus can be dramatically wrong, as with Galileo and the Catholic Church. The progress of science is famously marked by radical insights that are at first dismissed by the majority of scientists but eventually become major components of a science. One of the most famous and relevant for us is the theory of plate tectonics. In the modern scientific era, Alfred Wegener first proposed continental drift in 1915 in his book *The Origins of Continents and Oceans*. He based his idea on the shape of the west coast of Africa and the east coast of South America, which looked as if they would fit together like a jigsaw puzzle, as though they had once been together but had drifted apart, and also on the fact that fossils of identical animals and plants had been found on these two coasts. Since nobody could come up with a mechanism that would make the continents move, the idea was rejected pretty much out of hand. But after World War II, geological research revealed forces that moved continents, and Wegener's theory became the accepted norm.

This kind of scientific revolution is common enough to have its own name—a *paradigm shift*. When a science is changing rapidly, the conventional wisdom becomes less and less trustworthy. This is happening with global ecology right now.

On the other hand, there are always crazies and oddballs who claim to have explanations and solutions to everything. The British Petroleum oil spill in 2010 in the Gulf of Mexico gave rise to many proposed solutions, including the suggestion that the whole thing could be cured with an atomic bomb. One proponent of this said there was no danger of any kind in doing this. Now *that's* wacky.

People can also be easily fooled, as with the classic hoax of the mysterious crop circles in England, which I described in chapter 1. To save you the trouble of searching back, here is the story. In the late 1970s, circular patterns began to appear "mysteriously" in grain fields in southern England. Several explanations were suggested, the most popular of which were aliens and pranksters. A journal and a research organization headed by a scientist arose, as well as a number of books, magazines, and clubs devoted solely to crop circles. Scientists from Great Britain and Japan brought in scientific equipment to study the strange patterns. Then, in September 1991, 13 years after the first crop circles, two men confessed to creating them. They showed how they entered the fields along paths made by tractors and dragged planks through the crops to make the circles. They demonstrated their technique to reporters and some crop-circle experts.

Despite their confession, some people continue to believe that the crop circles were caused by something else, and crop-circle organizations still exist—some have Web sites. One report published on the World Wide Web in 2003 stated that "strange orange lightning" was seen one evening and that crop circles appeared the next day.[17] It seems that some people love a mystery and a story about aliens.

So what can an ordinary citizen do? Whom can one believe? I've tried to cut through this problem by taking a different approach to seeking explanations. Rather than gathering a large number of scientists who already agree on the same thing, I've brought together scientists who disagreed with each other strongly and got them arguing together until a solution emerged or a statement acceptable to all was formulated despite their differences.

Insights arise out of this process, and I'll take a good insight any day over a conventional-wisdom consensus. I did this in the 1970s when the National Science Foundation asked me to bring scientists together to discuss whether there should be long-term monitoring of ecological variables, supported by a government program. This led to the Long-Term Ecological Research Program. I did this again in 1980 when NASA asked me to run workshops to find out if there really was such a thing as a global ecology. Most recently, as I said in chapter 13, I used this approach in 2004 with Henrik Saxe to discuss whether, and to what extent, we could make useful forecasts about the ecological effects of global warming.

This process becomes a kind of scientific mediation. I used to tell friends and colleagues that such a meeting was a failure if by the middle of the workshop most of the participants weren't furious at the leader (me) and if they weren't convinced that the entire thing was a failure. Only after all the vigorous arguing could something useful develop. Otherwise all one got was the conventional wisdom.

With such environmental problems as salmon in Oregon, effects of water diversion from Mono Lake's watershed by the city of Los Angeles, land-use planning in northern Minnesota, and environmental effects of a mining road through Canadian Tlingit First Nation lands, I used another approach: creating a very small steering committee of scientists, each well established in his or her discipline, each representing a discipline essential to solving the problem, and each relatively free of prior political and ideological assertions about the problem. These panels were small enough for vigorous arguments to develop, and the members spent many hours in energetic discussions, learning a lot from each other. The panel also had a budget that could be used to support new research (not by them) to obtain information that was necessary and lacking. And, as I said in chapter 13, in the Oregon study of salmon, meetings open to the public were added so that citizens could involve themselves directly.

My experience persuaded me that these approaches worked much better in breaking through ill-conceived conventional wisdom, but not everybody likes this approach. One of the things that has disappointed me about the global-warming debate in this new century is that the two sides (if one can simplify the entire complex argument that way)—scientists convinced that human actions are causing a global warming that will be disastrous, and those who aren't—have tended to meet independently. Each year the leaders of major scientific conferences tell the media that the truth about global warming will be made clear by the end of the meeting. But the other side disagrees with that "truth" and then holds its own meeting with the same promise. I wish they would get together and argue things out in public before a live audience.

One can also list questions for citizens to ask the representatives of a panel of like-thinkers who publish a consensus report. You start with the simple ones connected to the scientific method: What are the data the panel used? Do the data have statistical validity? Were different ways of thinking about the problem integrated into the discussions? (How citizens can involve themselves constructively in finding out the reliability of scientific consensus statements is a book-length topic on its own, beyond what I can get into here.) Arriving at a list of questions takes work, and it's not all that common to find a group of citizens who are open to changing their ideas and also interested enough in an issue to learn to ask these questions. It happened with salmon in part because many livelihoods were at stake, but also because there were people willing to invest their time and effort to learn and think through the information.

There's no free lunch in making a judgment about what a group of scientists tell you is the consensus. Unlike a carefully done scientific experiment, there is never any guarantee that we will arrive at a truth through the social process of meeting and arguing. But we can do better while we wait and hope for the Newton, Galileo, James Clerk Maxwell, Max Planck, and Albert Einstein of ecology to come along. It's an odd contradiction of our modern times that we have more data and more information about just about everything than any previous civilization, and we are informed continually that problems are difficult, but we still want easy answers. I can't count the number of times when, at a speech or meeting or in informal conversation, someone will agree about how complicated the entire problem is and in the next breath say, "Give me two reasons why I should [or shouldn't] believe nuclear energy is the solution to all our energy needs," or whatever the environmental topic is.

The media feed on this, as is well known—the 20-second sound bite that people are told to be ready to say when they prep for a TV or radio interview is taken as words of wisdom. We get used to slogans substituting for thoughts, and catchphrases masquerading as answers. Many government bureaucrats love this kind of thing—"If you get a lemon, make lemonade" is a "solution" I've often heard, and "Well, that's just the camel's nose under the tent," said with assurance, as a substitute for an analysis.

The Future of Nature

Nature in the twenty-first century is fast becoming a nature that we make. Indeed, that is a primary lesson, perhaps the most important result, of the vastly popular concern about global warming, but it extends also to our actions that are leading to major declines in the abundance of many forms of life, destroying habitats important to many species, including ones we tend to love, or love to harvest. The question is the degree to which our molding of nature will be intentional or unintentional, desirable or undesirable. What is the likely outcome of our modern role in nature? We can envision several specific futures. The worst, nature after a nuclear war, might be like nature 2 billion years ago, a biosphere of only bacteria, which we would not want—we couldn't be there.

A more likely future lies on the path we have followed, in which we continue to treat natural history as a hobby, not to be taken seriously, and deal with environmental problems after they have arisen, using whatever tools and knowledge we happen to have at the moment, on the assumption that nature can be taken apart and repaired and put back together again, like a machine.

Another future is one that we might achieve if we began a massive effort today to make up for what we have not done in the past: to obtain the information, knowledge, and understanding to manage nature wisely and prudently. To this end, we must have land available for baseline surveys and continued monitoring—available in a scientific sense, not necessarily in a political or legal sense—enough land so that we have baselines from which to measure our actions and conserve as much of the remaining biological diversity as possible. We must train professionals and allocate large sums of money for the right kinds of research and management. And we have to do this within democracies. Although the actions necessary would take place over large areas, it is not clear at this time whether the more successful approach will be government and international organizations and agreements, or whether, following more along the lines of the leasing of land-use rights, nongovernmental agreements will work better.

Wilderness in the Twenty-First Century

Since there is no longer any part of Earth that is untouched by our actions in some way, either directly or indirectly, there are no wildernesses in the sense of places completely unaffected by people. But there are three kinds of natural areas that we must maintain in the future: no-action wilderness, pre-agriculture wilderness, and conservation areas. The first two we can regard as true wilderness and designate legally as protected wilderness areas.

The first is an area untouched by direct human actions, no matter what happens. This kind of wilderness is necessary for observation as a baseline from which scientists can measure the effects of human actions elsewhere; it is an essential calibration of the dials we should set up to monitor the state of nature. Such areas are also important because they will help us maintain biological diversity. Some of them may be pleasant for recreation, but some may not be, and some may become a nature never seen before. As in Hutcheson Memorial Forest, this kind of wilderness might be occupied by introduced species and native species in novel combinations.

The second kind, pre-agricultural wilderness, is an area that has the appearance of landscape or seascape that most closely matches the ideal of wilderness as it has been thought about in recent decades. In North and South America, Australia, New Zealand, and other places in which the time of arrival of modern technological man is readily dated, the idea is to create natural areas that appear as they did when first viewed by European explorers. In the Americas, this would be the landscape of the seventeenth century. It is necessary to choose a time period that has the desired appearance; if we don't, then we face the situation discussed earlier for the Boundary Waters Canoe Area, which, from the end of the last ice age until the time of European colonization, passed from ice and tundra to spruce and jack-pine forest. If *natural* means simply before human intervention, then all these habitats could be claimed as natural, contrary to what people really mean and really want. What people want in the Boundary Waters Canoe Area is wilderness as seen by the voyageurs and a landscape that gives the feeling of being untouched by people.

Conservation areas, the third type of natural region, are set aside to conserve biological diversity, either for a specific species, such as Kirtland's warbler, or for a kind of ecological community. Because we have so altered the landscape and have allowed inadvertently only small patches of former habitats to remain, most of these areas require active intervention on our part if they are to persist. For example, to manage the habitat of Kirtland's warbler in a way that allows the species to survive, we must pay attention to the frequency of fires and increase or reduce the rate so that it best suits the needs of that species. If the global climate changes, whether or not because of our actions, we may have to relocate the natural area for the warbler and also learn how to persuade it to move there. Then again, male Kirtland's warblers have recently been heard singing in Wisconsin, far from where we believed they would nest.

It is important to understand the distinctions among these three kinds of natural areas, each of which represents a different aspect of the older meanings attached to *wilderness*. Each is quite different from the others, and generally it won't be possible to manage the same area to be all three at once: truly undisturbed; appearing as in a presettlement landscape; and functioning to conserve endangered species or biological diversity. Under the old perspective on nature, one could assume that all three goals would be accomplished in any area simply by removing all human actions. To be viable, each kind of natural area must be a certain size (as yet generally undetermined). For example, Kirtland's warbler conservation areas must be large enough to support the breeding territories of hundreds of males.

The smaller the conservation area, the more diverse and more intense our actions must be. The amount of intervention required increases as the size of any specific preserve decreases.[18] The smallest area is simply a zoo, in which we provide all the necessities and remove all the wastes for the forms of life that we maintain there. At the opposite extreme is nature before technological civilization, where vast areas unaffected by human beings existed. The amount of effort required to maintain a preserve of any size depends also on the characteristics of the species found there, including life-history characteristics, such as size and longevity.

As a general rule, large and longer-lived organisms require large habitats. Tsavo, the biggest national park mentioned in this book, was not extensive enough to function as an independent preserve for the African elephant without active intervention by people. Of the examples discussed in this book, the Boundary Waters Canoe Area is perhaps the area that could persist with the least direct human action. The largest mammal within the Boundary Waters is the moose, which is much smaller and shorter-lived than the elephant. (A moose weighs about 1,000 pounds and lives for about 17 years; elephants weigh as much as 6 tons and can live for 60 years.)

Maintaining wilderness areas in the future will require that we develop ways to secure them from undesirable uses. As resources become limited and the human population continues to grow, there will be increasing pressure on natural areas for extracting timber, harvesting wildlife, and mining minerals. As an example, the poaching of elephants, a crisis requiring strenuous countermeasures when Tsavo National Park was established, remains a serious problem. Elephant populations are undergoing a severe decline because poaching has continued widely, and it is unclear whether the African elephant will be able to survive in the wild in the twenty-first century unless new approaches are found for its security. Elsewhere, as resources such as firewood and valuable furniture timber become scarce, there will be more and more pressure for people to simply take them even from an area set aside as a preserve. How to ensure that large natural areas are physically secure is an issue that has received very little attention and is not a simple problem. A nature preserve surrounded by police with weapons would seem to violate the idea of a "nature" preserve and would require funds that might prove impossible to obtain.

To these kinds of formally designated wilderness, we should add the kind of landscape that Thoreau sought, a place where he could experience *wildness*, a spiritual state existing between a person and nature, which he distinguished from *wilderness*, which was land or water unused at present by people and thus a state of nature.[19] As I discuss in *No Man's Garden: Thoreau and A New Vision for Civilization and Nature*, wildness meant so much to Thoreau that elsewhere he wrote: "I caught a glimpse of a woodchuck stealing across my path, and felt a strange thrill of savage delight, and was strongly tempted to seize and devour him raw; not that I was hungry then, except for that wildness he represented."[20] For Thoreau it was possible to find this wildness in places quite close to home and civilization, like Walden. It is much like the idea behind Japanese gardens, meant for reflection and meditation.

Many places could evoke this sense of wildness, depending on one's sensitivities and experiences, but we can do much on the same scale to create some kinds of parks with this purpose in mind, including a city park that is isolated from the city's noises.

Biological Diversity in the Twenty-First Century

One of the best forecasts anyone can make about nature in the twenty-first century is that at least one species is very likely to go extinct. The fossil record suggests that on average about one species out of a million goes extinct every year, and there are about 1.5 million named (and therefore known to exist) species on Earth. The average duration of a mammal species is about 750,000 to 1 million years. Just stating this ought to make clear that what has been the average for life on Earth is not necessarily the best goal from our human point of view. People love the great diversity of life on Earth, at least the diversity of those that add to the beauty of our surroundings and to our pleasure. It also tells us that biological diversity in the twenty-first century is largely going to be up to us, whether we like it or not, unless there's some major physical catastrophe, such as a rogue asteroid crashing into Earth.

There are nine reasons that people value biological diversity: utilitarian; public service; ecological; moral; theological; aesthetic; recreational; spiritual; and creative.[21] *Utilitarian* means that a species or group of species provides a product that is of direct value to people. *Public service* means that nature and its diversity provide some service—such as the taking up of carbon dioxide by photosynthetic organisms or the pollination of flowers by bees, birds, and bats—that is essential or valuable to us and would be expensive or impossible to do ourselves. *Ecological* refers to the idea that species have roles in their ecosystems, and that some of these are necessary for the persistence of their ecosystems, perhaps even for the persistence of all life. Scientific research tells us which species have such ecosystem roles. *Moral* refers to the belief that species have a right to exist, independent of their value to us. *Theological* means that some religions value nature and its diversity directly, and a person who subscribes to that religion supports this belief.

Aesthetic, recreational, spiritual, and creative reasons for valuing biodiversity have to do with the intangible (nonmaterial) ways that nature and its diversity benefit people. These are often lumped together, but we separate them here. *Aesthetic* refers to the beauty of nature, including the variety of life. *Recreational* is what it seems—that people enjoy getting out into nature not just because it is beautiful to look at but because it provides us with activities that are enjoyable and healthy. *Spiritual* refers to the way that contact with nature and its diversity has moved people ever since nature and its diversity have been written about, an uplifting often perceived as a religious experience. *Creative* refers to the fact that artists, writers, and musicians find stimulation for their creativity in nature and its diversity.

Science helps us directly in determining what are utilitarian, public-service, and ecosystem functions of biological diversity. Scientific research can lead to discoveries that provide new utilitarian benefits from biological diversity. For example, medical research led to the discovery of the chemotherapy properties of paclitaxel (trade name Taxol), a chemical found in western cedar and used widely today in the treatment of certain cancers. (Ironically, until the compound could be made artificially, its discovery led to the harvest of cedar, an endangered species, creating an environmental controversy.)

We must remember that life is the focus, and that it is worthwhile to keep in mind what Tom Lovejoy said in 1989: "In the last analysis, even when we have learned to manage other aspects of the global environment, even if a population reaches a stable level, even if we reach a time when environmental crises have become history, even if most wastes have gone except the most long-lived, even if global cycles have settled back into more normal modes, then the best measure of how we have managed the global environment will be how much biological diversity has survived."[22]

A nice idea, if a single-factor one. Unfortunately, misinterpreted, this idea leads some to believe that the goal must be to maximize current biological diversity, prevent the extinction of *any* species now present on Earth. On the surface this seems simply environmental common sense, but it should be clear to you by now that it is yet another call for a steady state, so often called for in one form or another, sometimes as today's maximum; sometimes that of 1491, the year before Columbus arrived in the New World, as Charles Mann described it in his book *1491: New Revelations of the Americas Before Columbus.*[23]

Zoologist Paul S. Martin has called for resurrecting the biological diversity of some time before 10,000 years ago, before American Indians arrived in the New World.[24] He was asked in an interview published in *American Scientist*, "If you were suddenly put in charge of an effort to repopulate North America with species lost during the late Quaternary, where would you begin? What animals would you most like to see roaming the continent again?"

Martin replied, "I would like to see free-ranging elephants in secondary tropical forests of the Americas. Until the end of the Pleistocene, forest and savanna in the New World tropics supported three families of elephants."[25]

Given the vast problems that arise from invasive species, this is an extremely odd goal for someone concerned about conserving biological diversity.

To believe one of these fixed times is best and necessary is to believe in an imaginary world quite different from the history of life on Earth as so well revealed by twentieth-century studies of fossils. As I have discussed, and many paleontologists have shown, our planet has gone through many changes in life-forms, degrees of diversity, and dominant species, and each stage has been consistent with the persistence of life past its time.

Only when we impose a view of ourselves on this picture of nature, see ourselves as within and part of nature, do we become sensitive and fearful about the exact species array that accompanies us. Our fear of losing any of these appears, from this viewpoint, a fear for ourselves, a kind of helplessness—where will we be if any of these species are lost? It is part of a passive stance, in which we stand helpless, in fear of, in awe of, our environment.

We need not and cannot remain helpless and fearful. We must become active and do our best for humanity and nature. Admittedly this is an idealistic goal that will never be completely attainable given human failings, or (to put our failings in ecological and Darwinian evolutionary perspectives) what we are because we have evolved that way.

The City, Civilization, and Nature

One autumn afternoon on a visit to Venice, members of the World Wildlife Fund took me on a pleasant boat trip to see the work they were doing on conservation of the lagoon. We traveled to the north of the lagoon, to the marshes out of sight of the Grand Canal and Santa Maria della Salute, the scene with which I began this book. There in the salt marshes, nature seemed to rule. Along the shore, in that hazy Venetian sunlight so often portrayed by artists over the centuries, shorebirds lingered and breezes brushed the salt grass. In the declining sunlight, we could look away from the city and believe that civilization was far distant, even though we knew that the shape of the shoreline, the chemistry of the waters, and the quality of the air were strongly influenced by the city and civilization.

While I have chosen to illustrate the central qualities of a new management of the environment through examples that have to do with the forest and the ocean— the nature of the "out-there," seemingly away from our homes and our cities, away from the apparent center of our civilization—the management approaches I've discussed apply broadly to the total environment, including the environment of our cities and the city as an environment. As the great thinkers about cities and civilization, such as Lewis Mumford, have made us realize, the city has been both the center of civilization in the past and a part of, not separate from, its surroundings.[26] For civilization to survive, we cannot believe that the management of nature means only the management of wilderness preserves.

Most of us live in cities, and in most cities, unlike Venice, architecture and nature seem separate; we tend not to perceive so readily the close relationship between the city and its environment. However, as urban areas continue to spread over the landscape with the growth of human populations, it becomes imperative that we understand that cities must be managed as local environments, both for their own sake and because of their effects on regional and global environments.

Venice epitomizes the urban dilemma. Of all cities, Venice is perhaps most visibly set within and dependent on a fragile, changing, and human-influenced environment. Venice attracts us because of the blend of its architecture (the technological) with the surrounding lagoon (naturalistic, providing an appearance of the natural). Each, the intentional artifice and the intended natural, strongly influences the other. The lagoon has been much altered by human technologies over the centuries, beginning with the millions of saplings driven into the shifting muds to provide foundations for the architecture itself, and continuing with alterations of watercourses that led into the lagoon and of the lagoon itself.

These human actions have had both positive and negative results. In the twentieth century, removal of groundwater on the nearby mainland for use by modern industry increased the rate at which the city was sinking, making the architecture increasingly vulnerable to the *aqua alta*, the floodwaters that occur with high tides during winter storms. Air pollution from the city of Mestre, on the mainland, erodes the statues in St. Mark's Square and at La Salute. Sewage from the city flows into the lagoon, making the waters dangerous to fall into and making the city completely dependent on the flushing action of the tides and storms, which move the polluted waters out into the Adriatic.

Solutions to Venice's environmental problems, like the solutions to managing the habitats of Kirtland's warbler, the giant sequoia, and the sea otter, involve acceptance of nature's ambiguities and variability, the necessity for new knowledge, and recognition that our engineering must enter into modification of the environment as a constructive power. In 1988 such an approach was attempted in Venice. Newly designed floodgates were being put into place at the outer edge of the lagoon. The gates lie underwater, allowing the movement of the tides to flush the city's waters, until a storm comes. Then they are inflated, rise to block the high waters, modulate the action of the storms, and prevent damage to the city's buildings. We will have to wait to see if this specific technology is successful, but the approach takes into account risk, uncertainty, variability, and complexity. The conservation of Venice requires facts, knowledge, and a change in myth and metaphor about nature. Others whose expertise about cities is greater than mine will know how to translate the new management perspective downtown.

The largest bird sanctuary in the northeastern United States is—surprise!—in New York City. It is the Jamaica Bay Wildlife Refuge, covering more than 9,000 acres, 14 square miles, within view of Manhattan's Empire State Building and reachable by city subway and bus. More than 300 bird species have been seen there, including the glossy ibis, common farther south, and the curlew sandpiper, which breeds in northern Siberia. Thus the Jamaica Bay Wildlife Refuge is one of the major stopovers on the Atlantic bird migration flyway and plays an important role in biodiversity conservation.

Across the continent, the vicinity of the city of Los Angles is home to some animals listed as endangered by the state of California. These include the Least Bell's Vireo, a summer resident in riparian habitats of Southern California. It was once found from interior Northern California to the Sierra Nevada foothills and then to the coast ranges into Baja, California. But by the late 1980s its breeding range had been restricted to the Amargosa River in Inyo County on the east slope of the Sierra, and there were small scattered populations in Santa Barbara, Riverside, and San Diego counties. It is endangered because of habitat loss, much of it due to urbanization, as well as nest parasitism by cowbirds. Protection and restoration of riparian habitats in Southern California, as might occur through the development of urban greenways, which I'll talk about later, could help save this species.

In the past, environmental action has most often focused on wilderness, wildlife, endangered species, and the impact of pollution on natural landscapes outside cities. In the development of the modern environmental movement in the 1960s and 1970s, it was fashionable to consider everything about cities bad and everything about wilderness good. Cities were viewed as polluted, dirty, lacking in wildlife and native plants, and artificial—therefore bad. Wilderness was viewed as unpolluted, clean, teeming with wildlife and native plants, and natural—therefore good. Now it is time to turn more of our attention to city environments.

Kansas City, Missouri, at the confluence of the Kansas and Missouri rivers, illustrates the common disconnect between a city and its river. The Missouri River's floodplain provides a convenient transportation corridor, so the south shore is dominated by railroads, the north shore by the city's airport. Except for a small riverfront park, the river has little place in this city for recreation and relief for its citizens or for the conservation of nature.

One of the reasons cities are important to biodiversity conservation is that they are rarely located at random; instead, cities arise at crucial transportation sites or other important areas. Many cities, for example, are either at the confluence of major rivers or where a river flows into a major ocean harbor, like that of New York City. Saint Louis lies at the confluence of the Missouri and Mississippi rivers; Manaus, Brazil; Pittsburgh, Pennsylvania; Koblenz, Germany; and Khartoum, Sudan are at the confluence of several rivers. Other cities, such as Geneva, Switzerland, are located where a river enters or leaves a major lake. These are also locations important to many species—sometimes important parts of migration routes, as with Jamaica Bay; other times as rare and endangered habitats, as with the Least Bell's Vireo.

Prairie vegetation, which once occupied more land area than any other vegetation type in the United States, is rare today. Almost all prairies have been plowed and converted to farmland. Some of the most valuable remaining unplowed prairies are in cities, including city graveyards and railroad rights-of-way. Some important prairie restorations are within major midwestern and western U.S. cities. One is in Omaha, Nebraska; another in Lincoln, Nebraska, to name just a few. Cities can provide habitat for other endangered vegetation as well. For example, Lakeland, Florida, uses endangered plants in local landscaping with considerable success.

Some animal species typically considered residents only of wilderness or undeveloped wide-open landscapes are also doing well in cities, and urban habitats are helping to save them from extinction. Cooper's hawks are doing pretty well in Tucson, Arizona, a city of more than a half-million people. Although this hawk is a native of the surrounding Sonoran Desert, some of them are nesting in groves of trees within the city. Nest success in 2005 was 84%, between two-thirds and three-quarters of the juvenile hawks that left the nest were still alive six months later, and the population was increasing. Scientists studying the hawk in Tucson concluded that "urbanized landscape can provide high-quality habitat."

Peregrine falcons hunt pigeons above the streets of Manhattan. They nest on the ledges of skyscrapers, unnoticed by most New Yorkers until they dive on their prey in an impressive display of predation. The falcons disappeared from the city in the 1960s when DDT and other organic pollutants caused a thinning of their eggshells and a failure in reproduction But they were reintroduced into New York City in 1982, and today 32 falcons live there. The reintroduction of peregrine falcons illustrates an important recent trend: the growing understanding that city environments can help to conserve nature, including endangered species.

City residents can play a direct role in biodiversity conservation. Urban kitchen gardens—backyard and community gardens that provide vegetables and decorative plants—can be designed to provide biodiversity habitats. For instance, these gardens can include flowers that provide nectar for threatened or endangered hummingbirds.

Urban drainage structures can also be designed as wildlife habitats. A typical urban runoff design depends on concrete-lined ditches that speed the flow of water from city streets to lakes, rivers, or the ocean. These structures can be designed to create stream and marsh habitats for fish and mammals, with meandering waterways and storage areas that don't interfere with city processes.

Cities have had major destructive effects on biodiversity outside of city limits. They export waste products to the countryside, including polluted water, air, and solids. The average city resident in an industrial nation annually produces 1,660,000 kg (1,826 tons) of sewage, 660 kg (0.8 ton) of solid wastes, and 200 kg (440 lb) of air pollutants. If these are exported without care, they pollute the countryside, affecting biodiversity broadly.

Until the 1960s, the major cities on the Missouri River—the river that drains one-sixth of America's land—simply dumped untreated sewage into the river. Graffiti scribbled above a public toilet in Omaha, Nebraska, said, "Please flush the toilet; St. Louis needs our water." This waste disposal, along with channelization, construction of levees to protect cities from floods, and urban construction on floodplains and river shores, destroyed many riparian habitats. Restoration of the natural complex of river channels and riparian habitats in and near major cities along the Missouri River are doing much to improve biological conservation of fish, like the endangered pallid sturgeon, and aquatic invertebrates.

Biological conservationists have tended to ignore cities, but urban planners, park planners, and landscape architects have long focused on improving city environments.

As the city of Boston grew in the 19th century, demand increased for more land for buildings, a larger area for docking ships, and a better water supply. The need to control ocean floods and to dispose of solid and liquid wastes grew as well. Much of the original tidal flats area, which had been too wet to build on and too shallow to navigate, had been filled in and built on. Hills had been leveled and the marshes filled with soil. The largest project had been the filling of Back Bay, which began in 1858 and continued for decades. Once filled, however, the area suffered from flooding and water pollution.

The great landscape planner Frederick Law Olmsted, who also designed New York City's Central Park, solved all these problems with a water-control project called the "fens." His goal was to "abate existing nuisances" by keeping sewage out of the streams and ponds and building artificial banks for the streams to prevent flooding, and to do this in a natural-looking way.

Olmsted's solution included creating artificial watercourses by digging shallow depressions in the tidal flats, following meandering patterns like natural streams; setting aside other artificial depressions as holding ponds for tidal flooding; restoring a natural salt marsh planted with vegetation tolerant of brackish water; and planting the entire area to serve as a recreational park when not in flood. He put a tidal gate on the Charles River—Boston's major river—and had two major streams diverted directly through culverts into the Charles so that they flooded the fens only during flood periods. He reconstructed the Muddy River primarily to create new, accessible landscape. The result of Olmsted's vision was that control of water became an aesthetic addition to the city and a benefit to biodiversity. The blending of several goals made the development of the fens a landmark in city planning.

In 1902, Ebenezer Howard coined the term "garden city." He believed that city and countryside should be planned together. A "garden city" was one surrounded by a greenbelt—a belt of parkways, parks, or farmland. The idea was to locate garden cities in a set connected by greenbelts, forming a system of countryside and urban landscapes. The idea caught on, and garden cities were planned and developed in Great Britain and the United States. Greenbelt, Maryland, just outside Washington, DC, is one of these cities, as is Letchworth, England. Howard's garden city concept benefits biodiversity and continues to be a part of city planning more than a hundred years later.

Today approximately 45% of the world's population live in cities, and it is projected that 62% of the population will live in cities by the year 2025. In the United States, about 75% of the population live in urban areas. Megacities—huge metropolitan areas with more than 8 million residents—are cropping up more and more. In 1950 the world had only two: the New York City and nearby urban New Jersey metropolitan area (12.2 million residents altogether) and greater London (12.4 million). By 1975, Mexico City, Los Angeles, Tokyo, Shanghai, and São Paulo, Brazil, had joined this list. By 2002, the most recent date for which data are available, 30 urban areas had more than 8 million people.

With the growing human population, we can imagine two futures. In one, cities are pleasing and livable, foster biodiversity within their boundaries and beyond, use resources from outside the city in a sustainable way, and minimize pollution of the surrounding countryside. In the other future, cities continue to be seen as environmental negatives and are allowed to decay from the inside. People flee to grander and more expansive, and expensive, suburbs that occupy much land, and the poor who remain in the city live in an unhealthy and unpleasant environment. Which shall it be? Shall The Nature Conservancy continue to focus only or nonurban landscapes, indirectly promoting the second option? Or will it join the modern urban biodiversity-conservation movement, and in a new way extend its incredibly powerful conservation efforts to deal with urbanization of the modern world?

Without the recognition that cities are of and within the environment, the wilderness of the wolf and the moose, the nature that most of us think of as natural, cannot survive, and our own survival on the planet will come into question. The management of nature, although perhaps most easily illustrated by a wolf, a moose, and some trees far distant from our everyday lives, is not restricted to things from which most of us are separated. The management of nature lies just outside the door, and affects not only ourselves and our neighbors but also nature in its largest sense.

How We Must Change Our Way of Thinking About Nature

In discussing the organic view of nature, and the transition that took place with the Romantic poets of the nineteenth century, I wrote that the important point was a shift from an argument that God must exist because His world looked perfectly ordered, to the argument that God is evident in the power of natural forces—a shift from an explanation based simply on structural characteristics to an explanation based on processes and dynamic qualities. This is a transition that the science of ecology had to make and in many ways still has to make, from a perception of nature as fixed structure to nature as dynamic systems, with an emphasis on *dynamic*.

Limits to Our Actions

People ask me if I am confident that if we were to follow all the suggestions made here, we could and would solve all our environmental problems. In reply I tell them the story of "The Monkey's Dilemma," told to me by a person who founded, built, and owned a nature preserve in the wilderness mountains of Costa Rica.

In those rain forests lives a species of bat that makes its nest under the leaves of a small palm tree. The palm lives in the understory and has very large leaves. The two sides of each leaf droop down from the center vein like a small pitched roof with a central beam. A pair of bats come along and cut partway through the leaf on each side of that vein, so that the leaf becomes a four-sided roof, even better as an umbrella. The bats then build a nest that they glue to the underside of the leaf.

A species of monkeys in those forests like to eat the bats. When a monkey sees one of the specially cut leaves, he knows that a bat has made a nest there. But there's one problem: The bats use a nest for only a short while, then abandon it and make a new one. A species of wasps that also live in the forest tend to build their paperlike nests at the base of old bat nests on the undersides of the leaves. But if a monkey takes the time to look under a leaf to see whether it's still inhabited by bats or is now a wasp nest, he might scare the bats away and never get one to eat.

The monkey's dilemma is that, knowing this, there is only one approach— what we would call a management policy—and it's risky. He has to make a grab under the leaf knowing that if he's lucky a bat will be there and he'll get a meal, but if there's a wasp nest he'll get stung.

So it is with us and nature: No matter how much we know and how well we understand nature, our decisions are always going to be, at least to some extent, another example of the monkey's dilemma.

NOTES

Introduction

1. G. M. Woodwell, G. J. MacDonald, R. Revelle, and C. David Keeling (1979). *The Carbon Dioxide Problem: Implications for Policy in the Management of Energy and Other Resources.* Submitted to the Council on Environmental Quality, with no publisher listed.

2. I asked my colleague and good friend, Matthew J. Sobel, an expert on stochastic processes and on business, and at the time a faculty member at Yale I Operations Research, to join me in that work. The result was published as D. B. Botkin and M. J. Sobel (1977). *Optimum Sustainable Marine Mammal Populations.* Report to the U.S. Federal Marine Mammal Commission, 126 pp. This report is available on the Web at www.naturestudy.org.

3. Joseph Campbell (1959). *The Masks of God: Primitive Mythology.* (New York: Viking, 2nd ed.), pp. 3–4.

4. Ibid., p. 5.

5. C. Mackay (1932). *Extraordinary Popular Delusions and the Madness of Crowds* (published in 1841. There were included certain passages and chapters that were omitted from the edition of 1852, of which this book is a verbatim reprint.) (Boston: L. C. Page & Company Publishers), p. 99.

6. J. Lovelock (2006). *The Revenge of Gaia: Why the Earth Is Fighting Back—and How We Can Still Save Humanity* (New York: Perseus).

7. D. H. Fischer (1989). *Albion's Seed: Four British Folkways in America* (New York: Oxford University Press).

8. Fischer's complete list of folkways is:

". . . they always include the following things:—Speech ways, conventional patterns of written and spoken language: pronunciation, vocabulary, syntax and grammar—Building ways, prevailing forms of vernacular architecture and high architecture, which tend to be relation to one another—Family ways, the structure and function of the household and family, both in ideal and actuality,—Marriage ways, ideas of the marriage-bond, and cultural processes of courtship, marriage, and divorce—Gender ways, customs that regulate social relations between men and women—Sex ways . . .—Child-rearing ways . . ."
—Naming ways
—Age ways
—Death ways
—Religious ways, patterns of religious worship, theology, ecclesiology and church architecture
—Magic ways, normative beliefs and practices concerning the supernatural
—Learning ways
—Food ways
—Dress ways
—Sport ways
—Work ways, work ethics and work experiences
—Time ways
—Wealth ways
—Rank ways

—Social ways, conventional patterns of migration, settlement, association and affiliation
—Order ways, ideas of order, ordering institutions, forms of disorder, and treatment of the disorderly
—Power ways, attitudes toward authority . . .
—Freedom ways, prevailing ideas of liberty and restraint, and libertarian customs and institutions

9. Isabella L. Bird (1875). *The Hawaiian Archipelago: Six Months among the Palm Groves, Coral Reefs, and Volcanoes of the Sandwich Islands* (London: J. Murray). This book has been reprinted many times, most recently in 2007 as *Six Months in the Sandwich Islands* (Mutual Publishing).

10. John Masefield (1913). *Salt-water Poems and Ballads* (New York: Macmillan Co.), p. 55. The poem was first published in 1902 in *Salt-water Ballads*.

Chapter 1

1. The number of pilings used as a foundation for the church of Santa Maria della Salute is given in *Venezia, città nobilissima del Sansovino* (Venice, 1663), cited in A. Storti. (1981). *Venice: A Practical Guide* (Venice: Storti, p. 5).

2. Accounts differ on the actual date of the founding of Venice. Some authorities give a date as late as AD 811, but John Julian Norwich attributes the first settlers of Venice to the attack by Goths on Aquileia in AD 402 (*A History of Venice* [New York: Knopf, 1982], p. 4). He notes that Venetian histories give Friday, March 25, 421, as the date when the city of Venice was "formally brought into being," but he states that this standard story is not validated and perhaps marks the date of the establishment of a Paduan trading post on one of the islands in the lagoon. In 466, a meeting at Grado, 60 miles south of Venice, established a government for the islands, but the site of the modern city was apparently not yet under a formal government. By the mid-sixth century, Venice seems to have been established; in 568, the Lombards invaded Italy and whole communities fled to Venice (before this, individuals or families fled to the lagoon). As a sidenote to illustrate the obscurity of the actual founding of Venice, Norwich states that the "first" doge of Venice never existed, even though a painting of this purported doge hangs in Venice.

A remarkable and fascinating discussion of the decline and fall of the Roman Empire, including why Germanic tribes moved into Italy and attacked the Roman Empire, is J.J. O'Donnell's 2009 book, *The Ruin of the Roman Empire*. New York Ecco (Harper Collins).

3. Some of the Proceedings from this conference are published in A. A. Orio and D. B. Botkin (Eds.) (1986). Man's Role in Changing the Global Environment, Proceedings of an International Conference, Venice, Italy, October 21–26, 1985. *The Science of the Total Environment*, vol. 55: 1–399; and vol. 56: 1–415.

4. The term "spaceship Earth" has been attributed to Buckminster Fuller, and to Kenneth Boulding; it is said that Adlai Stevenson was the first to popularize it. However, a much earlier and possibly first use of the term was written by the political philosopher Henry George, in *Progress and Poverty*: "It is a well-provisioned ship, this on which we sail through space." (Kenneth Boulding, at the Sixth Resources for the Future Forum on Environmental Quality in a Growing Economy in Washington, D.C. on March 8,1966; Fuller, Buckminster (1963). *Operating Manual for Spaceship Earth*. New York: E.P. Dutton & Co; Stevenson, A. E. (1966). Speech given as U.S. Ambassador to the United Nations, Economic and Social Council, Geneva, Switzerland, July 9, 1965.

5. Among the ecologists leading the movement to accept the non-equilibrium concept of ecological systems are Stuart Pimm (See for example *The Balance of Nature?*, University of Chicago Press,1991); Klaus Rohde (his book. *Nonequilibrium Ecology* New York,

Cambridge University Press, (2006) provides a thorough scholarly review of evidence supporting the dynamic characteristics of ecological systems). Stephen Hubbell's "neutral theory" is non-equilibrium (See for example, Hubbell, S. P. (2001). *The Unified Neutral Theory of Biodiversity and Biogeography.* Princeton, NJ, Princeton Univ. Press.). Donald Strong's "density-vague" concept,—that regulation is usually not very strong—supports this view (See, for example, Strong, D. (1986). "Density-vague population change, " *Trends in Ecology & Evolution* 1(2): 39–42). The classic papers by Joe Connell about "intermediate disturbance" are also consistent with this view. (Among his many papers, see, for example Connell, J. H. and R. O. Slatyer. (1977). " Mechanisms of Succession in Natural Communities and Their Role in Community Stability and Organization." *The American Naturalist* 111(982): 1110–1144.)

6. L. J. Henderson (1913/1966). *The Fitness of the Environment* (New York: Macmillan; Boston: Beacon Press).

7. Ibid., pp. 108–109. Among the key passages in Henderson's book concerning the remarkable properties of water are the following: "water shares the characteristic of very high specific heat with a very small number of substances, among which hydrogen and ammonia are probably the only important chemical individuals" (Ibid., p. 84); "water possesses certain nearly unique qualifications which are largely responsible for making the earth habitable. . . . The most obvious effect of the high specific heat of water is the tendency of the ocean and of all lakes and streams to maintain a nearly constant temperature. . . . A second effect . . . is the moderation of both summer and winter temperatures of the earth" (pp. 85–86). He also lists as important the high latent heat of melting and evaporation of water (the amount of heat required to convert a unit of ice to water and a unit of water to water vapor). These tend to stabilize the temperatures of the planet, for "so long as water and ice exist in contact, the system constitutes a thermostat. . . . Heating serves merely to melt the ice, cooling to freeze the water" (p. 93). In addition, water has an unusually high freezing point, which Henderson calls "one of the most important facts with which we are concerned, for while a very large number of chemical processes take place quite freely at 0°, the conditions are very different at the freezing point of ammonia, for instance," which is much lower and reactions therefore take place much more slowly (pp. 93–94). The latent heat of evaporation "is by far the highest known," which is "one of the most important regulatory factors at present known to meteorologists" (p. 98).

8. H. J. Morowitz (1979). *Energy Flow in Biology* (Woodbridge, Conn.: Oxbow Press).

9. Personal communication from my friend and colleague Bassett Maguire, with whom I published scientific papers about the underlying theoretical possibilities of closed ecological systems.

10. E. O. Wilson (1988). *Biodiversity* (Washington, D.C.: National Academy Press).

11. C. L. Glacken (1967). *Traces on the Rhodian Shore: Nature and Culture in Western Thought from Ancient Times to the End of the Eighteenth Century* (Berkeley: University of California Press).

12. G. P. Marsh (1864/1967). *Man and Nature* (ed. D. Lowenthal). (Cambridge, Mass.: Harvard University Press), pp. 29–30.

13. D. Lowenthal (1958). *George Perkins Marsh, Versatile Vermonter* (New York: Columbia University Press).

14. D. B. Botkin and M. J. Sobel (1975). Stability in Time-Varying Ecosystems. *American Naturalist, 109,* 624–646.

More discussion of the importance of change in ecological systems can be found in D. B. Botkin, S. Golubeck, B. Maguire, B. Moore III, H. J. Morowitz, and L. B. Slobodkin (). Closed Regenerative Life Support Systems for Space Travel: Their Development Poses Fundamental Questions for Ecological Science. In R. Homquist (Ed.), *Life Sciences and*

Space Research XVII (Elmsford, N.Y.: Pergamon Press); D. B. Botkin (1980). A Grandfather Clock Down the Staircase: Stability and Disturbance in Natural Ecosystems. In R. H. Waring (Ed.), *Forests: Fresh Perspectives from Ecosystem Analysis*, Proceedings of the Fortieth Annual Biology Colloquium (Corvallis: Oregon State University Press).

15. Galileo and Kepler were born in the sixteenth century and lived into the seventeenth. Newton was born in the seventeenth century and lived into the eighteenth. Their dates are: Kepler 1571–1630; Galileo 1564–1642; Newton 1643–1727.

16. M. H. Nicolson (1959). *Mountain Gloom and Mountain Glory: The Development of the Aesthetics of the Infinite* (Ithaca, N.Y.: Cornell University Press). [Reprinted with an introduction by historian William Cronon as: Nicolson, M. H. (2011). *Mountain Gloom and Mountain Glory*. Seattle, University of Washington Press (paperback).

17. M. H. Nicolson (1959). Work previously cited.

18. The digital analysis of human DNA was reported in N. Wade. (2011). Decoding DNA With Semiconductors. *The New York Times*.

19. L. Ferry. (1995). *The New Ecological Order* (Chicago: University of Chicago Press).

20. Quotes are from L. Ferry.

21. J. Hooper. (2006). Population decline set to turn Venice into Italy's Disneyland. *The Guardian*. Manchester, guardian.co.uk © Guardian News and Media Limited 2011. It is also worth noting that discussions continue about how to best forecast *aqua alta*. See, for example, P. Teatini, N. Castelletto, M. Ferronato, G. Gambolati and L. Tosi. Water Resources Research, Vol. 47, W12507, Pg. 17, Dec. 7, 2011. DOI:10.1029/2011WR010900.

Chapter 2

1. A. J. Nicolson (1933). The Balance of Animal Populations. *Journal of Animal Ecology* 2:133.

2. Charles Elton (1930). *Animal Ecology and Evolution* (New York: Oxford University Press).

3. The word *nature* has been used to mean (1) the natural world on the Earth as it exists without human beings or civilization—that is, the environment, including mountains, plains, rivers, lakes, oceans, air, rocks, and all nonhuman, nondomesticated living things; and (2) the universe, with all its phenomena, including objects and forces (*The Random House Dictionary of the English Language*, 2nd ed. [New York: Random House, 1988]). Throughout this book, unless otherwise noted, the word *nature* stands for the first definition. The first and second definitions are not so clearly distinguished in modern discussions of the classical Greek and Roman philosophers; nature is seen as encompassing both concepts: nature as the universe, and nature as the Earth.

In the science of ecology and in most discussions of natural history, the term *environment* generally means the nonliving aspects of nature, but in popular writing it is sometimes synonymous with the first definition of nature. In the science of ecology, the word *ecosystem* is the closest term to the first definition of nature. Ecologists do not speak of the universe as nature except in specific contexts, as in dealing with ecological life-support systems that might be used in space travel.

4. This paragraph is based on Daphne Sheldrick. (1973). *The Tsavo Story* (London: Collins and Harvill Press). The quote is on page 113. Additional information on the Tsavo elephant population, historical and recent, is from http://www.tsavoelephants.org/.

5. Ibid., p. 113.

6. Ibid., pp. 186–187.

7. Ibid., p. 283.

8. Ibid., p. 277–278.

9. Ibid., p. 190.

10. Barbara McKnight, 2009 Tsavo Elephant Research, www.tsavoelephants.org Accessed November 28, 2009. An area count between January 28 and February 2, 2008, with 11 aircraft flying 10 hours a day to count all the wildlife gave an estimate of 11,600 individuals (www.tsavoelephants.org/fieldnotes.html).

11. Information about fisheries catch is from D. B. Botkin and E. A. Keller (1987). *Environmental Studies: Earth as a Living Planet* (Columbus: Merrill Pub. Co., p. 220 and Table 8.2).

12. Specific data obtained from National Marine Fisheries Service. http://swr.nmfs.noaa. gov/biologic.htm; Pacific Fishery Management Council. Preseason Report I, Stock Abundance Analysis for 2008 Ocean Salmon Fisheries, Chapter 2, February 2008.

13. D. Jolly. (2011). European Union Proposes Overhaul of Fisheries Policy. *The New York Times.*

14. D. B. Botkin and E. A. Keller (2009). *Environmental Sciences: The Earth as a Living Planet* (7th ed.). New York: John Wiley, Chapter 13.

15. Pierre-François Verhulst was born in Brussels on October 28, 1804, and died on February 15, 1849. G. E. Hutchinson has written a brief and interesting biography of him. Verhulst taught at the University of Ghent and became interested in mathematical problems related to the repayment of public debt by state lotteries. Hutchinson suggests that this was the "beginning of his interest in social and demographic problems." He went to Italy during the Belgian revolution of 1830 and "became unsuccessfully involved in politics and also published one original historical study. By 1834 he had returned to science and began giving instruction in mathematics at the *École militaire*, where he later became a professor." Hutchinson notes that "though his contemporaries ignored the logistic, his interest in population studies led him into sociological problems . . . and the Belgian government appointed him to commissions on the relief of poverty and on state insurance policy" (*An Introduction to Population Ecology* [New Haven, Conn.: Yale University Press, 1978], p. 20).

16. A. Lotka (1925). *Elements of Physical Biology*. Baltimore: Williams and Wilkins, Baltimore. Reprinted as A. J. Lotka (1956). *Elements of Mathematical Biology*. New York: Dover Publications, Inc.

17. K. D. Young and R. J. Van Aarde (2010). Density as an explanatory variable of movements and calf survival in savanna elephants across southern Africa. *Journal of Animal Ecology, 79*(3), 662–673.

18. A. J. Lotka (1925). *Elements of Physical Biology*, reprinted in 1956 as *Elements of Mathematical Biology* (New York: Dover). Influential works by the others referred in this paragraph are: G. F. Gause (1934). *The Struggle for Existence* (Baltimore: Williams and Wilkins); V. J. Vernadsky (1929). *La Biosphère* (Paris: Félix Alcan); and V. Volterra (1926). Variazioni e fluttuazioni del numero d'individui in specie animali conviventi. *Mem. R. Acad. Naz. dei Lincei*, 6th Ser., 2, 31–113.

19. Occasionally there have been scientific papers that suggest a logistic growth has taken place over a short time, as in J. H. Connell (1961). The effects of competition, predation by *Thais lapillus* and other factors on natural populations of the barnacle, *Balanus blanoides*. *Ecological Monographs, 31*, 61–104. He shows a logistic-like growth over a period of a few weeks from the initial settlement of an area by barnacles.

20. For the teaching of the logistic growth curve in current ecology textbooks, see M. Molles et al. (2008). *Ecology: Concepts Applications* (New York: McGraw Hill, p. 259ff.); R. E. Ricklefs and Gary Miller (2000). *Ecology* (W. H. Freeman, p. 314ff.); and T. M. Smith and R. L. Smith (2006). *Elements of Ecology* (New York: Benjamin Cummings, Pearson International Edition, p. 224ff.).

The best discussion in a current ecology textbook of the logistic is in M. Begon, C. R. Townsend, and J. Harper (2006). *Ecology: From Individuals to Ecosystems* (Malden,

MA: Blackwell Publishing). This book presents the logistic as a beginning theoretical idea, and tells the student that it has all the shortcomings described for a mathematically discrete version of the same concept, and that "the logistic is therefore doomed to be a model of perfectly compensating density dependence. Nevertheless, in spite of these limitations, the equations will be an integral component of models in Chapters 8 and 10, and it has played a central role in the development of ecology" (p. 151). This is how the equation *should* be taught, as a beginning way to approach the entire question of how populations grow and are limited, and then to proceed beyond it to realistic situations and actual observations.

21. Publications resulting from the salmon studies I directed include:

D. B. Botkin, K. Cummins, T. Dunne, H. Regier, M. J. Sobel, and L. M. Talbot (1993). *Status and Future of Anadromous Fish of Western Oregon and Northern California, Rationale for a New Approach* (Santa Barbara, CA: Center for the Study of the Environment, 40 pp.).

M. J. Sobel and D. B. Botkin (1994). *Status and Future of Salmon of Western Oregon and Northern California, Forecasting Spring Chinook Runs* (Santa Barbara, CA: Center for the Study of the Environment, 42 pp.).

D. B. Botkin, K. Cummins, T. Dunne, H. Regier, M. J. Sobel, and L. M. Talbot (1995). *Status and Future of Anadromous Fish of Western Oregon and Northern California: Findings and Options* (Santa Barbara, CA: Center for the Study of the Environment).

22. Public Law 92–522, H.R. 10420, 92nd Cong., October 21, 1972. See also, for 21st century definitions, NOAA's website, http://www.nmfs.noaa.gov/pr/glossary.htm, which states the following: "**Optimum Sustainable Population:** defined by the MMPA section 3(9), with respect to any population stock, the number of animals which will result in the maximum productivity of the population or the species, keeping in mind the carrying capacity of the habitat and the health of the ecosystem of which they form a constituent element. (16 U.S.C. 1362(3)(9)). Optimum Sustainable Population is further defined by Federal regulations (50 CFR 216.3) as is a population size which falls within a range from the population level of a given species or stock which is the largest supportable within the ecosystem to the population level that results in maximum net productivity. Maximum net productivity is the greatest net annual increment in population numbers or biomass resulting from additions to the population due to reproduction and/or growth less losses due to natural mortality." Note also that the most recent available annual report of the Marine Mammal Commission (http://mmc.gov/reports/annual/welcome.shtml) uses the term "optimum sustainable population" a number of times, but does not contain a definition of the term. (Marine Mammal Commission Annual Report to Congress, 2009, Marine Mammal Commsion, 4340 East-West Highway, Room 700, Bethesda, MD, 20814

23. D. B. Botkin and M. J. Sobel. *Optimum Sustainable Marine Mammal Populations.* Report prepared for the Marine Mammal Commission, 1977.

24. Bockstoce's and my work on the bowhead whale include the following publications: J. R. Bockstoce and D. B. Botkin (1983). The History of the Reduction of the Bowhead Whale (*Balaena mysticetus*) Population by the Pelagic Whaling Industry, 1848–1914. In: M. F. Tillman and G. P. Donovan (Eds.), *International Whaling Commission Reports (Special Issue 5 on Historical Whaling Records).*

J. R. Bockstoce, D. B. Botkin, A. Philp, B. W. Collins, and J. C. George (2007). The Geographic Distribution of Bowhead Whales in the Bering, Chukchi, and Beaufort Seas: Evidence from Whaleship Records, 1849–1914. *Marine Fisheries Review, 67*(3), 1–25.

25. Robards, M. D., J. J. Burns, et al. (2009). "Limitations of an optimum sustainable population or potential biological removal approach for conserving marine mammals: Pacific walrus case study." Journal of Environmental Management **xxx**: 1–10.

25. C. J. Glacken (1967). *Traces on the Rhodian Shore: Nature and Culture in Western Thought from Ancient Times to the End of the Eighteenth Century* (Berkeley: University of California Press).

26. R. Nash (1967). *Wilderness and the American Mind* (New Haven: Yale Univ. Press).

27. Plotinus (1956). *The Enneads* (trans. S. Mackenna, revised by B. S. Page, 3rd ed.). (London: Faber and Faber Ltd.).

Chapter 3

1. R. H. MacArthur (1972). *Geographical Ecology* (New York: Harper & Row).

2. http://www.coppercountry.com/article_42.php

3. W. S. Cooper (1913). The Climax Forest of Isle Royale, Lake Superior, And Its Development. *Botanical Gazette*, 55, 1–44, 115–140, 189–234.

4. A. Murie (1934). *The Moose of Isle Royale* (Univ. of Michigan Museum Zoology Misc. Publ. 25, 44 pp.).

5. P. S. Martin and D. A. Burney (1999). Bring back the elephants. *Wild Earth* (Spring issue): 57–64; P. S. Martin and Christine R. Szuter (1999). War Zones and Game Sinks in Lewis and Clark's West. *Conservation Biology*, 13(1), 36–45; A. Esty. An interview with Paul S. Martin. *American Scientist*.

6. The complex hunting behavior of wolves is described by D. Mech. *The Wolves of Isle Royale* (U. S. National Park Service Fauna Series no. 7; Washington, D.C.: Government Printing Office, 1966) and *The Wolf: The Ecology and Behavior of an Endangered Species* (Garden City, N.Y.: Natural History Press, 1970).

7. D. B. Botkin, P. A. Jordan, A. S. Dominski, H. D. Lowendorf, and G. E. Hutchinson (1973). Sodium Dynamics in a Northern Terrestrial Ecosystem. *Proceedings of the National Academy of Sciences*, 2745–2748. At the time of this writing, the wolves of Isle Royale have undergone a drastic decline, suffering, I am told, from a viral disease introduced on the island by domestic dogs of tourists. In such ways even what appears to be an excellent example of wilderness is now indirectly touched by our hands.

8. Vucetich, J. A. a. R. O. P. (2010). *Ecological Studies of Wolves on Isle Royale Annual Report 2009–10*. (Houghton, MI, School of Forest Resources and Environmental Science, Michigan Technological University). See also Nelson, M. P., R. O. Peterson, et al. (2008). "The Isle Royale Wolf-Moose Project: Fifty Years of Challenge and Insight, "The George Wright Forum 25 of Number 2 (2008)(2): 98–113, and Vucetich, J. A., M. P. Nelson, et al. (2012). "Should Isle Royale Wolves be Reintroduced? A Case Study on Wilderness Management in a Changing World." The George Forum 29(1): 126–147.

9. The books referred to here are: C. R. Darwin (1859). *The Origin of the Species by Means of Natural Selection or the Preservation of Favored Races in the Struggle for Life* (London: Murray); G. P. Marsh (1864/1967). *Man and Nature* (ed. D. Lowenthal; Cambridge, Mass.: Harvard University Press); and E. Haeckel (1866). *Generelle Morphologie der Organismen: Allgemeine Gründz—ge der organischen Formen-wissenschaft, mechanisch begründet durch die von Charles Darwin reformirte Desendenz-Theorie* (2 vols.; Berlin: Reimer).

The word "ecology" is derived from the Greek word *oikos*, meaning "a dwelling place." Haeckel coined the term as *Oekologie*, but the spelling was anglicized at the Madison, Wisconsin, Botanical Congress of 1893 into its present spelling, according to R. P. McIntosh. Ernst Heinrich Haeckel was a German biologist who lived from 1834 to 1919. Within his field, he was known for his studies of marine organisms, but was also well known, according to McIntosh, as a leading proponent of Darwinian ideas in Germany ("Ecology Since 1900," pp. 33–372, in B. J. Taylor and T. J. White [Eds.]. [1976]. *Issues and Ideas in America* [Norman: University of Oklahoma Press]).

10. Excellent examples of machines as objects of art are shown in R. G. Wilson, D. H. Pilgrim, and D. Tashjian (1986). *The Machine Age in America: 1918–1941* (New York: Abrams).

11. For more information on the history of ecology, see McIntosh, "Ecology Since 1900" and *The Background of Ecology: Concept and Theory* (New York: Cambridge University Press, 1985); D. Worster (1977). *Nature's Economy: The Roots of Ecology* (San Francisco: Sierra Club Books).

12. S. A. Forbes. (1925). The Lake as a Microcosm. *Illinois Natural History Survey Bulletin, 15,* 549. Also of interest are additional comments by Forbes: "Although every species has to fight its way inch by inch from the egg to maturity, yet no species is exterminated, but each is maintained at a regular average number which we shall find good reason to believe is the greatest for which there is, year after year, a sufficient supply of food" (Ibid., p. 549). This becomes of interest in regard to assumptions about the expected yield of animals such as fish or wild game under management regimes.

The use of the term "balance of nature" by ecologists is discussed in F. N. Egerton (1973). Changing Concepts of the Balance of Nature. *Quarterly Review of Biology, 48,* 322–350.

13. B. J. Le Boeuf and Richard M. Laws. (1994). *Elephant Seals: Population Ecology, Behavior, and Physiology.* (Berkeley, CA, University of California Press.)

14. B. J. Le Beoeuf (1981). The Elephant Seal. In P. A. Jewel, S. J. Holt, and D. Hart (Eds.), *Problems in the Management of Locally Abundant Wild Mammals* (New York: Academic Press). The scientific name of the northern elephant seal is *Mirounga angustirostris.*

15. NOAA Office of Protected Resources. http://www.nmfs.noaa.gov/pr/species/mammals/pinnipeds/northernelephantseal.htm. Accessed March 10, 2010. According to this site, "Though a complete population count of elephant seals is not possible because all age classes are not ashore at the same time, the most recent estimate of the California breeding stock was approximately 124,000 individuals."

16. Marine Mammal Commission, 2010, list of species of special concern, http://mmc.gov/species/specialconcern.html. Accessed March 10, 2010.

17. The scientific name of the sandhill crane is *Grus canadensis.*

18. R. S. Miller, G. S. Hochbaum, and D. B. Botkin (1972). A simulation model for management of sandhill cranes. *Yale Univ. School of Forestry and Environmental Studies Bulletin* No. 80, 49 pp.; R. S. Miller and D. B. Botkin (1974). Endangered species: models and predictions. *Amer. Sci., 62,* 172–181.

19. North Prairie Wildlife Research Center (2006). The Cranes Status Survey and Conservation Action Plan Sandhill Crane (*Grus canadensis*). USGS,.

20. F. N. Egerton. (1975). Aristotle's Population Biology. *Arethusa, 8,* 307–330.

21. Gause, *The Struggle for Existence,* p. 25. That the comments came from Volterra's son-in-law is stated by G. E. Hutchinson in *An Introduction to Population Ecology* (New Haven: Yale University Press, 1978), p. 120.

22. Oddly enough, this is still true in 2012. See the list of ecology textbooks in an earlier note. Those still teach the Lotka-Volterra equations just as they still teach the logistic. Once again, this would be okay if the approach was to follow Occam's razor and try to find the simplest explanation consistent with observations. Then one would start with the simplest explanation of predator–prey interactions (i.e., the Lotka Volterra equations) and move onward to something the could reproduce actual predator–prey data. But that still is not typical.

23. See, for example, in the use of variations of the Lotka-Volterra model W. W. Murdoch, B. E. Kendall, R. M. Nisbet, C. J. Briggs, E. McCauley, and R. Bolser. (2002). Few-species models for many-species communities. *Nature, 417,* 541–543.

The abstract of this paper (minus reference citations) is: "*Most species live in species-rich food webs; yet, for a century, most mathematical models for population dynamics have included only one or two species. We ask whether such models are relevant to the real world. Two-species population models of an interacting consumer and resource collapse to one-species dynamics when recruitment to the resource population is unrelated to resource abundance, thereby weakening the coupling between consumer and resource. We predict that, in nature, generalist consumers that feed on many species should similarly show one-species dynamics. We test this prediction using cyclic populations, in which it is easier to infer underlying mechanisms, and which are widespread in nature. Here we show that one-species cycles can be distinguished from consumer-resource cycles by their periods. We then analyze a large number of time series from cyclic populations in nature and show that almost all cycling, generalist consumers examined have periods that are consistent with one-species dynamics. Thus generalist consumers indeed behave as if they were one-species populations, and a one-species model is a valid representation for generalist population dynamics in many-species food webs."*

And the paper discusses the kind of cyclic (oscillatory) periodicity just as in the original Lotka-Volterra model. For those interested in delving more deeply into this, in these contemporary approaches, and in related papers of the past 50 years, the basic Lotka-Volterra model for a single predator and single prey has been generalized in two ways: one to involve many species and the second to replace the predator with a competitor or parasite or some other kind of interacting species. The discussion of this goes beyond what I can cover in this book. The primary point here is that the models of this school are deterministic differential equations that require and lead to the kind of stability characteristic of the classic balance of nature, and are therefore philosophically consistent with that ancient view. This holds whether or not these modern formulations account for observations (as the authors claim) or not. My point here is that this continues to restrict our viewpoint of what nature really is and how nature actually works, but Murdoch et al. and others will no doubt strongly disagree.

Also Lord Robert May, whose work in theoretical ecology has been influential since the latter part of the twentieth century, made considerable use of the Lotka-Volterra model and the logistic early in his career, as in May, R. M. (1973). *Stability and Complexity in model ecosystems.* Princeton, NJ, Princeton University Press.) His book, *Theoretical Ecology, Principles and Applications* (originally published in 1976, now in its third edition, 2007), said by its publisher, Oxford University Press to be "the single most influential book in Theoretical Ecology," continues these uses.

24. This simple experiment demonstrates that under certain circumstances a predator can cause the extinction of its prey, after which the predator also suffers extinction. If this occurred consistently in nature, there would be few living predators and prey. The contrary is true: a great many organisms, including ourselves, are predators or prey or both, and certain predator and prey pairs are known from the fossil record to have persisted for a very long time. Gause's experiment raised a number of questions, one of which is: If a predator can completely eliminate its prey in a laboratory experiment, how can predator and prey coexist in nature? Gause found one answer to this question in a second experiment, in which he provided the paramecia with a refuge that consisted of sediment containing food. A paramecium covered by sediment was protected from attack by its predator. Some paramecia spent some of the time in the refuge and were missed by the predator. With the refuge, the outcome was quite different. The predator declined and became extinct, after which the prey continued to increase, undergoing an exponential rise until the end of the experiment. The refuge provided protection for the prey, but did not lead to the coexistence of predator and prey. The second experiment suggests that complexity in the environment may increase the chance of persistence of the prey.

A Lotka-Volterra predator regulates its prey. Without the predator, the prey would increase exponentially, a situation that, as we have seen, cannot be sustained in the real world indefinitely and that leads inevitably to a population crash, when the population exceeds its own resources. The predator, in turn, depends on the prey. Without the prey, the predator declines exponentially, eventually becoming extinct. Whether the predator and prey persist depends on the relative values of the intrinsic net rate of increase of each (the rate of increase in the absence of the other species), and the relative impact of predator on prey and prey on predator.

25. Gause, *Struggle for Existence*, p. 140.

26. R. J. Beverton and S. J. Holt (1957). *On the Dynamics of Exploited Fish Populations*. (London: Chapman and Hall. Reprinted by Blackburn Press in 2004.)

27. See R. O. Peterson. (1999). Wolf–Moose Interaction on Isle Royale: The End of Natural Regulation? *Ecological Applications, 9*, 10–16.

Long-term population fluctuations of wolves and moose in Isle Royale National Park, Michigan, are used to evaluate a central tenet of the "natural regulation" concept commonly applied by the National Park Service (NPS) in the United States, namely that wild cervid populations exhibit density dependence that, even in the absence of large predators, will stabilize population growth. This tenet, restated as a hypothesis, is rejected based on moose population response to a chronic wolf decline. In 1980–1996 with wolf numbers down, partly due to introduced disease, moose numbers increased to a historic high level. There was insufficient density dependence in moose reproduction and mortality to stabilize moose numbers. In 1996 moose suffered a crash; 80% died, primarily from starvation. These fluctuations, along with the possibility that the highly inbred wolf population may become extinct, will challenge NPS policy. The longstanding NPS management tradition of nonintervention may not be compatible with the current policy that stresses maintenance of natural ecological processes, such as a predator–prey system.

28. K. E. F. Watt. (1962). Use of Mathematics in Population Ecology. *Annual Review of Entomology, 7*, 243–252.

29. This discussion is based on D. B. Botkin and M. J. Sobel (1975). Stability in Time-Varying Ecosystems. *American Naturalist, 109*, 625–646.

30. For those of you interested in more details, Buzz Holling's first explanation of resiliency of ecological system was explicitly a statement about a system with classical static stability. C. S. Holling (1974). Resilience and Stability of Ecological Systems. *Annual Review of Ecology and Systematics, 4*, 1–23. In engineering, this kind of stability is formulated exactly mathematically, but that goes beyond what I hope to cover in this book.

31. R. M. May (1973). *Stability and Complexity in Model Ecosystem* (Princeton, NJ: Princeton University Press); R. M. May (1981). *Theoretical Ecology, Principles and Applications* (Sunderland, MA: Sinauer Assoc., Inc.).

32. H. G. Andrewartha and L. C. Birch (1954). *The Distribution and Abundance of Animals* (Chicago: University of Chicago Press).

33. D. Lack (1954). *The Natural Regulation of Animal Numbers* (London: Oxford University Press).

34. C. Elton (1924). Periodic fluctuations in the numbers of animals: their causes and effects. *J. Exper. Biol., 2*, 119–163; C. S. Elton (1942). *Voles, Mice and Lemmings: Problems in Population Dynamics*. (Oxford: Clarendon Press.)

35. C. S. Elton (1942). *Voles, Mice, and Lemmings* (Oxford: Clarendon Press). These records owe their origin to Charles II of England, who in 1679 granted ownership and exclusive trading rights to all the land, including "countries, coasts, and confines of the seas, bays, lakes, rivers, creeks, and sounds," along Hudson Bay in Canada to a firm later known as the Hudson's Bay Company.

36. Most long-term records for animal populations come from commercial harvesting—as for example the records of haddock from Icelandic fishing grounds and whales in the Pacific. These records also show variation, rather than constancy, over time. But they are even more likely to be confounded by the effects of variations in effort—the number of boats and the market for the fish—than the Hudson's Bay Company's records of animals caught in traps.

This is shown dramatically in the catches for both haddock and whales from 1915 to 1919 and 1939 to 1945, when fishing was halted by world wars.

37. D. Lack (1954). *The Natural Regulation of Animal Numbers* (London: Oxford University Press.)

38. Lack, *The Natural Regulation of Animal Numbers*, p. 1.

39. There has been extensive writing in the ecological scientific literature about the regulation of populations, especially for animal populations. A general distinction is between density-dependent and density-independent population regulation. Density-dependent population regulation is when a population regulates itself, as with a hypothetical logistic population—as density increases, death rates rise and birth rates decline, exactly as Lotka explained the logistic in terms of a laboratory container of fruit flies.

Given the amount that has been written in the scientific literature about population regulation, some standard ecology textbooks spend very little time on the concept, some not really discussing it at all or enough to explain it to the point I just did here. See, for example, C. R. Townsend, Michael Begon, and John L. Harper (2003). *Essentials of Ecology.* (Malden, MA: Blackwell).

Chapter 4

1. Thomas Pownall. (1784/1949). *A Topographical Description of the Dominion of the United States* (Pittsburgh: Lois Mulkean).

2. P. Kalm (1963). *Travels in North America: The America of 1750* (2 vols., trans. A. B. Benson) (New York: Dover).

3. Ibid., p. 175.

4. J. T. Cunningham (1954). Woodland Treasure. *Audubon* July–August.

5. For some stories about Murray Buell that reveal him as a wonderful person, see my 2003 book *Strange Encounters: Adventures of a Renegade Naturalist* (New York: Penguin [Tarcher] Books).

6. L. Barnett (1954, Nov. 8). The Woods of Home. *Life.*

7. T. Pownall (1776, republished 1949). *Topographical Description of the Dominions of the United States of America [Being a Revised and Enlarged Edition of a Topographical Description of Such Parts of North America as Are Obtained in the (Annexed) Map of the Middle British Colonies &C. In North America].* (Pittsburgh: University of Pittsburgh Press.)

8. Among the earliest uses of the term "succession" are in: H. D. Thoreau (1860). The Succession of Forest Trees; An Address Read to the Middlesex Agricultural Society in Concord, September, 1860 (Extracted from the 8th Annual Report of the Massachusetts Board of Agriculture); and G. P. Marsh (1864/1967). *Man and Nature* (D. Lowenthal; Cambridge, Mass.: Harvard Univ. Press).

9. M. F. Buell, H. F. Buell, and J. A. Small (1954). Fire in the History of Mettler's Woods. *Torreya, 81,* 253–255.

10. There has been considerable discussion of the role of American Indians in the burning of forests, especially in the Atlantic coastal states. In a classic and often-cited paper, G. M. Day suggests that fires were common and set on purpose to clear the forest in order to make traveling and hunting easier and to drive game (G. M. Day [1953]. The Indian as an Ecological Factor in the Northeastern Forest. *Ecology, 34,* 329–346). A number of historical sources mention large fires and suggest that they were set on purpose. For example, Adriaen Van der Donck, a Dutch settler, wrote in 1655 that "a yearly custom" of the Indians was an autumn burning of the woods and that this burning was done "to render hunting easier . . . to thin out and clear the woods of all dead substances and grass, which grow better the ensuing spring . . . to circumscribe and enclose the game within the lines of fire,

when it is more easily taken" (A Description of the New Netherlands. In T. F. O'Donnell [1968]. *Collections of the New York Historical Society* [2nd series] [Syracuse, N. Y.: Syracuse University Press], 1:125–242). E. W. B. Russell provides one of the most thorough analyses of the role of American Indians in setting fires and concludes that "the frequent use of fires by Indians to burn the forests was probably at most a local occurrence," but "their use of fire for many purposes did, however, increase the frequency of fires above the low levels caused by lightning." She argues that the activity of the Indians did not lead to wholesale burning of the entire forests of New Jersey every year (E. W. B. Russell [1983]. Indian-Set Fires in Northeastern USA. *Ecology, 64*, 78–88). However, the alteration of the forests that I have described required a much less frequent recurrence in any one area. A frequency of once a decade, consistent with the evidence of fire scars on the trees in Hutcheson Memorial Forest, would be sufficient to favor oaks over maples without preventing oak reproduction.

11. The information about the forests of the Maryland shore, including historical observations, is from N. Sampson et al. (2000). *Chesapeake Forests Plan: Resource Assessment* (The Sampson Group, prepared for the Conservation Fund).

12. The interpretation of forests as part of a changing landscape described here is mine; other ecologists still oppose it and hold on to the older idea. For example, Russell states that "the similarity of contemporary forests in North America to those that predated European colonization indicates the resilience, and thus the static stability of the forest types" (Russell [1980]. Vegetation Change in Northern New Jersey from Precolonization to the Present: A Palynological Interpretation. *Bulletin of the Torrey Botanical Club, 107*, 432) using a term from one of my papers (D. B. Botkin and M. J. Sobel [1975]. Stability in Time-Varying Ecosystems. *American Naturalist, 109*, 625–646).

She states that three periods are observable in the pollen records: pre-colonial, when forests dominated; 1740 to 1850, when there was extensive land clearing by European colonists, and reforestation after 1850, when much of the land was allowed to return to forests. According to her analysis, chestnut and oaks dominated the pre-colonial upland forests. Thus, Russell concludes, "The common denominator in the original as well as in the present forests is oak. . . . Differences are primarily in the less dominant taxa" (p. 444). In part, the point that I am making is revealed in the changes in the less dominant taxa, since they represent the beginnings of a change delayed by the longevity of oaks. Oak as a genus could dominate the pollen found in lakes because oaks are abundant on the settled and manipulated landscape of which Hutcheson Memorial Forest is only a tiny fraction.

Although pollen deposits provide a wonderful kind of information not available any other way, they offer only a crude index of abundance. Species differ greatly in their production of pollen, and since the size of the grains differs among species, the distance that pollen travels varies with species. The percentage of pollen from oaks might vary by 10% or 20%, and oaks would still appear as the dominant species. The abundance of oaks could vary as much, and the percentage of pollen falling into the ponds under study could still appear relatively constant. Moreover, the area from which pollen reaches the ponds includes land that has been directly altered by human activities since pre-settlement times, whereas Hutcheson Memorial Forest is a tiny remnant that was not cut. Finally, the changes that were beginning to appear in Hutcheson Forest, such as the shift in the understory saplings toward maples (too young to produce pollen) and the gradual influx of exotic species, would not show up in the pollen records of the ponds that Russell studied. Chestnut has been essentially eliminated except as root sprouts. Native maples have increased in abundance for too short a time, and introduced species have grown in the state for too short a time and are too small a factor in the overall heavily manipulated landscape to show up as a large percentage of the pollen found in lakes. The pollen deposits are simply not sensitive to the changes that I have described.

Russell's comments are therefore valuable to compare with mine to see the importance of point of view and interpretation of data in what is usually regarded as a subject of "objective" facts. To put this another way, while pollen deposits provide us with a wonderful window from which to view the ancient past, in the case of the history of Hutcheson Memorial Forest, a tiny spot of natural forest on a large and settled landscape, the information in pollen percentages collected from muds in a distant lake is filtered many times, by the mathematical calculations of percentages, by the differential flux of pollen through lake water, by the variations in the aerodynamic characteristics of the pollen of several species as the grains are wafted by the winds, by the differences among species in their production of their male element in reproduction. Without pursuing the details of the scientific analysis further, we can say that it is difficult to discover what has happened to less than 100 acres in less than 200 years from the pollen deposits.

13. Marsh, *Man and Nature*, p. 29 and p. 35.

14. W. S. Cooper. (1913). The Climax Forest of Isle Royale, Lake Superior, and Its Development. *Botanical Gazette, 55*, 1–44, 115–140, 189–234. There is no single, universal division of vegetation communities, but the number of different communities on the landscape is large. The United States, for example, has been divided into 114 vegetation communities occupying the lower 48 states (A. W. Küchler [1964]. *Potential Natural Vegetation of the Conterminous United States*. American Geographical Society Special Publication, no. 36. [New York: American Geographical Society]).

15. R. H. Whittaker (1970). *Communities and Ecosystems* (London: Macmillan), p. 73.

16. E. P. Odum (1969). The Strategy of Ecosystem Development. *Science, 164*, 262.

17. A. D. Barnosky (2009). *Heatstroke: Nature in an Age of Global Warming*. (Washington, D.C.: Island Press), p. 30.

18. The discovery that there have been major ice ages can be traced back to the observations of the farmers and herders of the Swiss Alps. As a glacier expands, it pushes ahead of itself a great mound of boulders, rock, and soil, including particles of all sizes, from clays to sands. Some of this debris is pushed to the sides and to the front ends of the ice.

At the leading edge of the glacier, the melting ice creates a rapidly flowing stream, colored military gray by the many small clay-like particles known as "glacial flour" that are suspended in the rushing waters. As a glacier melts and recedes, the flow of water becomes more intense, and the glacial streams move fast enough to carry larger particles of sand and gravel. They are deposited in beds according to their size, the heavier gravel dropping from the stream when the waters first begin to slow down, the sands depositing in slightly quieter waters, and so forth.

The study of the history of the continental glaciations has continued with geologists, climatologists, geographers, and biologists seeking the details of the duration and extent of these ice sheets.

19. For another discussion of the history of the discovery of continental glaciations, see my 2001 book *No Man's Garden: Thoreau and a New Vision for Nature and Civilization* (Washington, D.C.: Island Press.)

20. K. W. Butzer (1971). *Environment and Archeology* (Chicago and New York: Aldine Atherton).

21. Again I note that more discussion about the origin of knowledge about ice ages in the nineteenth century is in my 2001 book *No Man's Garden: Thoreau and a New Vision for Civilization and Nature* (Washington, D.C.: Island Press).

22. The evolution of our species and its upright-walking, bipedal close ancesters is a study undergoing rapid changes as new fossil evidence is found, and there are major controversies about which of our ancestral species evolved when. But the general picture is one of our genus, *Homo* arising during the Pleistocene and genus *Australiopithecus* including

species evolving as early as 4 million years ago and others evolving between 3 and 1.1 million years ago. Among the best current summaries are the website http://www.talkorigins.org/ faqs/homs/ Fossil Hominids

The Evidence for Human Evolution, Copyright © 1996-2011 by Jim Foley, [Last Update: May 31, 2011]. One of the best current summaries is Potts, R. and C. Sloan (2010). *What Does It Mean To Be Human?* Washington, DC, National Geographic.

23. H. M. Heinselman (1973). Fire in the Virgin Forests of the Boundary Waters Canoe Area, Minnesota. *Journal of Quaternary Research, 3,* 329–382. The Boundary Waters Canoe Area has also been relatively undisturbed. It has been subject to some human influences, though, both before and after European colonization, especially the suppression of fires from 1900 until quite recently, and logging and trapping before its establishment as a protected wilderness. See also Heinselman, M. L., 1981, pp. 374–405 in (1981). Fire and succession in the conifer forests of Northern North America. In D. C. West, H. H. Shugart, and D. B. Botkin (Eds.), 1981, *Forest Succession: Concepts and Applications* (New York: Springer-Verlag): 517.

24. Flint, *Glacial and Quaternary Geology.*

25. M. B. Davis (1976). Pleistocene Biogeography of Temperate Deciduous Forests. *GeoScience and Man, 13,* 13–26. In the discussion that follows, the information about rates of movement of trees also comes from this reference.

26. Davis, Pleistocene Biogeography of Temperate Deciduous Forests. Where they overlap, beech occurs on slightly warmer, better-drained soils, and hemlock on cooler, moister valleys, stream sides, and ridges. Neither hemlock nor beech occurs on Isle Royale, although both species grow farther east in forests that are otherwise typical of the warmer areas of the island, forests that on Isle Royale contain yellow birch and sugar maple and that in New York, New England, and Pennsylvania contain yellow birch, sugar maple, hemlock, and beech. Given enough time, these species might also reach Isle Royale; they would grow there if they were planted.

27. Hall, F. G., D. B. Botkin, D. E. Strebel, K. D. Woods, and S. J. Goetz, 1991, Large Scale Patterns in Forest Succession As Determined by Remote Sensing, *Ecology* 72: 628–640.

28. Hansen makes this point both directly and indirectly in many of his papers, as well as his writings and speeches for the general public, whenever he writes or talks about "tipping points" and related ideas. E. g. J. Hansen (2008). Tipping Point: Perspective of a Climatologist. *2008–2009 State of the Wild.* For specific scientific papers that reinforce the idea that forest communities are constant in relation to the climate, see both the following paper and the references it cites: D. B. Botkin, Henrik Saxe, Miguel B. Araújo, Richard Betts, Richard H. W. Bradshaw, Tomas Cedhagen, Peter Chesson, Terry P. Dawson, Julie Etterson, Daniel P. Faith, Simon Ferrier, Antoine Guisan, Anja Skjoldborg Hansen, David W. Hilbert, Craig Loehle, Chris Margules, Mark New, Matthew J. Sobel, and David R. B. Stockwell (2007). Forecasting Effects of Global Warming on Biodiversity. *BioScience, 57*(3), 227–236.

29. Rinderpest is a disease produced by virus of the genus Morbillivirus; the symptoms are lesions of the skin, diarrhea, and high fever. The word comes from the German *rinder,* which means "cattle." A. R. E. Sinclair (1977). *The African Buffalo* (Chicago: University of Chicago Press).

30. D. A. Livingstone (1975). Late Quarternary Climatic Change in Africa. *Annual Review of Ecology and Systematics, 6,* 275.

31. Mario de Vivo and Ana Paula Carmignotto (2004). Holocene vegetation change and the mammal faunas of South America and Africa. *Journal of Biogeography, 31,* 941–957.

32. The New Guinea sediment cores were obtained from fire sites stretched out over almost a 60-mile distance between Laiagam and Mount Itagen Town (approximately latitude 5° longitude 145°). The information discussed here is from D. Walker, The Changing Vegetation of the Montane Tropics.

33. J. M. Bowler, G. S. Hope, J. N. Jennings, G. Singh, and D. Walker (1976). Late Quaternary Climates of Australia and New Guinea. *Quaternary Research, 6*, 359–394; also D. Walker, Changing Vegetation of the Montane Tropics.

The evidence of past glaciers in New Guinea is similar to that in East Africa. There are large moraines ending at elevations of 9,000 to 11,000 feet on Mount Carstensz and Mount Wilhelm in New Guinea. Dating of the organic debris on the top of the moraines sets a minimum age and suggests that the ice retreated before 13,000 years ago. But there were several advances afterward, one after 11,600 and another sometime before 3,000 years ago. In more recent periods, between 2,500 and 1,300 years ago, there were three glacial advances.

34. A. P. Kershaw. (1976). A Late Pleistocene and Holocene Pollen Diagram from Lynch's Crater, Northeastern Queensland, Australia. *New Phytologist, 77*, 469–498. The history of forests near Lynch's Crater in Australia is clouded by the possible influence of human beings. The change from vegetation characteristic of high-rainfall areas to that of lower-rainfall regions could also result from an increase in the frequency of fire. The drier-site vegetation can survive more frequent fires than the rain-forest vegetation. It is unlikely that the rate of natural (lightning-caused) fire would have varied sufficiently to produce the modification in the vegetation. But the change in fire frequency could have resulted from an increase in the number of fires set by human beings. The earliest radiocarbon date for *Homo sapiens* in Australia is 38,000 years ago, coincident with the beginning of the long but definite change in the vegetation. Thus we cannot be certain whether the temporal patterns in vegetation near Lynch's Crater, whatever the cause, were primarily the result of human activities or the result of climatic change.

35. J. M. Bowler, G. S. Hope, J. N. Jennings, G. Singh, and D. Walker (1976). Late Quaternary Climates of Australia and New Guinea. *Quaternary Research, 6*, 359–394. And more recently, the following confirms the dynamics: Hopkins, M. S., J. Ash, et al. (1993). "Charcoal Evidence of the Spatial Extent of the Eucalyptus Woodland Expansions and Rainforest Contractions in North Queensland During the Late Pleistocene." Journal of Biogeography **20**(4): 357-372. They wrote "Recent geoscientific and biogeo- graphic studies suggest that most, if not all, the tropical rainforest areas of the world underwent a series of marked changes in climate associated with Quaternary glaciations, and that both tropical rainforest composition and distribu- tion were affected markedly by these."

36. Walker, The Changing Vegetation of the Montane Tropics, *Search* p. 217 and p. 220.

37. E. K. Faison, D. R. Foster, W. W. Oswald, B. C. S. Hansen, and A. E. Doughty (2006). Early Holocene openlands in southern New England. *Ecology, 87*(10), 2537–2547.

38. M. B. Bush and H. Hooghiemstra (2005). Tropical Biotic Responses to Climate Change. In T. E. Lovejoy and L. Hannah (Eds.), *Climate Change and Biodiversity*. (New Haven, Yale University Press), p. 129.

39. M. B. Bush and H. Hooghiemstra (2005). Tropical Biotic Responses to Climate Change. In T. E. Lovejoy and L. Hannah (Eds.), *Climate Change and Biodiversity* (New Haven: Yale University Press), p. 129.

40. Ibid., pp. 125–141.

40. Ibid., pp. 128–129.

41. J. Walker, C. H. Thompson, I. F. Fergus, and B. R. Tunstall (1981). Plant Succession and Soil Development in Coastal Sand Dunes of Subtropical Eastern Australia. In D. C. West, H. H. Shugart, and D. B. Botkin (Eds.), *Forest Succession: Concepts and Applications* (New York: Springer-Verlag).

42. W. A. Reiners, I. A. Worley, and D. B. Lawrence (1971). Plant Diversity in a Chronosequence at Glacier Bay, Alaska. *Ecology, 52*, 55–69.

43. J. Byelich, et al. (1985). *Kirtland's Warbler Recovery Plan*. U.S. Dept. of Interior and Fish and Wildlife Service, p. 1.

44. M. F. Heinselman. (1981). Fire and Succession in the Conifer Forests of Northern North America. In D. C. West, H. H. Shugart, and D. B. Botkin (Eds.), *Forest Succession: Concepts and Applications* (New York: Springer-Verlag).

45. Information about the Kirtland's warbler and its habitat is from H. Mayfield (1969). *The Kirtland's Warbler* (Bloomfield Hills, Mich.: Cranbrook Institute of Science), pp. 24–25; and Byelich et al., *Kirtland's Warbler Recovery Plan*, p. 12.

46. Norman A. Wood, quoted in Mayfield, *The Kirtland's Warbler*, p. 23.

47. Byelich et al., *Kirtland's Warbler Recovery Plan*, p. 22.

48. L. Line (1964). The Bird Worth a Forest Fire. *Audubon*, pp. 371–375.

Chapter 5

1. L. J. Henderson (1913/1966). *The Fitness of the Environment* (New York: Macmillan; and Boston, Beacon Press).

2. In regard to a teleological view of nature and the universe, the "anthropic principle," was proposed in the late twentieth century. This is the idea that the fundamental laws of the universe are "tuned" to permit the evolution of life and consciousness. For example, see J. D. Barrow, and F. J. Tipler (1986). *The Anthropic Cosmological Principle* (New York: Oxford University Press). This idea is discussed by scientists interested in the origin of life and in the exploration of space, as indicated by G. Wald (1974). Fitness in the Universe: Choices and Necessities. In J. Oró, S. L. Miller, C. Ponnamperuma, and R. S. Young (Eds.), *Cosmochemical Evolution and the Origins of Life* (Dordrecht: Reidel). The editors of this book have been active in the study of the origin of life and in biological issues that make use of satellite technology. Other relevant references include B. J. Carr and M. J. Rees (1979). The Anthropic Principle and the Structure of the Physical World. *Nature, 278*, 605–612; G. Gale (1981). The Anthropic Principle. *Scientific American, 245*, 154–171; B. Carter (1974). Large Number Coincidences and the Anthropic Principle in Cosmology. In M. S. Longair (Ed.), *Confrontation of Cosmological Theories with Observational Data* (Dordrecht: Reidel). I thank B. Callicott for these references.

3. A. Leopold, "Deer Irruptions." Reprinted in *Wisconsin Conservation Department Publication 321*(1943):3–11. This is the source usually quoted as initiating the mountain lion/ Kaibab deer story. See also A. Leopold, L. K. Sowls, and D. L. Spencer (1947). A Survey of Over-populated Deer Ranges in the United States. *Journal of Wildlife Management, 11*, 162–177. Accounts based on Leopold's can be found in W. S. Allee, A. E. Emerson, O. Park, T. Park, and K. P. Schmidt (1949). *Principles of Animal Ecology* (Philadelphia: Saunders); D. Lack (1954). *The Natural Regulation of Animal Numbers* (London: Oxford University Press); H. G. Andrewartha and L. C. Birch (1954). *The Distribution and Abundance of Animals* (Chicago: University of Chicago Press); and E. P. Odum (1971). *Fundamentals of Ecology* (Philadelphia: Saunders).

4. D. I. Rasmussen (1941). Biotic Communities of Kaibab Plateau, Arizona. *Ecological Monographs, 3*, 229–275.

5. A. Leopold, L. K. Sowls, and D. L. Spencer (1947). A Survey of Over-populated Deer Ranges in the United States. *Journal of Wildlife Management, 11*, 162–177. The following information is based on the excellent article by G. Caughley (1970). Eruption of Ungulate Populations, with Emphasis on Himalayan Thar in New Zealand. *Ecology, 49*, 54–72.

6. A. Leopold (1949). *A Sand County Almanac and Sketches Here and There* (New York: Oxford University Press), pp. 130–132. Those familiar with Aldo Leopold's work know that the point of view toward predators expressed in this book was a major change from the viewpoint that Leopold had held earlier in his career, when, during his years with the United States Forest Service, he was a public advocate of predator eradication from the

southwestern ranges. According to B. Callicott, who is an expert on Leopold's career, it was very difficult for Leopold to admit that he was mistaken, and his article "Thinking Like a Mountain" was written as a kind of confession of past misdeeds. This is discussed in D. Ribbens (1987). The Making of *A Sand County Almanac*. In B. Callicott (Ed.), *Companion to A Sand County Almanac* (Madison: University of Wisconsin Press). The discussion, begun by Caughley, suggesting that there may be little evidence that the mountain lions actually had a beneficial effect may seem ironic, since it would seem to deny the point of view with which Leopold ended his career and which he arrived at only after a long and apparently introspective consideration. But the main thrust of my discussion is that we must seek a view that is consistent with our observations, a point of view clearly shared by Leopold. The fundamental purpose of my discussion is to help us achieve a better way to live with our environment, which was Leopold's desire as well. That mountain lions may not regulate the abundance of their prey is not an argument in favor of hunting lions. The conservation of endangered species and of biological diversity has, in recent years, expanded greatly to a much larger scientific and philosophical basis, as will be discussed later. Since science is a process and knowledge continually changes, so our interpretations and our understanding of how to achieve that goal must change.

7. This discussion is based on the excellent analysis by Caughley, "Eruption of Ungulate Populations."

8. Rasmussen, "Biotic Communities of Kaibab Plateau, Arizona."

9. G. Caughley (1970). Eruption of Ungulate Populations, with Emphasis on Himalayan Thar in New Zealand. *Ecology, 51*(1), 53–72.

10. D. Binkley, Margaret M. Moore, William H. Romme, and Peter M. Brown (2006). Was Aldo Leopold Right about the Kaibab Deer Herd? *Ecosystems, 9*(2), 227–241.

11. A. C. Martin, Herbert S. Zim, and Arnold L. Nelson (1951). *American Wildlife & Plants, A Guide to Wildlife Food Habits: The Use of Trees, Shrubs, Weeds, and Herbs by Birds and Mammals of the United States.* (New York: Dover).

12. D. Binkley, Margaret M. Moore, William H. Romme, and Peter M. Brown (2006). Was Aldo Leopold Right about the Kaibab Deer Herd? *Ecosystems, 9*, 227–241. These authors argue that the forest supervisor and others later, after the fact, changed their minds and said that Leopold must have been right and the Kaibab deer population must have gotten very high. I have made the decision to stay with their original observation, not what they decided they ought to have believed after the fact. But the reader can see an opposing interpretation to mine by reading the above journal article. These authors also argue that the history of the aspen stands, following their interpretation, is consistent with Leopold's story.

13. G. Chaucer (1957). *The Works of Geoffrey Chaucer* (F. N. Robinson, ed.) (Boston: Houghton Mifflin), p. 271. "The Book of the Duchess" was one of Chaucer's earliest works. It was written as an eulogy to and upon the death of Blanche, the first wife of John of Gaunt (whom you will remember turns up as a character in Shakespeare). Another of Chaucer's works, *The Parliament of Fowls*, is written for John of Gaunt during his deep mourning over the death of Blanche. Robinson writes that in this work Chaucer was heavily influenced by French writing of the time, suggesting that this kind of description of the wondrous balance of nature in what ecologists call a climax forest was characteristic of the time. The (loose) translation from Middle English to modern American English was done by myself (DBB).

14. Quoted in L. P. Coonen and C. M. Porter (1976). Thomas Jefferson and American Biology. *BioScience, 26*, 747.

15. William Derham (1798). *Physico-Theology: or, A Demonstration of the Being and Attributes of God, from His Work of Creation* (London: A. Strahan et al.). The original edition included the statement that this work was "the substance of sixteen sermons, preached in St. Mary-le-Bow Church, London; at the Honourable Mr. Boyle's lectures, in the years 1711 and 1712."

16. These issues are thoroughly discussed by C. J. Glacken in his excellent 1967 book *Traces on the Rhodian Shore: Nature and Culture in Western Thought from Ancient Times to the end of the Eighteenth Century* (Berkeley: University of California Press). Another classic and important book on this topic is A. O. Lovejoy (1942). *The Great Chain of Being* (Cambridge, Mass.: Harvard University Press). Other interesting analyses can be found in F. N. Egerton (1973). Changing concepts of the balance of nature. *Quarterly Review of Biology, 48*, 322–350. The discussion that follows merely outlines the history of these ideas; a reader interested in this history should refer to these references especially.

17. Derham, *Physico-Theology*, Vol. I, p. 257.

18. Ibid., pp. 257–259.

19. Quoted in F. Egerton (1973). Changing Concepts of the Balance of Nature. *Quarterly Review of Biology, 48*, 338.

20. Glacken, *Traces on the Rhodian Shore*, p. 36.

21. Ibid., p. 62.

22. Herodotus (1964). *The History of Herodotus* (trans. George Rawlinson, ed. E. H. Blakeney) (New York: Dutton), p. 148. See Glacken (1967), op. cit., for further discussion of the Greek and Roman ideas about the character of nature.

23. Cicero (1972). *The Nature of the Gods* (trans. H. C. P. McGregor) (Aylesbury: Penguin). The quotations that follow are from pp. 172–173.

24. This quote from Cicero is from a slightly different translation, which uses terminology more familiar to us. C. D. Young (trans.) (1877). *Cicero's Tusculan Disputations, also treatises on The Nature of the Gods and on The Commonwealth.* (New York: Harper and Brothers) (available online).

25. Quoted in A.O. Lovejoy, *The Great Chain of Being*, p. 50.

26. Psalm 104 (King James Version) includes the following:

> 10 He sendeth the springs into the valleys,
> which run among the hills.
> 11 They give drink to every beast of the field:
> the wild asses quench their thirst.
> 12 By them shall the fowls of the heaven have their habitation,
> which sing among the branches.
> 13 He watereth the hills from his chambers:
> the earth is satisfied with the fruit of thy works.
> 14 He causeth the grass to grow for the cattle,
> and herb for the service of man:
> that he may bring forth food out of the earth;
> . . .
> 16 The trees of the LORD are full of sap;
> the cedars of Lebanon, which he hath planted;
> 17 where the birds make their nests:
> as for the stork, the fir trees are her house.
> 18 The high hills are a refuge for the wild goats;
> and the rocks for the conies.

27. Botkin, D. B. 2004. *Beyond the Stony Mountains: Nature in the American West from Lewis and Clark to Today*, Oxford University Press, N. Y. How beliefs about nature affected the expedition are explained in greater detail than is possible here.

28. A. Pope (1734). *An Essay on Man*, quoted in Lovejoy, *The Great Chain of Being*, p. 60.

29. G. L. Leclerc (1812). *Natural History, General and Particular* (vol. 3, trans. W. Smellie) (London: C. Wood). The quotations that follow are from pp. 455–457.

30. Cicero, p. 176.

31. Cicero, p. 177.

32. Lucretius (1968). *De Rerum Natura* (trans. R. Humphries) (Bloomington: Indiana University Press), Book 5, lines 200–237, pp. 164–165.

Chapter 6

1. Lucretius (Titus Lucretius Carus) (1968). *De Re Natura*, trans. R. Humphries (Bloomington: Indiana University Press).

2. M. H. Nicolson (1959). *Mountain Gloom and Mountain Glory: The Development of the Aesthetics of the Infinite* (Ithaca, N.Y.: Cornell University Press), pp. 160–161. Reprinted in 2009 with a foreword by William Cronon as *Mountain Gloom and Mountain Glory* (Seattle: University of Washington Press). A wonderful book.

3. A. Kircher (1665). *Mundus Subterraneus*, parts of which were published as *The Vulcanoes or Burning and Fire-Vomiting Mountains, Famous in the World: With Their Remarkables* (no trans. given) (London: John Allen, 1669), p. 35 and Preface (n.p.).

4. Ibid. Quotations in this paragraph are from pp. 56, 43, and 3.

5. Ibid. Quotes in this paragraph are from pp. 55 and 58.

6. A. E. Nevala (2006). Into the 'Mouth of Hell'. *Oceanus* http://www.whoi.edu/oceanus/viewArticle.do?id=13350

7. The idea that change is intrinsic in nature can be found in the poetry of Patrick Casey, especially "Fallay et Instabilis," a poem of the "Caroline" period: "'Tis a strange thing," he wrote, "Nothing but change I see" with the sea never in rest and "mountains do sink down." Quoted in Nicolson, *Mountain Gloom and Mountain Glory*, p. 155.

It is also worth pointing out that Alfred North Whitehead distinguished an organic theory from a materialistic (read "mechanistic") theory in a more abstract way: "On the materialistic theory, there is material—such as matter or electricity—which endures. On the organic theory, the only endurances are structures of activity, and the structures are evolved.

"Enduring things are thus the outcome of a temporal process; whereas eternal things are the elements required for the very being of the process." (From Whitehead [1925/1959]. *Science and the Modern World* [New York: Mentor Books], pp. 101–102).

8. I would like to acknowledge my debt to Nicolson for her excellent discussion in *Mountain Gloom and Mountain Glory* of the history of the aesthetics of nature and the change in the perceptions of nature from the Renaissance to the Romantics, and for her general insights tying this discussion to the classical Greeks and Romans. My discussions extend hers, and hers provide a key background for this chapter.

9. Quoted in Nicolson, *Mountain Gloom and Mountain Glory*, pp. 198 and 200.

10. Ibid., pp. 199–200.

11. From *Nicolai Stenonis de Solido intra Solidum naturaliter contento dissertation in prodromus* (Florence, 1669). English translation published in 1671 as *The Prodromus to a Dissertation concerning Solids Naturally Contained within Solids* (London: Henry Oldenburg). Quoted in Nicolson, *Mountain Gloom and Mountain Glory*, p. 156.

12. P. Fletcher (1633). *The Purple Island, with the Piscatory Eclogues and Poeticall Miscellenie*. The original edition states that it was "Printed by the Printers to the Universitie of Cambridge. 1633." It was reprinted by Renascence Editions; the text was transcribed by Daniel Gustav Anderson, July 2003. It reproduces the 1633 publication. The copyright for this edition is held by the University of Oregon. Available at http://www.luminarium.org/renascence-editions/island/pi1.html#poem

13. Lucretius (1968). *De Re Natura*, trans. R. Humphries (Bloomington: University Press), Book V: quotations are selected from lines 784–836.

14. Ibid. Quotations selected from Book V, lines 247–260.

15. Nicolson, *Mountain Gloom and Mountain Glory*, pp. 160–161.

16. Moulton, G. E. (1986). *The Journals of the Lewis and Clark Expedition: August 30, 1803 to August 24, 1804*. Lincoln, NE, University of Nebraska Press.

17. T. Pownall (1949). *A Topographical Description of the Dominion of the United States*, ed. Lois Mulkean (Pittsburgh: University of Pittsburgh Press), p. 24.

18. Both the quotations about the mountains and the oceans are from Nicolson, *Mountain Gloom and Mountain Glory*, pp. 305–306.

19. *The Folly and Unreasonableness of Atheism Demonstrated from the Origin and Frame of the World* was originally published as Boyle lectures—1692, no. 1, Early English books, 1641–1700—807:23

20. The quote from E. Warren is from his poem, *Geologia*, quoted in Nicolson, *Mountain Gloom and Mountain Glory*, p. 267; the quote from R. Bentley is also from Nicolson, p. 262. Bentley's book was originally published in London and the quotations are from pp. 35–38 of the 1693 edition.

21. As a personal note, some of Andrew Marvell's poems are among my favorites, including these lines from *To His Coy Mistress*:

> Had we but world enough, and time,
> This coyness, lady, were no crime.
> We would sit down and think which way
> To walk, and pass our long love's day;
> . . .
> But at my back I always hear
> Time's winged chariot hurrying near;
> And yonder all before us lie
> Deserts of vast eternity.

22. The quote from Andrew Marvell is from his poem, "Upon the Hill and Grove at Bill-borow," *Poems and Letters*, ed. H. M. Margoliouth, (Oxford: Clarendon Press, 1927) I, p. 56. The quote from John Dennis is from Nicolson, *Mountain Gloom and Mountain Glory*, p. 277, while the quote from James Thomson is also from Nicolson, p. 335.

23. Quoted in Ibid., p. 388.

24. I discuss Thoreau and nature in much greater detail in my 2001 book *No Man's Garden: Thoreau and a New Vision for Civilization and Nature* (Island Press).

25. The nature he sought to confront is what we refer to today as biological nature, as distinct from the cosmos, as I noted earlier with different terminology. The word "nature" is applied to both terms. In this book, "nature" will always refer to biological nature, including what scientists call ecosystems, landscapes, the biosphere, populations, and species of wild organisms.

26. Thoreau, *The Maine Woods* (Moldenhauer edition), p. 63.

27. Ibid., p. 64.

28. Ibid.

29. Ibid.

30. Ibid., p. 70.

31. Thoreau had a considerable background and lifelong interest in the detailed study of nature and the collection of specimens. About 1845, when he was starting his sojourn at Walden, Thoreau began to study vegetation. He obtained a basic library of botanical guides. He learned plant taxonomy—the relatively new system of the naming and description of plants as developed by Linnaeus. He kept up with the work of his contemporaries who studied vegetation, including Asa Gray and Agassiz. Thoreau also

studied fish in the rivers, streams, and lakes of Concord. He read books about insects and became acquainted with Thaddeus Harris, an entomologist at Harvard. According to Sattelmeyer, Thoreau "had been interested in ornithology since boyhood (a family album of bird sightings survives, dating from the 1830s and containing entries by Henry, his brother John, and his sister Sophia), and he compiled a large collection of birds' nests and eggs. In 1850 he was elected a corresponding member of the Boston Society of Natural History, to which he contributed specimens and various written accounts over the years, and whose library and collections he used regularly in pursuing his studies. (His own extensive collections of Indian artifacts, birds' nests and eggs, and pressed plants went to the Society after his death.) In his work as a surveyor, he made a more intimate acquaintance with the farms, swamps, and woodlots of Concord. Gradually his townsmen, who had generally looked askance at his activities, began to come around for help in identifying plants and animals, and to bring him new items for his collections." From the introduction by R Sattelmeyer, who also was the editor of the Peregrine Smith Books (Salt Lake City), of *The Natural History Essays*, by Henry David Thoreau, 1980. pp. xix–xx. See also the excellent book compiled by Bradley Dean from Thoreau's manuscripts, Thoreau, H. D. and B. P. Dean (editor) (1996). *Faith in a Seed: The Dispersion Of Seeds And Other Late Natural History Writings*. Washington, DC, Island Press.

32. Thoreau, *Faith in a Seed* (Dean edition), Foreword, p. xiv. Dean also states that Thoreau's "observations on dispersal distances of fleshy fruits carried by animals remain important a century and a half later for their precision and anticipation of plant-animal mutualism."

33. Ibid., Foreword, p. xv.

34. Ibid., pp. 28–29.

35. Ibid., p. xiv.

36. Thoreau, *The Maine Woods* (Moldenhauer edition), p. 60.

37. F. E. Clements (1936). Nature and Structure of the Climax. *Journal of Ecology*, 22, 257. See Chapter 4 for direct quotations from Clements's writings.

38. R. H. Whittaker (1957). Recent Evolution of Ecological Concepts in Relation to the Eastern Forests of North America. *American Journal of Botany*, 44, 197–205.

39. J. E. Lovelock (1979). *Gaia: A New Look at Life on Earth* (New York: Oxford University Press). The first quote is from p. xii, the second from p. 127.

40. Ibid., p. 57. See also Lovelock, J. (2006). *The Revenge of Gaia: Why the Earth Is Fighting Back – and How We Can Still Save Humanity*. New York, Perseus.

41. Ibid., p. 62.

42. Ibid., pp. ix–x.

43. Ibid., p. 53.

Chapter 7

1. Edward Alden Jewell (Dec. 11, 1927). 'Machines, Machines!' The Futurist's Cry. *New York Times*, Section V, p. 13. Jewell was an art critic for the *New York Times*.

2. David Foster (June 17, 1987). *Santa Barbara News-Press*, p. B5.

3. The rise of the mechanical paradigm led to a change in the perception of God, beginning with Descartes's argument that God created nature and the human mind according to the same rational laws; thus it is possible to know nature through reason. The philosopher B. Callicott has written to me that later scientists-theologians described their activity as "thinking God's thoughts after Him." In such ways, the mechanical view of nature began to change the perception of God from the divine patriarch, which is a more organic image, to the cosmic engineer.

An interesting discussion of a major change in our ideas about machinery, mechanics, and technology can be found in Frederick Turner (November 1984). Escape from Modernism. *Harper's*, pp. 47–55. Turner refers to a present-day "crisis of materialism," analogous to the "crisis of Christianity" that occurred 400 years ago.

4. M. H. Nicolson (1959). *Mountain Gloom and Mountain Glory: The Development of the Aesthetics of the Infinite* (Ithaca, N.Y.: Cornell University Press), p. 161.

5. Ibid., p. 161. In *Science and the Modern World*, Alfred North Whitehead provided one of the classic discussions of the transition of ideas from the pre-scientific to the scientific, and the impact of the new physics on philosophy; he wrote that "the nature-poetry of the romantic revival (in the nineteenth century) was a protest on behalf of the organic view of nature, and also a protest against the exclusion of value from the essence of matter of fact" (Whitehead [1925/1959]. *Science and the Modern World* [New York: Mentor Books], p. 90).

6. R. Powers (2005). *Mark Twain: A Life* (New York: Free Press), p. 521.

7. J. Mokyr (1990). *The Lever of Riches* (New York: Oxford University Press).

8. M. Hale (1677). *The Primitive Origination of Mankind Considered and Examined According to the Light of Nature* (London: W. Godbid), p. 211. Quoted in F. N. Egerton (1973). Changing Concepts of the Balance of Nature. *Quarterly Review of Biology, 48,* 331.

9. See, for example, the excellent historical analyses of the changes in the extent of Alpine glaciers in E. Le Roy Ladurie (1971). *Times of Feast, Times of Famine: A History of Climate Since the Year 1000* (Garden City, N.Y.: Doubleday).

10. Ibid.

11. A. von Humboldt (1851). *Cosmos*, vol. 3 (New York: Harper & Brothers), pp. 9–10.

12. A. von Humboldt (1886). *Views of Nature* (London: Bell), p. 288.

13. G. P. Marsh (1864/1967). *Man and Nature*, ed. D. Lowenthal (Cambridge, Mass.: Harvard University Press), pp. 29–30.

14. G. W. Wetherill and C. L. Drake (1980). The Earth and Planetary Sciences. *Science, 209,* 96–104.

15. C. D. Steward (1907). A Race on the Missouri. *The Century Magazine,* quoted in B. A. Botkin (1944). *A Treasury of American Folklore* (New York: Crown).

16. I discuss, with more stories and analysis, the dynamics of the Missouri River and attempts to control the river, in Botkin, D. B. 2004. *Beyond the Stony Mountains: Nature in the American West from Lewis and Clark to Today*, Oxford University Press, N. Y.

16. William H. Allen (1993). The Great Flood of 1993. *BioScience, 43*(11), 732–733.

17. George L. Church (July 26, 1993). Untitled. *Time Magazine,* p. 27.

18. Richard Price "'Some never will' get over the flood," *USA Today*, August 9, 1993, p. 1.

19. Mumford, L. (1968). *The City in History: Its Origins, Its Transformations, and Its Prospects.* New York, Harvest Books; and Mumford, L. (1974). *Pentagon Of Power: The Myth Of The Machine, Vol. II.* New York, Harvest Books.

20. Saylor, Steven, 1993, *Catilina's Riddle.* New York: St. Martin's Minotaur.

21. Antoine de Saint-Exupéry (2002). *Wind, Sand and Stars* (Harvest Books), p. 43. Since I first read this when I was in high school, it has been one of my favorite books. It and Joseph Conrad's novels were the greatest influence on me about how I wanted to live my life, and have tried to do so ever since high school.

22. Ibid., p. 46.

23. Hale, *Primitive Origination of Mankind,* p. 211.

24. Quoted in Nicolson, *Mountain Gloom and Mountain Glory,* p. 153. As another example, she cites Joseph Blancanus, *Sphaera Mundi* (1620).

Chapter 8

1. Antoine de Saint-Exupéry (2002). *Wind, Sand and Stars* (Harvest Books), p. 45.
2. Calvin Tomkins (Jan. 8, 1966). In the Outlaw Area [a profile of Buckminster Fuller]. *The New Yorker.*
3. F. Turner (Nov. 1984). Escape from Modernism. *Harper's*, p. 48.
4. *The Columbia Viking Desk Encyclopedia*, S. V. "bacteria."
5. S. Sonea and M. Panisset (1983). *A New Bacteriology* (Boston: Jones and Bartlett), pp. 8–9.
6. L. Margulis (1983). Preface to S. Sonea and M. Panisset, *A New Bacteriology* (Boston: Jones and Bartlett), p. viii.
7. Sonea and Panisset, *A New Bacteriology*, pp. 1–8.
8. Kalm, *Travels in North America*, pp. 369–370. I originally recounted this story in D. B. Botkin (1977). Life and Death in a Forest: The Computer as an Aid to Understanding. In *Ecosystem Modeling in Theory and Practice*, ed. C. A. S. Hall and J. W. Day, Jr. (New York: Wiley).
9. The first papers we published on the JABOWA computer model of forest growth, establishing it as original with us, were: Botkin, D.B., J.F. Janak and J.R. Wallis. 1970, A simulator for northeastern forest growth: a contribution of the Hubbard Brook Ecosystem Study and IBM Research, *IBM Research Report 3140*, Yorktown Heights, NY. 21 pp; Botkin, D.B., J.R. Janak and J.R. Wallis, 1971, A simulation of forest growth, pp. 812–819, In: Proceedings of the Summer Computer Simulation Conference, Board of Simulation Conferences, Denver, CO.; Botkin, D.B., J.R. Janak and J.R. Wallis, 1972A, Rationale, limitations and assumptions of a northeast forest growth simulator, *IBM J. of Research and Development 16*: 101–116; and Botkin, D.B., J.F. Janak and J.R. Wallis, 1972B, Some ecological consequences of a computer model of forest growth, *J. Ecology 60*: 849–872.

The model has been used, and continues to be used, widely around the world, sometimes renamed, but the algorithms are, in most cases, identical to those of the original model. One example is: Shugart, H.H., 1984, A Theory of Forest Dynamics, Springer - Verlag, N.Y. For the record, Shugart has claimed that his forest model is original with him, but it is not.

It's an interesting story about how the cooperative research with Janak and Wallis came about. My first faculty appointment was at the Yale School of Forestry (later Forestry and Environmental Studies), and I was beginning to think through how I would create a computer model of forest growth. A Yale colleague, Tom Siccama, and I began to think about various approaches. At that time, IBM was being sued by the federal government for bundling software and hardware, which was considered a violation of anti-monopoly laws. IBM decided it needed to improve its public image and set up eight summer programs that would involve socially beneficial applications of computers. The Dean of the Yale Forestry School got a call from IBM asking if anyone had any ideas of what might be done. The information came to me, and I thought, perfect! I went over to the IBM lab and gave a presentation of my ideas about how to develop the computer model of forests. Janak and Wallis had selected that topic to attend, and immediately said that it sounded good and feasible. We worked together for several years, from my point of view a delightful cooperation both personally and professionally.

As a side objective, IBM let us develop a real-time video display of the forest growth with touch-screen capabilities—if you wanted to cut down a tree, you touched it and it was cut and disappeared from the screen and the computer program. My talented graduate student, Al Doolittle, was hired to do the video—touch screen program. It took two computers to do this, an IBM 360 (mainframe) to run the model and a smaller computer that took the output and created the visual display. Today, all of this runs on a laptop, but to my knowledge this was the first use of interactive display in ecology and probably one of the first applications of interactive displays ever.

10. K. E. F. Watt (1962). Use of mathematics in population ecology. *Annual Review of Entomology, 7,* 243–252. Watt was concerned with some of the things that concern me and are still lacking in ecological theory. Here are a few relevant quotes from this paper (minus the reference citations):

> The historical development of physics has been characterized by a high degree of integration between theory and experiment. In quantum mechanics, for example, the closeness of this integration in the last few decades has been amazing. On the other hand, many writers have noticed the lack of such integration in population ecology. This is an extremely important point, since the single factor most necessary for rapid evolution of a branch of science is this close integration of theory and empirical work. It is clear that sophisticated theory by itself is not adequate since population ecology has had such theories for about four decades. These theories have never had an impact on the rate of scientific evolution like that of the physical theories which have often been tested weeks and even hours after they were published. It is also clear that elaborate experimental and observational studies, not directed and interpreted by theories, do not greatly advance a science. The dead files of ecologists abound with vast sets of unused data which were obtained from studies fundamentally populational in intent. This author knows of several dozen such sets.
>
> We must conclude that no matter how active a theoretical branch of science becomes, it will ultimately run into a cul-de-sac if unrelated to the here-now world of our sense impressions. On the other hand, a field program will most certainly also be of limited usefulness if unrelated to theory. One aim of this review is to point out types of activity which can increase the degree of integration in theory and practice.

Perhaps most important to our discussions is that Watt notes the dependence on steady-state models, but believed ecologists would move beyond these.

> Many of the earlier ecological models were designed for steady-state, or equilibrium, conditions in which large-scale changes of the physicochemical environment did not occur. However, as more and more data become available which show that very large-order fluctuations are influenced by weather, steady-state models may decline in popularity.

11. D. B. Botkin (1993). *Forest Dynamics: An Ecological Model* (New York: Oxford University Press). (Out of print but available from me as a pdf file.)

12. R. Leemans and I. C. Prentice (1987). Description and Simulation of Tree-layer Composition and Size Distributions in a Primaeval *Picea-Pinus* Forest. *Vegetatio, 69,* 147–156; D. C. West, H. H. Shugart, and D. B. Botkin, eds. (1981). *Forest Succession: Concepts and Applications* (New York: Springer-Verlag). The idea that computer simulation can be used to mimic nature in any realistic way was not well known in the late twentieth century to the public or necessarily accepted among scientists in ecology. But the tables have turned in our century; it is more and more common for computer forecasts to be treated as the truth, to replace observations. In some situations, funds are available for computer simulation but not for the empirical research necessary to validate those simulation programs.

13. Ngugi, Michael R. and Daniel B. Botkin, 2011, "Validation of a multispecies forest dynamics model using 50-year growth from Eucalyptus forests in eastern Australia," *Ecological Modelling. 222:* 3261– 3270.

14. Two of his books about system dynamics are J. W. Forrester (1961). *Industrial Dynamics* (Cambridge, MA: MIT Press) and J. W. Forrester (1973). *World Dynamics* (Cambridge: MIT [Wright-Allen] Press).

15. D. H. Meadows, Dennis L. Meadows, Jorgen Randers, and William W. Behrens (1972). *Limits to Growth* (New York: Universe Books).

16. G. E. Yule (1920). The Wind Bloweth Where It Listeth. *Cambridge Review, 41*, 184.

17. P. Frank (1947). *Einstein, His Life and Times* (G. Rosen, trans.) (New York: Alfred A. Knopf), p. 208.

18. As an example, radioactive isotopes will decay, but when any particular atom will undergo such decay seems to be a matter of chance whose probability can be determined. Einstein could not accept this interpretation. An alternative is that there are beneath these observations some other, completely deterministic rules that govern events such as the decay of an atom of a radioactive isotope, and therefore such events only appear random because of the level at which we observe them.

The uncertainty discussed in quantum theory has to do not only with the small size and high speed of particles, but more fundamentally with the idea that observations require the exchange of energy between the observer and the observed, that this exchange of energy affects the system being observed, and that subatomic particles are small enough for this energy exchange to have a significant effect. Thus, in an attempt to determine the position and velocity of a very small particle, the observer strikes the particle with a photon, which then changes either the position or the velocity or both. This now-fundamental idea, the Heisenberg uncertainty principle, is today familiar to students of physics. The amount of uncertainty in quantum mechanics can be calculated.

While this is an important idea in twentieth-century physics, it is not my intention to make an exact analogy with it. I am not trying to claim that at the level at which we observe wolves and trees, the process of observation necessarily has a significant effect on the state of the system we are observing. The Heisenberg uncertainty effect is too small for these middle-scale systems and for our means of observation. The point I wish to make is that quantum theory opened up a different perception of nature, nature as fundamentally uncertain, and thus provides a metaphor that may make it easier for us to accept a change in our perception of biological nature.

Clearly, my discussion is not meant to be a definitive treatment of any of the philosophical issues raised by quantum physics. My purpose is merely to show some of the connections between twentieth-century physics and the life sciences in regard to the fundamental ideas about nature. Other scholarly books deal with these aspects of the philosophy of science. One of the more interesting and easy to read is Erwin Schrödinger (1952). *Science and the Human Temperament* and *Science and Humanism* (London: Cambridge University Press).

19. There are recent attempts among philosophers to grapple with these difficult issues: for example, B. Callicott (1985). Intrinsic Value, Quantum Theory, and Environmental Ethics. *Environmental Ethics, 7*, 257–275. Again, I am not attempting in this book to make a definitive analysis of these philosophical issues, but more simply to suggest through examples the changes that are taking place in our ideas.

20. M. A. Kominz and N. G. Pisias (1979). Pleistocene Climate: Deterministic or Stochastic? *Science, 204*, 171–172.

21. The information on the whooping crane (*Grus americana*), is from R. S. Miller and D. B. Botkin (1974). Endangered Species: Models and Predictions. *American Scientist, 62*, 172–181; and R. P. Allen (1952). *The Whooping Crane: National Audubon Society Research Reports* no. 3 (New York: National Audubon Society).

22. T. Stehn (2009). *Whooping Crane Recovery Activities*. U. S. Fish and Wildlife Service. (Aransas, TX: U. S. Government), 27. I thank Tom Stehn for his excellent work. I understand he retired in 2012.

23. Caesar, *The Civil Wars*. 3, 68 quoted in A. Goldsworthy (2006). *Caesar: Life of a Colossus* (New Haven: Yale University Press), p. 405.

24. Branches of mathematics that deal with stochastic processes have advanced rapidly in the twentieth century and have, to a limited but growing extent, begun to influence the science of ecology. Thus they are also supporting the shift away from a mechanical metaphor of nature. There is an extensive scientific literature on stochastic processes in ecology. It is not my purpose here to provide a definitive review of that literature, but merely to introduce the general concepts.

As mentioned briefly in the chapter, there are curious deterministic systems that appear random in the sense that future states cannot be predicted simply from present states. For example, there are some mathematical equations whose future values can be determined only by calculating them one at a time. The relevance of this to ecological phenomena was discussed in the early 1970s by the mathematician J. A. Yorke (S. A. Woodin and J. A. Yorke [1975]. Disturbance, Fluctuating Rates of Resource Recruitment, and Increased Diversity. In *Ecosystem Analysis and Prediction*, ed. S. A. Levin [Philadelphia: Society for Industrial and Applied Mathematics]) and by R. May (R. May [1974]. Biological Populations with Nonoverlapping Generations. *Science*, 186, 645–647). Some of the methods used to generate "random numbers" by computers make use of such equations.

25. This is the work of Dr. Mel Manalis of the University of California, Santa Barbara.

26. As an example, I worked with a expert mathematician in the field of stochastic processes; we developed a model of a single population of elephants that took into account in an explicit manner that births and deaths are matters of chance. The analytic mathematics (the pencil-and-paper math) gave quite interesting results. The pencil-and-paper mathematics led to the framework of a model of the elephant populations, but only when this framework was translated into a computer program could the approach be used for realistic situations with an environment that changes over time. The pencil-and-paper part of the work was published as L. S. Wu and D. B. Botkin (1980). Of Elephants and Men: A Discrete, Stochastic Model for Long-lived Species with Complex Life Histories. *American Naturalist, 116*, 831–849.

Chapter 9

1. Plutarch, *Concerning the Face Which Appears in the Orb of the Moon*, Quoted in Glacken, *Traces on the Rhodian Shore*, p. 74.

2. G. E. Hutchinson (1954). The Biochemistry of the Terrestrial Atmosphere. *The Earth as a Planet*, ed. G. P. Kuiper (Chicago: University of Chicago Press), p. 372.

3. M. S. Gordon, G. A. Bartholomew, A. D. Ginnell, G. B. Jorgensen, and F. M. White (1977). *Animal Physiology* (New York: Macmillan).

4. G. E. Hutchinson (1950). The Biogeochemistry of Vertebrate Excretion. *Bulletin of the American Museum of Natural History, 96*, 1–554.

5. L. J. Henderson (1913). *The Fitness of the Environment* (Boston: Macmillan).

6. D. B. Botkin and E. A. Keller (1987). *Environmental Studies: The Earth as a Living Planet* (Columbus, Ohio: Merrill).

7. Quoted in C. Lyell (1832). *Principles of Geology; Being an Attempt to Explain the Former Changes of the Earth's Surface, by Reference to Causes Now in Operation*, Vol. II (London: John Murray), p. 190. Sir Charles Lyell lived from 1797 to 1875. His book is usually accepted as the first modern book on the science of geology. Sedgwick, by the way, although one of Darwin's professors and his companion on some natural history trips and

a correspondent when Darwin was on the voyage of the *Beagle*, never accepted Darwin's theory of biological evolution.

8. Ibid., pp. 190–191.

9. Ibid., pp. 191–192. An excellent modern discussion of the factors that determine erosion and the situations when vegetation can play an important role and when it cannot is R. C. Sidle, A. J. Pearce, and C. L. O'Loughlin (1985). *Hillslope Stability and Land Use*, Water Resources Monograph Series, no. 11 (Washington, D. C.: American Geophysical Union).

10. Lyell, *Principles of Geology*, p. 192.

11. Ibid.

12. Estimates of world organic matter are from various sources; all are rather rough estimates. In this update, I have used J-Y. Fang and Z. M. Wang (2001). Forest biomass estimation at regional and global levels, with special reference to China's forest biomass. *Ecological Research, 16*(3), 587–592, for a ballpark figure of forest biomass per square kilometer, and then the area of forests of the world from various sources. Since forests are the largest single storage of living carbon, this provides a starting point, giving an estimate of about 200 billion metric tons. The world total is unlikely to be more than five times that, and 1,000 billion metric tons is a typical estimate, which is published widely. It isn't especially accurate or precise but is good enough for our purposes here.

13. The mass of the Earth is 6×10^{24} kg, or 6 billion trillion metric tons; the total living biomass is estimated to be on the order of 1.2×10^{15} kg or 1,000 billion metric tons. 1.2×10^{15} kg divided by 6×10^{24} kg is 0.2×10^{-9}. The mass of the atmosphere is 5×10^{18} kg. Dividing total biomass (1.2×10^{15} kg) by the mass of the atmosphere (5×10^{18} kg) gives 0.24×10^{-3}.

14. L. K. Nash (1952). *Plants and the Atmosphere* (Cambridge, Mass.: Harvard University Press). Joseph Priestly (1733–1804) was a theologian as well as a scientist; born in England, he immigrated to the United States in 1794.

15. S. A. Tyler and E. S. Barghoorn (1954). Occurrence of Structurally Preserved Plants in Precambrian Rocks of the Canadian Shield. *Science, 119*, 606–608.

16. L. Margulis and J. E. Lovelock (1974). Biological Modulation of the Earth's Atmosphere. *Icarus, 21*, 471–489. Additional interesting information about the time when only prokaryotes lived on the Earth can be found in L. Margulis (1982). *Early Life* (Boston: Science Books International) and L. Margulis (1981). *Symbiosis in Cell Evolution* (San Francisco: W. H. Freeman).

17. S. M. Awramik, J. W. Schopf, and M. R. Walter (1983). Filamentous fossil bacteria from the Archean of Western Australia. *Precambrian Research, 20*, 357–374.

18. Whether the 3.5-billion-year-old stromatolites were produced by living things is still debated, but the weight of evidence supports a biological origin. Prof. Stan Awramik, the discoverer of the oldest fossils, writes that "stromatolites are composed of micrite or clay-size grains. The main stromatolite structures at Shark Bay, however, may have originally formed a few thousand years ago. Growth is still going on, but there may have been one or more hiatuses, and the microbiota is dominated by cyanobacteria AND diatoms. In general, however, the Shark Bay stromatolites are looked at as the modern analog for ALL ancient stromatolites." (Email, Sept. 9, 2010). See also: A. C. Allwood, John P. Grotzingerc, Andrew H. Knoll, Ian W. Burch, Mark S. Anderson, Max L. Coleman, and Isik Kanik (2009). Controls on development and diversity of Early Archean stromatolites. *Proceedings of the National Academy of Sciences USA, 106*(24), 9548–9555, which indicates a biological origin.

19. L. Margulis and R. Guerrero (1989). From Planetary Atmospheres to Microbial Communities: A Stroll Through Space and Time. In *Changing The Global Environment: Perspectives on Human Involvement*, ed. D. B. Botkin, M. Caswell, J. E. Estes, and A. Orio (Orlando, Fla.: Academic Press).

20. Except, of course, a lot of the free hydrogen would escape into space from Earth's orbit, because our planet does not have enough gravitational pull—it's too small a planet— to hold on to free hydrogen very well.

21. For a history of the Earth's atmosphere, see J. C. G. Walker (1977). *Evolution of the Atmosphere* (New York: Macmillan). The early concentration of oxygen and hydrogen is discussed in J. C. G. Walker (1977). Oxygen and Hydrogen in the Primitive Atmosphere. *Pure and Applied Geophysics, 116,* 222–231. The preceding two paragraphs are from D. B. Botkin and E. A. Keller (1982). *Environmental Studies: The Earth as a Living Planet* (Columbus: Merrill), p. 57.

22. D. M. Gates (1979). *Biophysical Ecology* (New York: Springer-Verlag).

23. Ibid.

24. The heading of this section is in deference to G. Evelyn Hutchinson's wonderful 1965 book, *The Ecological Theater and the Evolutionary Play* (Yale University Press).

25. P. Westbroek (1991). *Life as a Geological Force: Dynamics of the Earth* (New York: Norton [Commonwealth Fund Book Program]).

26. Taylor's contribution to the theory of plate tectonics is reported in B. Skinner, S. Porter, and D. B. Botkin (1999). *The Blue Planet* (New York: John Wiley & Sons).

27. Wegener's book was published in German in 1922, but he had proposed the idea in a 1915 scientific paper. His book has been variously reprinted, such as A. Wegener (1966). *The Origin of Continents and Oceans.* (New York: Dover; translated from the fourth revised German edition by John Biram).

28. G. G. Simpson (1943). Mammals and the Nature of Continents. *American Journal of Science, 241,* 1–31.

29. P. Westbroek (1989). The Impact of Life on the Planet Earth. Some General Considerations. In *Changing the Global Environment: Perspectives on Human Involvement,* ed D. B. Botkin, M. F. Caswell, J. E. Estes, and A. A. Orio (Boston: Academic Press, pp. 37–48). The material quoted here is from pp. 37–38.

30. Ibid., pp. 39–40.

31. D. B. Botkin, M. B. Davis, J. Estes, A. Knoll, R. V. O'Neill, L. Orgel, L. B. Slobodkin, J. C. G. Walker, J. Walsh, and D. C. White (1986). *Remote Sensing of the Biosphere* (Washington, D. C.: National Academy of Sciences).

32. The evolution of diatoms provides another example of a one-way process in the biosphere. Diatoms are single-celled algae that have a hard shell made of silicon. When diatoms evolved, they were the first major group of organisms to make use of the dissolved silicon in the oceans. Their hard silicon shells provided protection against enemies and in this way were a major evolutionary advance. The diatoms made use of a previously unused resource and gained an evolutionary advantage from it. There were two results: the local biological result was the evolution of many kinds of diatoms; the global (biospheric) result appears to have been a change in the cycling of carbon. Diatoms live at the surface of the ocean, where there is enough light for photosynthesis; but when they die, their shells sink to the ocean bottom. This transport of diatom shells from the ocean surface downward produced a new major storage of silicon and carbon on the ocean floor, creating diatomaceous earth, which we know as chalk.

The evolution that resulted in calcareous shells and internal skeletons containing calcium also resulted in major new opportunities for animals because the shells provide protection against predators, among other advantages. This biological use of calcium led to an increase in the production of limestone and in the storage of large quantities of carbon in limestone, which indirectly allowed an additional buildup in the atmosphere of a large amount of free oxygen. Just as in the case of the evolution of diatoms, the evolution of calcium-based shells and skeletons led to a new wave of biological evolution and to a major change in the biosphere.

33. For a fascinating discussion about how life could have evolved, consistent with the laws of thermodynamics and the chemistry of the universe, see J. Trefil, Harold J. Morowitz, and Eric Smith (2009). The Origin of Life. *American Scientist, 97*(May–June), 206–213, as well as H. J. Morowitz (1999). A theory of biochemical organization, metabolic pathways, and evolution. *Complexity, 4,* 39–53; E. Smith and H. J. Morowitz (2004). Universality in intermediary metabolism. *Proceedings of the National Ácademy of Sciences of the USA, 101,* 13168–13173; H. J. Morowitz and E. Smith (2007). Energy flow and the organization of life. *Complexity, 13,* 51–59.

34. Lyell, Principles of Geology, p. 189.

35. Gulliver, Austin F.; Hill, Graham; Adelman, Saul J. (1994), "Vega: A rapidly rotating pole-on star", *The Astrophysical Journal* **429** (2): L81–L84.

36. J. Hansen, P. K. Makiko Sato, David Beerling, Robert Berner, et al. (2008). Target Atmospheric CO_2: Where Should Humanity Aim? *The Open Atmospheric Science Journal, 2,* 217–231.

37. IPCC (2007). Summary for Policymakers. In: *Climate Change 2007: The Physical Science Basis. Contribution of Working Group I to the Fourth Assessment Report of the Intergovernmental Panel on Climate Change,* eds. S. Solomon, D. Qin, M. Manning, Z. Chen, M. Marquis, K. B. Avery, M. Tignor, and H. L. Miller (Cambridge and New York: Cambridge University Press).

38. Lewis Thomas (Nov. 25, 1913–Dec. 3, 1993). *Dictionary of Literary Biography: Volume 275: Twentieth-Century American Nature Writers: Prose, 2003.*
Ann Woodlief, Virginia Commonwealth University. http://www.vcu.edu/engweb/ LewisThomas.htm

39. L. Thomas (1975). *The Lives of a Cell: Notes of a Biology Watcher* (New York: Bantam Books), p. 4.

40. J. E. Lovelock (1979). *Gaia: A New Look at Life on Earth* (New York: Oxford University Press). The first quote is from p. xii, the second from p. 127.

Lovelock adds a third Gaian principle, which concerns a more technical matter: "Gaian responses to changes for the worse must obey the rules of cybernetics, where the time constant and the loop gain are important factors. Thus the regulation of oxygen has a time constant measured in thousands of years" (p. 127).

Lovelock's ideas are in one way a return to the use of the organic model for nature. His statement that "Gaia has vital organs at the core" is one example, but of even greater interest is his statement that salinity control "may be a key Gaian regulatory function," which he explains in terms of the human kidney (p. 57). However, his ideas are a blending of an organic and a mechanistic view, and they include a use of organic, mechanistic, and computer-based metaphors to explain the functionings of the biosphere. In his book, the homeostasis of the biosphere is explained in terms of the temperature regulation of an electric oven, of the functioning of a cybernetic system (that is, calculating device, a computer), and an analogy with the human body, as mentioned. In some places, there is a tendency to view the biosphere only mechanistically, as in "the only difference between non-living and living systems is in the scale of their intricacy, a distinction which fades all the time as the complexity and capacity of automated systems continue to evolve" (p. 62).

A problem in discussions of this kind is that it is difficult, if not impossible, to avoid presenting the ideas as though there were purposefulness on the part of life. This difficulty is a well-known flaw in scientific discussions, known as a "teleological argument"—giving a sense of intentionality to objects that cannot have consciousness, desire, and purpose. Lovelock writes that "occasionally it has been difficult, without excessive circumlocution, to avoid talking of Gaia as if she were known to be sentient. This is meant no more seriously than is the appellation 'she' when given to a ship by those who sail in her, as a recognition that even pieces of wood and metal when specifically designed and assembled may achieve a composite identity with its own characteristic signature, as distinct from being the mere sum of its parts" (pp. ix–x). The book is sometimes metaphysical, going beyond the issue of scientific observations (is the biosphere biologically regulated?) to the question of whether this regulation is purposeful, as in Lovelock's discussion of the possibility that "the Earth's surface temperature is actively maintained at an optimum by and for the complex entity which is Gaia, and has been so maintained for most of her existence" (p. 53).

41. A. J. Watson and J. E. Lovelock (1983). Biological homeostasis of the global environment: the parable of Daisyworld. *Tellus B International Meteorological Institute*, 35(4), 286–289.

42. *Los Angeles Times*, Sunday, March 7, 1993, p. A21.

43. Charles Elton (1930). *Animal Ecology and Evolution* (New York: Oxford University Press).

44. The African folksong, "Sambo Caesar," can be heard on the CD The Alan Lomax Collection Sampler, originally released in April 1997 by Rounder Records, and still available today. Alan Lomax (1915 – July 19, 2002) was one of the twentieth century's most important folklorist, including his many recordings of folk music from around the world.

45. D. B. Botkin, ed. (1986). *Remote Sensing of the Biosphere* (Washington, D. C.: National Academy Press).

The term *biosphere* is used in this book to mean the planetary system that includes and sustains life. The term has had other meanings; it was coined in the late nineteenth century by Edward Suess to refer to the total amount of organic matter on the Earth—which we now refer to as "total biomass"—and it has been used to mean simply the place where life is found on the Earth (that is, the extent of the distribution of life). The term *ecosphere* is synonymous with biosphere.

Chapter 10

1. P. Kalm (1966). *Travels in America: The America of 1750*, ed. A. B. Benson, English version 1770 (New York: Dover), pp. 300 and 309.

2. The first recorded observation of the giant sequoia (*Sequoiadendron giganteum* (Lindl.) *Buchholz*) was made in 1839 by Zenas Leonard of Clearfield, Pennsylvania, but his obscure publication was little known (R. J. Hartesveldt, J. T. Harvey, H. S. Shellhammer, and R. E. Stecker [1975]. *The Giant Sequoia of the Sierra Nevada* [Washington, D. C.: Government Printing Office]). The report published in 1852 in the *Sonora Herald* brought attention to the tree. Too big to move and display in their entirety, sequoia trees were cut apart and parts of them were displayed. The bark of one sequoia 30 feet in diameter was removed completely to a height of 120 feet (thereby killing the tree) and reconstructed at the Crystal Palace in England in the 1850s (p. 7). The sequoia were also actively logged for timber, even though much of the fragile wood of the tree was destroyed when the huge trunks crashed to the ground.

3. That national parks were America's answer to the cultural trappings of Europe is the idea of Alfred Runte, a historian of American national parks, as discussed in *National Parks: The American Experience* (First edition: Lincoln: University of Nebraska Press, 1979; 4th Edition [Paperback]. New York City, Taylor Trade Publishers, Inc.). The quotations from Clarence King and Horace Greeley in this paragraph are from the 1st edition, pp. 21–22. Readers interested in the national parks and their role in American society would enjoy another of Al Runte's books, Runte, A. (2011). *Trains of Discovery: Railroads and the Legacy of Our National Parks* (5th Edition [Paperback]. Boulder, CO, Roberts Rinehart.

4. Yosemite Valley became a California state park in 1864; a national park was established there in 1890 (Ibid., p. 16, plate 1).

5. Hartesveldt, Shellhammer, and Stecker, *Giant Sequoia of the Sierra Nevada;* and R. J. Hartesveldt (1968). Fire Ecology of the Giant Sequoia: Controlled Fires May Be One Solution to Survival of the Species, *Natural History, 73*, 12–19.

6. Jan Van Wagtendonk, Sequoia National Park scientist, personal communication.

7. Anonymous (2012). *Sequoia & Kings Canyon National Park Fire and Fuels Management Plan: 2011 Annual Update*. U. S. D. o. I. National Park Service. Washington, DC, National Park Service, U. S. Department of Interior.

8. C. G. Lorimer et al. (2009). Presettlement and modern disturbance regimes in coast redwood forests: Implications for the conservation of old-growth stands. *Forest Ecology and Management, 258*(7), 1038–1054.

9. Some people believe that doing nothing is the same as having no policy, but, on the contrary, a policy of intentional inaction is a definite policy, while the absence of any plan can lead to any sort of effect on the environment.

10. National Park Service (2008). Wildland Fire in Yellowstone. http://www.nps.gov/yell/naturescience/wildlandfire.htm

11. Ibid.

12. Julie Johnson (Sept. 14, 1988). Administration Rethinks Policy On Forest Fires. *New York Times*.

13. Philip Shabecoff (Sept. 15, 1988). Park Service Plans to Revise Fire Recovery Policy. *New York Times*.

14. Chase, A. (1988). A Voice From Yellowstone; 'Neither Fire Suppression Nor Natural Burn Is A Sound Scientific Option.' New York City: New York Times.

15. P. J. White and Robert A. Garrott (2005). Northern Yellowstone elk after wolf restoration. *Wildlife Society Bulletin, 33*(3), 942–955.

16. Elk populations in Yellowstone National Park in the twenty-first century so far have numbered between 5,000 and 10,000, with the peak reached in 2004–2005 and the 2009 population at 6,070. In late 2009, the bison population in the park numbered about 3,000 ([2010]. *Yellowstone Science, 18*(1), 3–4). D. B. Tyers (2008). Moose Population History on the Northern Yellowstone. *Yellowstone Science, 16*(1), 3–11.

17. Information on other species are from National Park Service, "Yellowstone Elk," http://www.nps.gov/yell/naturescience/elk.htm; "Yellowstone Bison" http://www.nps.gov/yell/naturescience/bison.htm;"Wolves of Yellowstone," http://www.nps.gov/yell/naturescience/wolves.htm

18. Covington's research has been reported on in a number of publications, including the following:

David Malakoff (2002). Arizona ecologist puts stamp on forest restoration debate. *Science, 297*, 2194–2196.

R. F. Noss, P. Beier, W. W. Covington, R. E. Grumbine, D. B. Lindenmayer, J. W. Prather, F. Schmiegelow, T. D. Sisk, and D. J. Vosick (2006). Recommendations for Integrating Restoration Ecology and Conservation Biology in Ponderosa Pine Forests of the Southwestern United States. *Restoration Ecology, 14*, 4–10.

P. Z. Fulé, J. E. Crouse, A. E. Cocke, M. M. Moore, and W. W. Covington (2004). Changes in canopy fuels and potential fire behavior 1880–2040: Grand Canyon, Arizona. *Ecological Modelling, 175*, 231–248.

W. W. Covington (2003). Restoring ecosystem health in frequent-fire forests of the American West. *Ecological Restoration, 21*(1), 7–11.

W. W. Covington, P. Z. Fulé, S. C. Hart, and R. P. Weaver (2001). Modeling ecological restoration effects on ponderosa pine forest structure. *Restoration Ecology, 9*(4), 421–431. W. W. Covington (2000). Helping western forests heal. *Nature, 408*, 135–136.

19. John C. Campbell et al. (2007). *Long-term Trends from Ecosystem Research at the Hubbard Brook Experimental Forest*. U. S. Dept. of Agriculture, Forestry Service. General Technical Report NRS-17.

20. The young pines growing under the old red oak tree near Pellston, Michigan, suggested that there is a considerable lag in the response of a mature forests to climatic change,

which is another way that forests are different from the simpler ideas that were dominant during the first half of the twentieth century, when it was assumed that natural vegetation would respond rapidly to changes in climate.

21. R. A. Sedjo and D. B. Botkin (1997). Using Forest Plantations to Spare the Natural Forest. *Environment, 39*(10), 14–20.

Chapter 11

1. Endangered Species Act of 1973. U. S. Congress.

2. D. H. Fischer (1989). *Albion's Seed: Four British Folkways in America* (New York: Oxford University Press).

3. Endangered Species Act of 1973. U.S. Congress

3. The work that I did for the Olympic Natural Resources Center was an outgrowth of the previous project I had directed for the state of Oregon, to provide an objective scientific analysis of the relative effects of forest practices on salmon. The final report of that study is Botkin, D.B., K. Cummins, T. Dunne, H. Regier, M. J. Sobel, and L. M. Talbot, 1995, *Status and Future of Anadromous Fish of Western Oregon and Northern California: Findings and Options*, Center for the Study of the Environment, Santa Barbara, CA. It is available online at www.naturestudy.org.

4. M. Norton-Griffiths (1978). *Counting Animals.* No. 1 of a series of handbooks on techniques currently used in African wildlife ecology, Ed. J. J. R. Grimsdel. (Nairobi: African Wildlife Foundation).

5. My books about the Lewis and Clark expedition are D. B. Botkin (1995). *Our Natural History: The Lessons of Lewis and Clark* (Putnam, N.Y., paperback published fall 1996, republished by Oxford University Press); D. B. Botkin (1999). *Passage of Discovery: The American Rivers Guide to The Missouri River of Lewis and Clark* (New York: Perigee Books [a Division of Penguin-Putnam]); D. B. Botkin (2004). *Beyond the Stony Mountains: Nature in the American West from Lewis and Clark to Today* (New York: Oxford University Press). New editions of *Passage of Discovery* and *Beyond the Stony Mountains* became available in 2012 in all ebook formats, from Croton River Publishers, distribution in Inscribe Digital Corporation.

6. The grizzly bear information and Lewis and Clark stories are from D. B. Botkin (2004). *Beyond the Stony Mountains: Nature in the American West from Lewis and Clark to Today* (New York: Oxford University Press).

7. I worked out how many grizzles there might have been in 1805 out of curiosity, but didn't consider the calculation accurate enough or reliable enough to report in a scientific journal. However, two scientists, Andrea S. Laliberte and William J. Ripple, read my book *Our Natural History: The Lessons of Lewis and Clark* and used the method I suggested to estimate early-nineteenth-century grizzly abundance from the Lewis and Clark journals. They published their results in the highly respected journal *BioScience.* Looking back, perhaps I underestimated the scientific value of such calculations based on unique, historical information. (Andrew S. Laliberte and William J. Ripple [2003]. Wildlife encounters by Lewis & Clark: a spatial analysis of Native American/wildlife interactions. *BioScience, 53*(10): 994–1003.)

8. The classic study by the Craighead brothers was reported in: Craighead, John, J., Joel R. Varney Frank C. Craighead , Jr., 1974, "A population analysis of the Yellowstone grizzly bears, "Montana Forest & Conversation Experiment Station, School of Forestry, University of Montana. Bulletin – 40, Missoula, MT. Then later, John Craighead summarized the brothers' work in: Craighead, J. J., Jay S. Sumner, et al.(1995). The Grizzly Bears of Yellowstone: Their Ecology in the Yellowstone Ecosystem, 1959–1992 Washington, D.C.,

Island Press. A review of this book was published by Mark Boyce, a wildlife ecologist: Boyce, M. S. (1997). "Review Essay: The Grizzly Bears of Yellowstone Their Ecology in the Yellowstone Ecosystem, 1959–1992. "Yellowstone Science:18-21. Another important publication regarding these bears is: McCullough, D. R.(1986). Bears: Their Biology and Management. Sixth International Conference on Bear Research and Management, Grand Canyon, Arizona, USA, February 1983(1986), 6:21–32. "Historical range of abundance" is also sometimes written as "historic range of abundance," but since "historic" means "of species' historical importance," it is inappropriate in this context. The use of the historical range of abundance to set the modern range is consistent with the geologic concept of uniformitarianism, the idea that the past explains the present and the present explains the past—that processes that operated in the past continue to operate the same way in the present, and vice versa. Understanding the past therefore provides a key to understanding future possibilities. This is how early geologists used modern rates of erosion to estimate how long it could have taken for mountains to be worn down near to sea level by water erosion.

9. Sevheen, C. (1982). Grizzly Bear Recovery Plan. U.S._Fish_and_Wildlife_Service. Missoula, MT. U.S. Fish and Wildlife **Service:** 181 pp.

10. For example, the following paper in one of the early ones using a genetic basis for estimating a minimum viable population size: J. F. Lehmkuhl (1984). Determining size and dispersion of minimum viable populations for land management planning and species conservation. *Environmental Management, 8*(2), 167–176.

11. M. E. Soulé (1985). What is Conservation Biology? *BioScience, 35,* 727–734.

12. A more thorough discussion of how to manage populations and ecosystems in a time-varying environment is discussed in Botkin, D.B. and M.J. Sobel, 1975, Stability in time-varying ecosystems, *Amer. Nat. 109*: 625–646.

13. Eric S. Menges (1991). The Application of Minimum Viable Population Theory to Plants. Chapter 3, pp. 45–62, in D. A. Falk and K. E. Holsinger, eds. *Genetics and Conservation of Rare Plants* (New York: Oxford University Press). More recently, the following extended the analysis of minimum viable populations: D. H. Reed, Julian J. O'Grady, Barry W. Brook, Jonathan D. Ballou, and Richard Frankham (2003). Estimates of minimum viable population sizes for vertebrates and factors influencing those estimates. *Biological Conservation, 113,* 23–34. But, unfortunately from the point of view of my book, the authors use the logistic model and start with a population, N(t) at K, so the method is useless for the real world as far as I am concerned, just as likely to mislead as to lead, and most likely to divert attention from realistic, empirically based estimates of minimum viable population sizes.

14. D. H. Reed, Julian J. O'Grady, Barry W. Brook, Jonathan D. Ballou, and Richard Frankham (2003). Estimates of minimum viable population sizes for vertebrates and factors influencing those estimates. *Biological Conservation, 113,* 23–34.

15. Larry Slobodkin, who died in 2009, was, along with G. Evelyn Hutcheson, another of the twentieth century's smartest and most imaginative and insightful ecologists, and also, from my experience, ecology's best standup comic. When I was working at Woods Hole, one year the Ecosystems Center invited Larry to teach the summer marine ecology course (which he had done many times in the past). While Larry was giving the course, we had another well-known and very successful ecologist visit, and he wanted to hear Larry's lecture. When Larry saw him in the classroom, he greeted him warmly and told the students that, when this visiting professor was a student, he had taken Larry's marine ecology course three times. Afterwards, I asked him if this was true. "Sure," he said, "of course, the material was the same. But the jokes were so good." Larry was one of my favorite colleagues, and the fact that he had a great sense of humor should only be remembered as part of his greatness, not instead of.

16. Lee Talbot, whom I'm mentioned before as one of our time's leading biological conservationists, was the primary author of the Marine Mammal Protection Act of 1972 and the Endangered Species Act of 1973. He said to me on a visit to Santa Barbara that he thought condors feeding on dead marine mammals was not only likely, but they might have been the major food for the condors.

17. This information about sea otters is from J. A. Estes and J. F. Palmisano (1974). Sea Otters: Their Role in Structuring Nearshore Communities. *Science, 185,* 1058–1060; and D. O. Duggins (1980). Kelp Beds and Sea Otters: An Experimental Approach. *Ecology, 61,* 447–453.

18. D. B. Lindenmayer, C. R. Margules, and D. B. Botkin (2000). Indicators of Biodiversity for Ecologically Sustainable Forest Management. *Conservation Biology, 14*(4), 941–950.

19. K. W. Kenyon (1969). *The Sea Otter in the Eastern Pacific Ocean.* North American Fauna, no. 68, Bureau of Sport Fisheries and Wildlife, United States Department of the Interior (Washington, D. C.: Government Printing Office).

20. Chapter 3, "Species of Special Concern," in *Marine Mammal Commission—Annual Report for 2002,* p. 177.

21. Public Law 92–522, H.R. 10420, 92nd Cong., Oct. 21, 1972.

22. *Marine Mammal Commission—Annual Report for 2008,* p. 122.

23. Chapter 3, "Species of Special Concern," in *Marine Mammal Commission—Annual Report for 2002,* p. 177.

24. Ibid.

25. *Marine Mammal Commission—Annual Report for 2008,* p. 127.

Chapter 12

1. Isabella Bird visited the Hawaiian Islands in 1873. Her experiences were published originally as *Six Months Among the Palm Groves, Coral Reefs and Volcanoes of the Sandwich Islands* (London: John Murray, 1890), reprinted as *Six Months In the Sandwich Islands* (Rutland, Vt.: Tuttle, 1983), p. 37.

2. B. Aldrin and K. Abraham (2009). *Magnificent Desolation: The Long Journey Home from the Moon* (New York: Crown Books), p. 34.

3. Naess, Arne, 1989, Ecology, Community and Lifestyle:Outline of an Ecolosophy, translated by D. Rothenberg, Cambridge University Press, New York, p.169. Naess at the time was at the Council of Environmental Studies, University of Oslo. He is considered to be primary philosopher of the deep ecology movement.

4. P. Krugman (June 29, 2009) Betraying the Planet [op-ed]. *New York Times.*

5. Huffington Post (Sept. 8, 2011). Rick Perry: Global Warming Based On Scientists Manipulating Data. Available at: http://www.huffingtonpost.com/2011/08/17/rick-perry-global-warming_n_929235.html

6. F. Zwiers and G. Hegerl (May 15, 2008). Climate Change: Attributing Cause and Effect. *Nature, 453,* 296–297.

7. The term "global warming" is usually used today in a context that implies the warming is caused by humans, but strictly speaking it is sometimes, and certainly can be, used to mean the present warming whether or not it is caused by human actions. See, for example, the Intergovernment Panel on Climate Change report IPCC AR4 SYR (2007). Core Writing Team; R. K. Pachauri and A. Reisinger, eds. *Climate Change 2007: Synthesis Report. Contribution of Working Groups I, II and III to the Fourth Assessment Report of the Intergovernmental Panel on Climate Change.* IPCC. ISBN ISBN 92-9169-122-4. http://www.ipcc.ch/publications_and_data/ar4/syr/en/contents.html.

8. Climatologists prefer that the temperature be represented as a difference between a long-term average and reconstructed values. This allows consistency in the methods. Ecologically, the actual temperature is preferable, because living things respond to temperature, not determined deviation. For example, a tree is affected much differently by a temperature rise from –1°C to +1°C—just below freezing to just above freezing—than to a 2° increase from 9°C to 11°C. The best long-term temperature reconstructions are from Antarctic ice cores. These are used by scientists to represent Earth's global temperature change. Scientists acknowledge that there are regional differences and the changes reconstructed near the South Pole may or may not be representative of the entire Earth, but they are the best we have.

The average global modern surface temperature is usually represented by the available measurements from 1960 to 1990, which yield a value of 14.0°C, as obtained by the Hadley Meteorological Center and used frequently in many of the publications by climatologists (P. Brohan, J. J. Kennedy, I. Harris, S. F. B. Tett, and P. D. Jones [2006]. Uncertainty estimates in regional and global observed temperature changes: a new dataset from 1850. *Journal of Geophysical Research*, 111, D12106, doi:10.1029/2005JD006548), which is harder for us ecologists to put into the context of effects on living things, since they respond to the temperature, unaware of an estimate of some modern average. National Research Council (2010). America's Climate Choices: Panel on Advancing the Science of Climate Change. *Advancing the Science of Climate Change* (Washington, D.C.: The National Academies Press. ISBN 0309145880). http://www.nap.edu/catalog.php?record_id=12782.

9. E. Le Roy Ladurie (1971). *Times of Feast, Times of Famine: A History of Climate Since the Year 1000* (Garden City, NY: Doubleday & Co.).

10. H. H. Lamb (1995). *Climate, History and the Modern World* (London: Routledge [paperback]). See also H. H. Lamb (2011, first published in 1966). *A History of Climate Changes* (London: Routledge, 4 vols., reprinted).

11. Isabella Bird, p. 35.

12. NOAA data. Available at: http://www.esrl.noaa.gov/gmd/ccgg/trends/#mlo_full This website contains the following information:

Monthly mean atmospheric carbon dioxide at Mauna Loa Observatory, Hawaii. The carbon dioxide data, measured as the mole fraction in dry air, on Mauna Loa constitute the longest record of direct measurements of CO_2 in the atmosphere. They were started by C. David Keeling of the Scripps Institution of Oceanography in March of 1958 at a facility of the National Oceanic and Atmospheric Administration (Keeling, 1976). NOAA started its own CO_2 measurements in May of 1974, and they have run in parallel with those made by Scripps since then (Thoning, 1989). The black curve represents the seasonally corrected data.

Data are reported as a dry mole fraction defined as the number of molecules of carbon dioxide divided by the number of molecules of dry air multiplied by one million (ppm).

13. Some of the first important discussions of these measurements took place at a conference in 1972 at the Brookhaven National Laboratory and were published as *Carbon and the Biosphere*, G. M. Woodwell and E. V. Pecan, eds., Brookhaven National Laboratory Symposium, no. 24 (Oak Ridge, Tenn.: Technical Information Service, 1973).

14. Source for Keeling's graph in Figure 12.3 is NOAA data: ftp://ftp.cmdl.noaa.gov/ccg/co2/trends/co2_mm_mlo.txt. See www.esrl.noaa.gov/gmd/ccgg/trends/for additional details. Data from March 1958 through April 1974 have been obtained by C. David Keeling of the Scripps Institutin of Oceanography [SIO] and were obtained from the Scripps website (scrippsco2.ucsd.edu).

15. Svante Arrhenius (1896). On the Influence of Carbonic Acid in the Air Upon the Temperature of the Ground. *Philosophical Magazine, 41*, 237–276. Scientists before

Arrhenius understood that gases in the atmosphere absorbed different wavelengths of light differently, and proposed that this could affect the surface temperature. In 1820, the famous mathematician Joseph Fourier recognized this possibility.

16. See G. S. Callendar (1938). The Artificial Production of Carbon Dioxide and Its Influence on Temperature. *Quarterly Journal of the Royal Meteorological Society, 64,* 223–237; G. S. Callendar (1949). Can Carbon Dioxide Influence Climate? *Weather, 4,* 310–314; G. S. Callendar. (1958). On the Amount of Carbon Dioxide in the Atmosphere. *Tellus, 10,* 243.

17. C. A. Ekdahl and Charles D. Keeling (1973). Atmospheric carbon dioxide and radiocarbon in the natural carbon cycle: 1. Quantitative deductions from records at Mauna Loa Observatory and at the South Pole. *Carbon and the Biosphere.* George M. Woodwell and Erene V. Pecan (eds.), U.S. Atomic Energy Commission, Technical Information Center, Office of Information Services. Brookhaven National Laboratory Symposium No. 24: 51–81. See also NOAA data available at ftp://ftp.cmdl.noaa.gov/ccg/co2/trends/co2_mm_mlo. txt. Source of Keeling's original CO_2 graph: Carl A. Ekdahl and Charles D. Keeling (1973). Atmospheric carbon dioxide and radiocarbon in the natural carbon cycle: 1. Quantitative deductions from records at Mauna Loa Observatory and at the South Pole. *Carbon and the Biosphere.* In George M. Woodwell and Erene V. Pecan (eds.), U.S. Atomic Energy Commission, Technical Information Center, Office of Information Services. Brookhaven National Laboratory Symposium No. 24:51–81.)

See www.esrl.noaa.gov/gmd/ccgg/trends for additional details. Data from March 1958 through April 1974 have been obtained by C. David Keeling of the Scripps Institution of Oceanography (SIO) and were obtained from the Scripps website (scrippsco2.ucsd. edu).

18. Keeling reported a 3 years moving average of from increase from 312 parts per million carbon dioxide to 323 over the time from 1957 to 1973. Carl A. Ekdahl and Charles D. Keeling (1973). Atmospheric carbon dioxide and radiocarbon in the natural carbon cycle: 1. Quantitative deductions from records at Mauna Loa Observatory and at the South Pole. *Carbon and the Biosphere.* George M. Woodwell and Erene V. Pecan (eds.), U.S. Atomic Energy Commission, Technical Information Center, Office of Information Services. Brookhaven National Laboratory Symposium No. 24: 51–81.

19. Figure 12.5 is redrawn from the Hadley Meteorological Center's <u>HadCRUT3 and as shown on</u> http://www.cru.uea.ac.uk/cru/info/warming/

20. C. J. Glacken (1967). *Traces on the Rhodian Shore: Nature and Culture in Western Thought from Ancient Times to the End of the Eighteenth Century* (Berkeley: University of California Press), p. 74. Glacken points out that writers from Plutarch through those of the seventeenth century, including John Ray, John Keill, and Edmund Halley, did argue that uninhabited parts of the Earth were useful. (The argument by Plutarch is quoted at the beginning of Chapter 9.) I thank Dr. Charles Beveridge, editor of the Frederick Law Olmsted papers, for acquainting me with Glacken's great book.

21. *Santa Barbara News-Press,* April 24, 1988, p. A25. The article describes an analysis done by Lori Heise of the Worldwatch Institute of Washington, D.C. A recent article made the same point, reporting an estimate that an area the size of Australia would be required to absorb the 5 billion metric tons of carbon added every year from the burning of fossil fuels (W. Booth [1988]. Johnny Appleseed and the Greenhouse. *Science, 242,* 19–20).

22. P. M. Sheehan (2001). The Late Ordovician Mass Extinction. *Annual Review of Earth and Planetary Sciences, 29,* 331–364. The abstract of this paper states: "Near the end of the Late Ordovician, in the first of five mass extinctions in the Phanerozoic, about 85% of marine species died. The cause was a brief glacial interval that produced two pulses of extinction. The first pulse was at the beginning of the glaciation, when sea-level decline drained epicontinental seaways, produced a harsh climate in low and mid-latitudes, and

initiated active, deep-oceanic currents that aerated the deep oceans and brought nutrients and possibly toxic material up from oceanic depths. Following that initial pulse of extinction, surviving faunas adapted to the new ecologic setting. The glaciation ended suddenly, and as sea level rose, the climate moderated, and oceanic circulation stagnated, another pulse of extinction occurred. The second extinction marked the end of a long interval of ecologic stasis (an Ecologic-Evolutionary Unit). Recovery from the event took several million years, but the resulting fauna had ecologic patterns similar to the fauna that had become extinct. Other extinction events that eliminated similar or even smaller percentages of species had greater long-term ecologic effects."

23. NOAA (2009). *Global Climate Change Impacts in the United States* (Washington, D.C.: Cambridge University Press). Editors: Thomas R. Karl, NOAA National Climatic Data Center; Jerry M. Melillo, Marine Biological Laboratory; Thomas C. Peterson, NOAA National Climatic Data Center.

24. W. M. Marsh and J. Dozier (1981). *Landscape.* (New York: John Wiley & Sons). T. J. Crowley (2000). Causes of Climate Change Over the Past 1000 Years. *Science, 289,* 270–277. J. Hansen et al. (2005). Efficiency of Climate Forcings. *Journal of Geophysical Research, 110* (D18104), 45P.

25. J. D. Macdougall (2006). *Frozen Earth: The Once and Future Story of Ice Ages.* (Berkeley: University of California Press.)

26. Jostein Bakke, Øyvind Lie, Einar Heegaard, Trond Dokken, Gerald H. Haug, Hilary H. Birks, Peter Dulski, and Trygve Nilsen (2009). Rapid oceanic and atmospheric changes during the Younger Dryas cold period. *Nature Geoscience, 2,* 202–205. Published online: 15 February 2009 | doi:10.1038/ngeo439

27. T. J. Crowley (2000). Causes of Climate Change Over the Past 1000 Years. *Science, 289,* 270–277.

28. M. S. Pelto (1996). Recent changes in glacier and alpine runoff in the North Cascades, Washington. *Hydrological Processes, 10,* 1173–1180. A. G. Fountain, K. Jackson, H. J. Basagic, and D. Sitts (2007). A century of glacier change on Mount Baker, Washington. *Geological Society of America Abstracts with Programs, 39*(4), 67. Available at: http://www.mb-vrc.wwu.edu/abstracts/abstractText.php?id=35, retrieved March 31, 2009. V. A. Mohnen, W. Goldstein, and W. Wang (1991). The conflict over global warming. *Global Environmental Change, 1,* 109–123.

29. P. W. Mote and Georg Kaser (2007). The shrinking glaciers of Kilimanjaro: can global warming be blamed? The Kibo ice cap, a 'poster child' of global climate change, is being starved of snowfall and depleted by solar radiation. *American Scientist, 95*(4), 318.

30. In 1911, the U. S. Navy forecast that the Arctic Ocean could be ice-free by 2035. "As early as 2035, the Arctic Ocean could be ice free for a month, according to the US Navy's US Rear Admiral Dave Titley, addressing an Arctic conference in Tromsø, Norway. As a result, reports the Financial Times, the Bering Strait between the US and Russia will become as prominent as the Strait of Malacca and the Persian Gulf to global shipping. The route from Rotterdam, Netherlands to Yokohama, Japan is 40% shorter via the North Pole compared with the traditional passage through the Suez Canal." http://www.arcticprogress.com/2011/01/us-navy-north-pole-ice-free-2035/

See also W. Gibbs (July 11, 2000). Research Predicts Summer Doom for Northern Icecap. *New York Times:*

> OSLO, July 10—The mythic ice scape that stretches south in all directions from the North Pole is melting so fast that Norwegian scientists say it could disappear entirely each summer beginning in just 50 years, radically altering the Earth's environment, the global economy and the human imagination.

Climatologists have warned for a decade that the northern icecap is retreating. But researchers at the University of Bergen's Nansen Environmental and Remote Sensing Center are apparently the first to predict the disorienting specter of a watery North Pole open to cruise ships and the Polar Bear Swim Club within the lives of today's young people.

"The changes we've seen have been much faster and more dramatic than most people imagine," said Tore Furevik, 31, a polar researcher and co-author of the article "Toward an Ice-Free Arctic?" in the latest issue of the Norwegian science journal *Cicerone*.

Not all the scientists at the conference agreed that this was necessarily going to happen. The same article continues: "Dr. Drew Rothrock, a University of Washington oceanographer, said he agreed that Arctic sea ice was on a trajectory to disappear in 50 years. But, he added, that did not mean it would. The ice is being expelled from the Arctic by abnormally strong winds before it could achieve its accustomed thickness, he said, and that could be a temporary phenomenon.

"I think it is quite possible that in the next 10 years we will see the winds revert to a more historical pattern, so that the ice begins to reside longer in the Arctic and thicken up again," he said. "I would be cautious about predicting doom."

31. P. Chylek, M. K. Dubey, and G. Lesins (2006). Greenland warming of 1920–1930 and 1995–2005. *Geophysical Research Letters, 33*(L11707), 1–5.

32. Andrew R. Mahoney, John R. Bockstoce, Daniel B. Botkin, Hajo Eicken, and Robert A. Nisbet (in press). Sea Ice Distribution in the Bering and Chukchi Seas: Information from Historical Whaleships' Logbooks and Journals. *Arctic.*

33. H. J. Morowitz (1992). *The Thermodynamics of Pizza: Essays on Science and Everyday Life* (New Brunswick: Rutgers University Press).

34. K. Bennett (1990). Milankovitch Cycles and Their Effects on Species in Ecological and Evolutionary Time. *Paleobiology, 16*(1), 11–21.

35. C. Brahic, David L. Chandler, Michael Le Page, Phil McKenna, and Fred Pearce (2007). Climate myths. *New Scientist, 194*(2604).

36. P. Foukal, C. Frohlich, H. Sprint, and T. M. L. Wigley (2006). Variations in Solar Luminosity and Their Effect on the Earth's Climate. *Nature, 443,* 151–166. W. Soon (2007). Implications of the Secondary Role of Carbon Dioxide and Methane Forcing in Climate Change: Past, Present, and Future. *Physical Geography, 28*(2), 97–125.

37. Michael Rowan-Robinson (1985). *Fire & Ice: The Nuclear Winter* (Burnt Mill, Harlow, Essex, England: Longman).

38. Carl Sagan and Richard Turco (1990). *A Path Where No Man Thought: Nuclear Winter and the End of the Arms Race* (New York: Random House).

39. *Watts Up With That?* Guest post by Lon Hocker (June 10, 2010). A study: The temperature rise has caused the CO2 increase, not the other way around. Available at: http://www.prisonplanet.com/a-study-the-temperature-rise-has-caused-the-co2-increase-not-the-other-way-around.html.

40. M. A. K. Khalil and R. A. Rasmussen (1985). Causes of Increasing Atmospheric Methane: Depletion of Hydroxyl Radicals and the Rise of Emissions. *Atmospheric Environment, 19,* 397.

41. J. Hansen, A. Lacis, and M. Prather (1989). Greenhouse Effect of Chlorofluorocarbons and Other Trace Gases. *Journal of Geophysical Research, 94*(D13), 16417–16421.

42. The discussion of El Niño is taken from D. B. Botkin and E. A. Keller (2009). *Environmental Sciences: The Earth as a Living Planet* (New York: John Wiley, 7th ed.).

43. Jet Propulsion Laboratory. El Niño: When the Pacific Ocean Speaks, Earth Listens. Accessed Sept. 25, 2008 at www.jpl.nasa.gov

44. Al Gore (Feb. 28, 2010). We Can't Wish Away Climate Change. *New York Times*.

45. A. D. Barnosky (2009). *Heatstroke: Nature in an Age of Global Warming* (Washington, D.C.: Island Press).

46. Mackay, 1841, pp. 265–266.

47. A well-thought-of history of attempts to understand and make calculations about climate can be found in S. R. Weart (2008). *The Discovery of Global Warming* (Cambridge: rev. and exp. ed.). The book is very good about the physicists and mathematicians who explored climate change and the increasing sophistication of their calculations. It is written by a physicist (trained in that field; his career was as the director of the Center for History of Physics of the American Institute of Physics). But his history is as if there were few geologists and fewer oceanographers who discovered anything significant. For example, his history of the understanding of the ice ages completely omits the stories I have told here that involve Swiss farmers and Agassiz.

48. Veerabhadran Ramanathan is the Victor Alderson Professor of Applied Ocean Sciences and director of the Center for Atmospheric Sciences at the Scripps Institution of Oceanography, University of California, San Diego. Among his important research works are V. Ramanathan (1975). Greenhouse Effect Due to Chlorofluorocarbons: Climatic Implications. *Science, 190*, 50–51. doi:10.1126/science.190.4209.50. V. Ramanathan et al. (1985). Trace Gas Trends and Their Potential Role in Climate Change. *J. Geophys. Res., 90*(D3), 5547–5566. Available at: http://www.agu.org/pubs/crossref/1985/JD090iD03p05547.shtml. V. Ramanathan et al. (1983). The Response of a Spectral General Circulation Model to Refinements in Radiative Processes. *J. Atmos. Sci., 40*, 605–630. Available at: http://ams. allenpress.com/perlserv/?request=get-abstract&doi=10.1175%2F1520-0469(1983)040%3C0 605%3ATROASG%3E2.0.CO%3B2.

49. S. R. Weart, on the American Institute of Physics website (http://www.aip.org/history/climate/arakawa.htm) summarizing his 2008 book *The Discovery of Global Warming* (Cambridge: rev. and exp. ed.).

50. American Institute of Physics website http://www.aip.org/history/climate/GCM. htm#L000 summarizing S. R. Weart (2008). *The Discovery of Global Warming* (Cambridge: rev. and exp. ed.).

51. James Hansen (March 30, 2008). Tipping Point: Perspective of a Climatologist. *2008-2009 State of the Wild: A Global Portrait of Wildlife, Wildlands, and Oceans*. (Washington, D.C.: Island Press).

52. Tyco Brahe's most famous work was *De nova stella* (1573). His work is famous today among astronomers, not only because of the accuracy and high quality of that work, but also because he was the last of the major astronomers to work without a telescope.

53. J. C. Bergengren, Starley L. Thompson, David Pollard, and Robert M. Deconto (2001). Modeling global climate–vegetation interactions in a doubled CO_2 world. *Climatic Change, 50*, 31–75. The abstract of this paper states that "Terrestrial vegetation is resolved into 110 plant life forms, which represent groups of species with similar physiognomic characteristics and migrational responses to climate change, thus preserving the spatial integrity of each life-form distribution as climate changes. EVE generates a quantitative description of plant community structure defined by total vegetation cover and the fractional covers of life forms as a function of climate. The equilibrium distribution of each life form is predicted from monthly mean temperature, precipitation, and relative humidity, based on observed correlations with the present climate. The fractional covers of the life forms at each site are determined by parameterizations of dynamic ecological processes: competition for sunlight, disturbances by fire and treefall. A second model

(LEAF) simulates the seasonal phenology of EVE's plant canopies, driven by the daily climate at each location, and provides the physical quantities needed for coupling vegetation and climate models."

54. J. S. Armstrong (2001). *Principles of Forecasting—A Handbook for Researchers and Practitioners* (New York: Springer).

55. The original of this discussion was first published as D. B. Botkin (Dec. 19, 2009). Global Warming and an Odd Bull Moose. *Wall Street Journal*.

Chapter 13

1. D. B. Botkin and E. A. Keller (2009). *Environmental Sciences: The Earth as a Living Planet* (New York: John Wiley, 7th ed., Chapter 21).

2. J. Kanter (2008). Europe May Ban Imports of Some Biofuel Crops. *New York Times*.

3. T. L. Friedman (Sept. 14, 2011). Is It Weird Enough Yet? *New York Times*.

4. National Research Council (2006). *Surface Temperature Reconstructions for the Last 2,000 Years* (Washington D.C.: The National Academy Press).

5. M. E. Mann et al. (2009). Global signatures and dynamical origins of the Little Ice Age and Medieval Climate Anomaly. *Science, 326*, 1256–1260; B. M. Fagan (2008). *The Great Warming: Climate Change and the Rise and Fall of Civilizations*. (New York: Bloomsbury Press).

6. B. M. Fagan (2008). *The Great Warming: Climate Change and the Rise and Fall of Civilizations*. (New York: Bloomsbury Press) and Brown, N. M. (2008). *The Far Traveler: Voyages of a Viking Woman*. New York, Mariner Books: Houghton Mifflin Harcourt.

7. Historical evidence of climate change and its effects on civilization during the Little Ice Age are well documented in the classic book E. Le Roy Ladurie (1971). *Times of Feast, Times of Famine: A History of Climate Since the Year 1000* (Garden City, N.Y: Doubleday & Co., translated from French). It was published originally in French as E. Le Roy Ladurie. *Histoire humaine et comparee du Climat: Canicules et Glaciers (XIIIe—XVIIIe Siecle)*, Fayard. There is a 2004 reprinted edition of this original version.

8. This and the following material about North America in the Little Ice Age are from D. B. Botkin (1998). Introduction (pp. xix—xxviii). In T. J. Stiles (ed.), *In Their Own Words: The Colonizers: Early European Settlers and the Shaping of North America* (New York: Perigree Press).

9. Discussion of the effects of the Little Ice Age beyond Europe is from D. B. Botkin and E. A. Keller (2009). *Environmental Sciences: The Earth as a Living Planet* (New York: John Wiley, 7th ed.).

10. T. J. Stiles (ed.) (1998). *In Their Own Words: The Colonizers: Early European Settlers and the Shaping of North America* (New York: Perigree Press).

11. R. T. Lackey (2011). *Science: Beacon of Reality*. Plenary Address, 141st Annual Meeting of the American Fisheries Society, Seattle, Washington, September 5, 2011.

12. The projected effect depends on which climate model is used. The models that we used were from Goddard Space Flight Center, New York, Princeton University, and Oregon State University.

13. D. B. Botkin, R. A. Nisbet, and T. E. Reynales (1989). Effects of Climate Change on Forests of the Great Lake States. In J. B. Smith and D. A. Tirpak (eds.), *The Potential Effects of Global Climate Change on the United States* (Washington, D. C.: U. S. Environmental Protection Agency, EPA—203-05-89-0), pp. 2–1 to 2–31; D. B. Botkin and R. A. Nisbet (1990). *Response of Forests to Global Warming and CO_2 Fertilization. Report to EPA*; D. B. Botkin (1991). *Global Warming and Forests of the Great Lakes States: An Example of the Use of Quantitative Projections in Policy Analysis* (essay submitted for the George and Cynthia Mitchell International Prize Competition, 1991, which won first prize and

was published by the Mitchell Foundation, Houston, TX); D. B. Botkin and R. A. Nisbet (1992). Forest response to climatic change: effects of parameter estimation and choice of weather patterns on the reliability of projections. *Climatic Change, 20,* 87–111.

14. D. E. Haugen and A. K. Weatherspoon (2009). *Michigan Timber Industry: An Assessment of Timber Product Output and Use* (Newtown Square, PA: U. S. Forest Service, USDA: 96 pp).

15. The discussion of the major change in forest ownership is from D. B. Botkin and E. A. Keller (2011). *Environmental Sciences: The Earth as a Living Planet* (New York: John Wiley, 8th ed.).

16. Suming Jin and Steven A. Sader (2006). Effects of forest ownership and change on forest harvest rates, types and trends in northern Maine. *Science Direct* Available online (accessed May 2, 2006) at:
http://www.sciencedirect.com/science?_ob=ArticleURL&_udi=B6T6X-4JVTCCJ-4&_
user=10&_rdoc=1&_fmt=&_orig=search&_sort=d&_docanchor=&view=c&_search
StrId=1050312920&_rerunOrigin=google&_acct=C000050221&_version=1&_urlVe
rsion=0&_userid=10&md5=5c36bcd396911d59aff4583aec8d9095

17. D. B. Botkin, D. A. Woodby, and R. A. Nisbet (1991). Kirtland's Warbler Habitats: A Possible Early Indicator of Climatic Warming. *Biological Conservation, 56*(1), 63–78.

18. This discussion about the lack of interest in our Kirtland's warbler forecast was originally published as D. B. Botkin (Dec. 28, 2007). Science and soothsaying. *International Herald Tribune.*

19. Chris D. Thomas, Alison Cameron, Rhys E. Green, Michel Bakkenes, Linda J. Beaumont, Yvonne C. Collingham, Barend F. N. Erasmus, Marinez Ferreira de Siqueira, Alan Grainger, Lee Hannah, Lesley Hughes, Brian Huntley, Albert S. van Jaarsveld, Guy F. Midgley, Lera Miles, Miguel A. Ortega-Huerta, A. Townsend Peterson, Oliver L. Phillips, and Stephen E. Williams (2004). Extinction risk from climate change. *Nature, 427,* 145–148.

20. Konza Prairie Biological Station, a 3,487 hectare area of native tallgrass prairie in the Flint Hills of northeastern Kansas (39°05'N and 96°35'W), is located approximately 13 km south of Manhattan, KS. KPBS is owned by the Nature Conservancy and Kansas State University and operated as a field research station by the K-State Division of Biology. KPBS is divided into more than 60 watersheds that incorporate different fire frequencies and the presence/absence of large ungulate grazers (bison or cattle) in a long-term experimental setting (http://www.konza.ksu.edu/). This biological station is part of the Long-term Ecology Research Program (LTER), which I helped set up, when I was asked by the National Science Foundation in the early 1970s to convene two workshops to discuss the need for, and how to accomplish, long-term ecological monitoring. S. M. Scheiner, S. B. Cox, Michael Willig, Gary G. Mittelbach, Craig Osenberg, and Michael Kaspari(2000), Species richness, species-area curves and Simpson's paradox. *Evolutionary Ecology Research, 2,* 791–802.

21. D. B. Botkin and L. Simpson (1990). Biomass of the North American Boreal Forest: A Step Toward Accurate Global Measures. *Biogeochemistry, 9,* 161–174.

22. Liem, Mita Valina, 2007 "Indonesian rainforests are disappearing fast," New York Times, Published: Monday, June 4, 2007.

23. Small Cap Network website. Available at: http://www.smallcapnetwork.com/Asian-Demand-for-American-Crops-lifting-Railcar-Stocks-RAIL-TRN-GMT-ARII-GBY/s/via/3420/article/view/p/mid/1/id/127/

24. U.S. Grain Council online, Global Update: September 8, 2011. Available at: http://www.grains.org/global-update/3375-global-update-september-8-2011

25. The information about land area in agriculture and food crises is from D. B. Botkin and E. A. Keller (2011). *Environmental Sciences: The Earth as a Living Planet* (New York:

John Wiley, 8th ed.). The information about biofuels in from D. B. Botkin (2001). *Powering the Future: A Scientist's Guide to Energy Independence* (FT Press).

26. D. B. Botkin, Henrik Saxe, Miguel B. Araújo, Richard Betts, Richard H. W. Bradshaw, Tomas Cedhagen, Peter Chesson, Terry P. Dawson, Julie Etterson, Daniel P. Faith, Simon Ferrier, Antoine Guisan, Anja Skjoldborg Hansen, David W. Hilbert, Craig Loehle, Chris Margules, Mark New, Matthew J. Sobel, and David R. B. Stockwell (2007). Forecasting Effects of Global Warming on Biodiversity. *BioScience, 57*(3), 227–236.

27. In Botkin et al. (2007), we discuss the famous megafauna extinctions at the end of the last ice age: "A long-standing controversy regarding the role of people in Quaternary extinctions of large mammals speaks to the difficulty of quantifying impacts of multiple factors on species loss. The high extinction rate of large mammals has been widely recognized since the 19th century, and extinctions of large mammals and island birds over the past 100,000 years have been the subject of much conjecture. Paul Martin has made the now well-known case that the timing of extinctions followed human dispersal from Afro-Asia to other parts of the globe and that these extinctions resulted from human "blitzkrieg" overkill (Martin and Steadman 1999). But careful analysis of well-documented extinctions in Beringia suggests that human hunting was superimposed on a preexisting trend of diminishing animal population density (Shapiro et al. 2004, Guthrie 2006). These data suggest that the interaction of environmental change and human resource use can have a larger negative impact on biodiversity than either factor alone." The references cited are B. Shapiro et al. (2004). Rise and fall of the Beringian Steppe Bison. *Science, 306*, 1561–1565; and D. Guthrie (2006). New carbon dates link climatic change with human colonization and Pleistocene extinctions. *Nature, 441*, 207–209.

28. IPCC (2007). *Climate Change 2007: Synthesis Report* (Valencia, Spain: IPCC).

29. James Hansen (2008). Tipping Point: Perspective of a Climatologist. In: *2008–2009 State of the Wild: A Global Portrait of Wildlife, Wildlands, and Oceans* (Washington, D.C.: Island Press).

30. J. P. Grime, Jason D. Fridley, Andrew P. Askew, Ken Thompson, John G. Hodgson, and Chris R. Bennett (2008). Long-term Resistance to Simulated Climate Change in an Infertile Grassland. *Proceedings of the National Academy of Sciences, 105*(29), 10028–10032; Earth System Science Committee (1988). *Earth System Science: A Preview* (Boulder, Colorado: University Corporation for Atmospheric Research).

31. W. E. Bradshaw and C. M. Holzapfel (2006). Evolutionary response to rapid climate change. *Science, 312*, 1477–1478.

32. R. B. Huey et al. (2000). Rapid evolution of a geographic cline in size in an introduced fly. *Science, 287*, 308–309.

33. Anthony D. Barnosky (2009). *Heatstroke: Nature in an Age of Global Warming* (Washington, D.C.: Island Press).

34. E. Martinez-Meyer, A. Townsend Peterson, and W. W. Hargrove (2004). Ecological niches as stable distributional constraints on mammal species, with implications for Pleistocene extinctions and climate change projections for biodiversity. *Global Ecology and Biogeography, 13*, 305–314.

35. G-R. Walther, S. Berger, and M. T. Sykes (2005). An ecological 'footprint' of climate change. *Proceedings of the Royal Society B, 272*, 1427–1432; M. B. Araújo et al. (2005). Validation of species-climate impact models under climate change. *Global Change Biology, 11*, 1504–1513; M. B. Araújo et al. (2005). Reducing uncertainty in projections of extinction risk from climate change. *Global Ecology and Biogeography, 14*, 529–538.

36. C. F. Randin et al. (2006). Are niche-based species distribution models transferable in space? *Journal of Biogeography, 33*, 1689–1703.

37. M. B. Araújo and M. New (2006). Ensemble forecasting of species distributions. *Trends in Ecology and Evolution* doi:10.1016/j.tree.2006.09.010.

38. W. Thuiller (2004). Patterns and uncertainties of species' range shifts under climate change. *Global Change Biology, 10*, 2020–2027; J. J. Lawler et al. (2006). Predicting climate-induced range shifts: model differences and model reliability. *Global Change Biology, 12*, 1568–1584; R. G. Pearson (2006). Climate Change and the Migration Capacity of Species. *Trends in Ecology & Evolution, 21*, 111–113. As we point out in Botkin et al. (2007), "An additional complication is that the relationship between the occurrence of a species and climatic variables is not always correlated with the mean. For example, amphibian declines due to outbreaks of a pathogenic chytrid fungus (*Batrachochytrium dendrobatidis*) are related to the annual range of temperatures, not to the mean temperature." This is discussed in detail in J. A. Pounds et al. (2006). Widespread amphibian extinctions from epidemic disease driven by global warming. *Nature, 439*, 161–167.

39. W. Cook (1991). *The Road to the 707* (Bellevue, WA: TYC Pub. Co.), p. 3.

40. E. A. Harris (2011). The Heat Starts Early, Then Breaks a Record. *New York Times.*

41. Katharine Q. Seelye (Aug. 13, 2011). A Long, Cold Summer at Mount Rainier. *New York Times.*

42. Fritz Maytag, personal communication.

43. (2011). Global warming not to blame for 2011 droughts. *New Scientist.* Available at: http://www.newscientist.com/article/mg21028173.100-global-warming-not-to-blame-for-2011-droughts.html.

44. The discussion of La Niña and El Niño are from D. B. Botkin and E. A. Keller (2009). *Environmental Sciences: The Earth as a Living Planet* (New York: John Wiley, 7th ed.), Chapter 23.

Chapter 14

1. Plotinus (1956). *The Enneads* (trans. S. Mackenna, revised by B. S. Page) (London: Faber and Faber Ltd., 3rd ed.).

2. Paul MacCready (1987). Driven to Extremes. *Los Angeles Times Magazine.* MacCready is the designer of the Sunraycer car and the Gossamer Condor, the first human-powered aircraft.

3. J. Campbell (1987). *The Masks of God: Primitive Mythology* (New York: Penguin Books), p. 468.

4. This discussion is based on P. G. K. Kahn and S. M. Pompea (1978). Nautiloid Growth Rhythms and Dynamical Evolution of the Earth-Moon System. *Nature, 275*, 606–611.

5. This assumes that the Earth's day length has changed slowly and measurably, and that the masses of the Earth and moon have been constant.

6. Kahn and Pompea (1978).

7. G. P. Marsh (1864/1967). *Man and Nature* (ed. D. Lowenthal) (Cambridge, Mass.: Harvard University Press), p. 36.

8. P. Kalm (1966). *Travels in North America* (trans. A. B. Benson) (New York: Dover Books), pp. 308–309.

9. J. Campbell, *The Masks of God: Primitive Mythology* (New York: Viking, 1959), p. 468.

10. P. Sears (1935). *Deserts on the March* (Norman: University of Oklahoma Press), p. 3.

11. A. Leopold (1949). *A Sand County Almanac and Sketches Here and There* (New York: Oxford University Press), p. 207.

Postscript

1. An article in Forbes Magazine reported the following in 2012:
"THURSDAY, APRIL 19, 2012
Warmist Steve Zwick in Forbes: "We know who the active denialists are – not the people who buy the lies, mind you, but the people who create the lies. Let's start keeping track of them now, and when the famines come, let's make them pay. Let's let their houses burn"
http://tomnelson.blogspot.com/2012/04/warmist-steve-zwick-in-forbes-k...

2. Lovejoy, T. E. (1988). "Will unexpectedly the top blow off?" BioScience **38**(10): 722-726. His career has included many major posts, such as first Biodiversity Chair of the H. John Heinz III Center for Science, Economics and the Environment; University Professor of Environmental Science and Policy at George Mason University; Chair of the Scientific Technical Advisory Panel (STAP) for the Global Environment Facility (GEF); From 1999 to 2002, he served as chief biodiversity adviser to the President of the World Bank. He is Senior Adviser to the President of the United Nations Foundation, chair of the Yale Institute for Biospheric Studies, and is past president of the American Institute of Biological Sciences, past chairman of the United States Man and Biosphere Program, and past president of the Society for Conservation Biology.

3. D. E. Harper, D. B. Botkin, R. A. Carpenter, and B. W. Mar (1987). *Applying Ecology to Land Management in Southeast Asia*. East-West Center, Occasional Paper No. 2, 148 pp.

4. http://www.freshfromflorida.com/pi/chrp/greening/citrus_disease-june-2010.pdf.

5. Jared Hardner and Richard Rice (May 2002). Rethinking Green Consumerism. *Scientific American*.

6. Richard Rice. *A Lion's Tale: In Search of Conservation on the African Plain (and Beyond)* (in press).

7. One of the early major articles about nuclear winter was: Sagan, C. (1983-84). "Nuclear war and climate catastrophe. Some policy implications." *Foreign Affairs* **62**(257–292). A general history and review of the concern about nuclear winter and its integration of biology and meteorology, and its connection between science and politics, can be found in Badash, L. (2009). *A Nuclear Winter's Tale: Science and Politics in the 1980s (Transformations: Studies in the History of Science and Technology)*. Cambridge, MA, The MIT Press.

8. G. Pullis La Rouche (2003). *Birding in the United States: a demographic and economic analysis*. Addendum to the 2001 National Survey of Fishing, Hunting and Wildlife-Associated Recreation. Report 2001–1. U.S. Fish and Wildlife Service, Arlington, Virginia.

9. http://newsgroups.derkeiler.com/Archive/Rec/rec.sport.tennis/2007-09/msg07543.html

10. For information about the annual backyard bird count see: http://www.birdcount.org The website of the Cornell Lab of Ornithology is http://www.allaboutbirds.org/NetCommunity/page.aspx?pid=1774

11. D. B. Botkin, K. Cummins, T. Dunne, H. Regier, M. J. Sobel, and L. M. Talbot (1995). *Status and Future of Anadromous Fish of Western Oregon and Northern California: Findings and Options* (Santa Barbara, CA: Center for the Study of the Environment).

12. http://www.neoninc.org/

13. Anonymous (2008). The Next Big Climate Challenge: Governments Should Work Together to Build the Supercomputers Needed for Future Predictions that Can Capture the Detail Required to Inform Policy. *Nature, 453*(7193), 257.

14. D. B. Botkin and M. J. Sobel (1975). Stability in time-varying ecosystems. *American Naturalist, 109*, 625–646.

15. B. Vickers (ed.). (1987). *English Science: Bacon to Newton* (Cambridge English Prose Texts, paperback).

16. This discussion of the scientific method is based on Chapter 2 in D. B. Botkin and E. A. Keller (2011). *Environmental Science: Earth as a Living Planet* (New York: John Wiley; first published in 1991; the 8th edition was published in 2011). Oddly, although the scientific method is fundamental to science, few science textbooks these days actually explain the scientific method. It is unusual enough that one of our colleagues in the Department of Geological Sciences, University of California, Santa Barbara, has his graduate students read this chapter, written for undergraduates with no special scientific training.

One would think that all scientists would be very much involved in thinking about, clarifying, and perhaps improving the scientific method itself. But in fact many scientists just practice science naturally. One of the scientists for whom I have the greatest respect said that during her career she never really thought about science, she just did it. Perhaps thinking about the method is something like asking a musician to think about the theory of music or a baseball player to think about the theory of the game. As Yogi Berra said, "Think! How the hell are you gonna think and hit at the same time?" The same can be said of scientists and the scientific method.

17. W. E. Schmidt (September 10, 1991). Jovial Con Men" Take Credit(?) for Crop Circles. *New York Times*, p. B1. This information about crop circles is from Crop Circle News: http://cropcirclenews.com

18. This point is made in L. B. Slobodkin, D. B. Botkin, B. Maguire, Jr., B. Moore III, and H. J. Morowitz (1980). On the Epistemology of Ecosystem Analysis. In *Estuarine Perspectives*, ed. V. S. Kennedy (New York: Academic Press).

19. For further discussion of the American idea of wilderness, see R Nash. *Wilderness and the American Mind.*

20. Thoreau, *Walden*, Shanley edition, p. 210.

21. D. B. Botkin (2001). *No Man's Garden: Thoreau and a New Vision for Civilization and Nature* (Island Press).

22. T. E. Lovejoy (1989). Deforestation and Extinction of Species. In *Changing the Global Environment: Perspectives on Human Involvement*, ed. D. B. Botkin, M. F. Caswell, J. E. Estes and A. A. Orio (Boston, Mass.: Academic Press), p. 97.

23. C. C. Mann (2006). *1491: New Revelations of the Americas Before Columbus.* (New York: Vintage).

24. P. S. Martin and D. A. Burney (1999). Bring back the elephants. *Wild Earth* (Spring issue), 57–64.

25. http://www.americanscientist.org/bookshelf/id.3538,content.true,css.print/bookshelf.aspx

26. One can read any of Lewis Mumford's many books for insights about the city and civilization within nature, from his earlier books, such as *Technics and Civilization* (New York: Harcourt Brace Jovanovich, 1934), to the more recent ones, including *The City in History* (New York: Harcourt Brace Jovanovich, 1961), and *The Pentagon of Power: The Myth of the Machine* (New York: Harcourt Brace Jovanovich, 1964). Of course, there is a large field that deals with the history of cities, cities as environment, and urban planning. Books on these topics include C. C. McLaughlin, C. E. Beveridge, D. Schuyler, and J. T. Censer (eds.) (1977–86). *The Papers of Frederick Law Olmsted*, 4 vols. (Baltimore: Johns Hopkins University Press); J. A. Burton (1977). *Worlds Apart: Nature in the City* (Garden City, N. Y.: Doubleday); and A. W. Spirn (1984). *The Granite Garden: Urban Nature and Human Design* (New York: Basic Books).

INDEX